# Regulation of the Upstream Petroleum Sector

NEW HORIZONS IN ENVIRONMENTAL AND ENERGY LAW

**Series Editors:** Kurt Deketelaere, *Professor of Law, University of Leuven, Belgium and University of Dundee, Scotland* and Zen Makuch, *Reader in Law, Barrister, Imperial College, London, UK*

Environmental law – including the pressing considerations of energy law and climate change – is an increasingly important area of legal research and practice. Given the growing interdependence of global society and the significant steps being made towards environmental protection and energy efficiency, there are few people untouched by environmental and energy lawmaking processes.

At the same time, environmental and energy law is at a crossroads. The command and control methodology that evolved in the 1960s and 1970s for air, land and water protection may have reached the limit of its environmental protection achievements. New life needs to be injected into our environmental protection regimes – perhaps through the concept of sustainability in its environmental, economic and social forms. The same goes for energy policy and law, where liberalization, environmental protection and security of supply are at the centre of attention. This important series seeks to press forward the boundaries of environmental and energy law through innovative research into environmental and energy law, doctrine and case law. Adopting a wide interpretation of environmental and energy law, it includes contributions from both leading and emerging international scholars.

Titles in the series include:

Energy for the 21st Century
Opportunities and Challenges for Liquefied Natural Gas (LNG)
*Susan L. Sakmar*

Emissions Trading Design
A Critical Overview
*Stefan E. Weishaar*

EU Environmental Legislation
Legal Perspectives on Regulatory Strategies
*Edited by Marjan Peeters and Rosa Uylenburg*

Polar Oceans Governance in an Era of Environmental Change
*Edited by Tim Stephens and David L. VanderZwaag*

The Fragmentation of Global Climate Governance
Consequences and Management of Regime Interactions
*Harro van Asselt*

Renewable Energy Law in the EU
Legal Perspectives on Bottom-up Approaches
*Edited by Marjan Peeters and Thomas Schomerus*

Environmental Enforcement Networks
Concepts, Implementation and Effectiveness
*Edited by Michael Faure, Peter De Smedt and An Stas*

Regulation of the Upstream Petroleum Sector
A Comparative Study of Licensing and Concession Systems
*Edited by Tina Hunter*

# Regulation of the Upstream Petroleum Sector

## A Comparative Study of Licensing and Concession Systems

*Edited by*

Tina Hunter

*Reader in Energy Law, University of Aberdeen, UK and Associate of the Aberdeen University Centre for Energy Law, UK*

NEW HORIZONS IN ENVIRONMENTAL AND ENERGY LAW

Edward Elgar
PUBLISHING

Cheltenham, UK • Northampton, MA, USA

Published by
Edward Elgar Publishing Limited
The Lypiatts
15 Lansdown Road
Cheltenham
Glos GL50 2JA
UK

Edward Elgar Publishing, Inc.
William Pratt House
9 Dewey Court
Northampton
Massachusetts 01060
USA

A catalogue record for this book
is available from the British Library

Library of Congress Control Number: 2014954943

This book is available electronically in the **Elgar**online
Law subject collection
DOI 10.4337/9781783470112

FSC www.fsc.org MIX Paper from responsible sources FSC® C013056

ISBN 978 1 78347 010 5 (cased)
ISBN 978 1 78347 011 2 (eBook)

Typeset by Columns Design XML Ltd, Reading
Printed and bound in Great Britain by T.J. International Ltd, Padstow

# Contents

## PART III COMPARATIVE PETROLEUM REGULATION IN DEVELOPING PETROLEUM PROVINCES



## PART IV LEGAL ISSUES IN PETROLEUM REGULATION

# Figures

# Tables

# Contributors

**Simon Warikiyei Amaduobogha**

Simon holds a Bachelor of Laws degree from the Rivers State University of Science and Technology, Port Harcourt, Nigeria and a Master of Laws degree in International Environmental Law from the University of Nottingham, United Kingdom. He was called to the Nigerian Bar in 1998 as a barrister and solicitor of the Supreme Court of Nigeria. Simon practised with top commercial law firms in Nigeria, including Punuka Chambers and Aluko & Oyebode, before joining Dougam Law Firm (Legal Practitioners & Arbitrators) as senior partner in 2004. He is also a chartered arbitrator and lecturer in the Faculty of Law, Niger Delta University, Wilberforce Island, Bayelsa State, Nigeria.

Simon is currently a doctoral candidate at the Centre for Energy, Petroleum and Mineral Law and Policy (CEPMLP), University of Dundee, United Kingdom.

**Owen L. Anderson**

Professor Anderson is the Eugene Kuntz Chair in Law in Oil, Gas and Natural Resources, the George Lynn Cross Research Professor, and the Academic Director of the John B. Turner LL.M. Program in Energy, Natural Resources & Indigenous Peoples Law at the University of Oklahoma College of Law. He also regularly teaches at the University of Texas, the University of Dundee, the University of Melbourne and the University of Sydney. He was recently appointed a Visiting Professor in the Centre for Commercial Law Studies at Queen Mary University of London. He has lectured at numerous other universities and venues in Angola, Australia, Brazil, Canada, China, France, Italy, Jordan, Korea, Mexico, Namibia, the Netherlands, Norway, Peru, Portugal, Qatar, Russia, Tanzania, Thailand and the United Kingdom.

Professor Anderson has authored over 100 articles and is a co-author of *International Petroleum Transactions*; *Hemingway's Oil and Gas Law and Taxation*; *Cases and Materials on Oil & Gas Law*; and *A Student's Guide to Estates in Land and Future Interests*. He co-authors the annual supplements to *Kuntz on Oil and Gas Law*, a multi-volume treatise, and is co-author of several chapters on the multi-volume treatise, *Waters and*

*Water Rights*. He is a contributing author to *International Petroleum Exploration and Exploitation Agreements*, an editor of the *Oil and Gas Reporter* and Editor in Chief of the *Texas Title Standards*.

**Kylie Fletcher-Johnson**
Kylie is an assistant professor and PhD candidate at Bond University. She joined Bond in 2008 and teaches principles of contractual liability, law of obligations and mining and natural resources law.

Kylie was employed in the legal profession for almost a decade prior to joining Bond. She commenced her legal career in 1998, and was soon working as a member of a large corporate and commercial law team. In 2001, Kylie accepted a position working at a national top-tier firm where she was later promoted to the position of senior associate. In this role, Kylie advised energy and resource sector clients on corporate, commercial, financial and regulatory issues. In 2005, Kylie took up a position in-house at a Queensland gas and electricity distributor and retailer. Much of her time in that role was spent advising the company on its large-scale downstream gas and electricity activities.

**Greg Gordon**
Greg Gordon practised as a solicitor qualified in Scotland and England and Wales before moving into academia. He is a senior lecturer in the Aberdeen University Centre for Energy Law, which he co-directs, and is also the Co-Director of the university's LLM in Oil and Gas Law and of Aberdeen's involvement in the North Sea Energy Law Programme. He is an editor of and contributing author to Gordon, Paterson and Usenmez's *Oil and Gas Law: Current Issues and Emerging Trends*. His research interests include petroleum licensing, risk allocation in oil and gas contracts, the protection of health, safety and the environment in offshore oil and gas operations and civil liability for oil spills from offshore installations.

**Tina Hunter**
Dr Hunter is a Reader in Energy Law at the University of Aberdeen. Prior to this appointment, she was the inaugural Director of the Centre for International Minerals and Energy Law. She teaches and researches in the areas of national and international petroleum law, unconventional petroleum regulation, Arctic petroleum law and governance, property law and resources law and policy. She has received academic qualifications in marine sediments, geology, political science, applied science, and law, completing her PhD at the University of Bergen, Norway.

In addition to teaching and researching in Energy Law at the University of Aberdeen, Dr Hunter has previous or ongoing teaching appointments in Norway, Australia and Russia. She was a visiting scholar at the University of Bergen, the University of Aberdeen and the University of Texas at Austin. Tina has presented at over 50 conferences on three continents. Her expertise has been sought by national and international governments, and she has analyzed petroleum laws, drafted petroleum legislation (both conventional and unconventional) and advised governments on policy and governance relating to unconventional petroleum. She has also consulted to governments, industry groups and NGOs in several jurisdictions including Australia, New Zealand, the UK, Russia and Canada.

In June 2014, Dr Hunter participated in the unconventional gas learning symposium at the World Bank, Washington DC. She is presently a member of the South Australian Roundtable for Unconventional Gas Projects and the working group for the Modernising Queensland Resources Acts Program. She is an associate editor of *OGEL*, the Oil, Gas and Energy Law Journal.

**Andrey Kompaniets**
Andrey graduated from Moscow State Institute of International Relations of the Ministry of Foreign Affairs in Russia (MGIMO-University) with an LLB in 2007 and an LLM in 2009 with specializations in international and oil and gas law. Andrey is currently working on his PhD in contractual forms of subsoil use regulations. He also maintains the role of Deputy Head of International Projects Support Division JSC Gazprom.

**Souichirou Kozuka**
Souichirou Kozuka (PhD, Tokyo) is Professor of Law at Gakushuin University, Tokyo. He specializes in commercial law, corporate law and maritime law and researches in comparative law studies. His recent English publications include: 'The Law Applicable on the Continental Shelf and in the Exclusive Economic Zone: The Japanese Perspective' (co-authored with Hideyuki Nakamura), *Ocean Yearbook* Vol.25, pp. 357–78 (2011); 'Insurance Law Issues Due to the Great East Japan Earthquake of 2011', *Zeitschrift für Japanisches Recht* No. 33, pp. 3–11 (2012) and No. 34, pp. 87–94 (2013); and 'Policy and Politics in Contract Law Reform in Japan' (co-authored with Luke Nottage), in: *The Method and Culture of Comparative Law* (Maurice Adams and Dirk Heirbaut (eds.)), pp. 235–53 (2014). He is a correspondent of UNIDROIT (International Institute for the Unification of Private Law) and an Associate Member of the International Academy of Comparative Law (IACL).

**Christopher Kulander**

Professor Kulander has received academic qualifications in geology (BSc), geophysics (MSc and PhD) and law (JD). He teaches at the South Texas College of Law and serves as Director of the college's new Oil & Gas Law Institute. He is also Of Counsel to Haynes and Boone, LLP. He teaches domestic energy law, basic property law, mining law, and beginning and advanced oil and gas law courses. In addition, he teaches international energy law at Vytautas Magnus University, Kaunas, Lithuania, as a permanent visiting professor. He is admitted to practice in Texas and New Mexico. Professor Kulander has published twenty law review articles, as well as many other articles with a more practical focus in the industry literature, on diverse topics including energy lending, finance, oil and gas law, land use control, American Indian law, as well as on geology, and petroleum seismology. He is a prolific speaker and expert witness.

Prior to teaching, Professor Kulander practised for four years in the Houston office of Haynes and Boone, LLP within the Energy Practice Group, focusing on energy lending, finance and bankruptcy. Prior to that, he practised for two years with Cotton & Bledsoe in Midland, Texas, focusing on oil and gas title and leasing. Before law school, Professor Kulander worked for the U.S. Geological Survey as a geophysicist for the various energy teams supporting the National Oil and Gas Assessment group.

**Ernst Nordtveit**

Ernst Nordtveit is a professor at the Faculty of Law, University of Bergen, Norway. He took his law degree at the University of Bergen in 1980 and his doctoral degree at the same university in 1990. He became a professor in 1993. He was Vice Dean for research from 1996–98 and Dean from 1999–2009. He wrote his doctoral thesis on joint ventures in the petroleum industry and has worked with natural resource law, especially petroleum law, environmental law and energy law in general. He has been one of the organizers of an annual symposium on petroleum law in Bergen since 1995. He has been teaching property law, company law, petroleum law and energy law in Bergen and has also taught courses in energy law in Australia (Bond University), New Zealand (University of Auckland), Russia (Northern (Arctic) Federal University, Arkhangelsk) and Belgium (University of Ghent).

Professor Nordtveit has been a member of numerous Norwegian national committees to prepare legislation and also a member of the committee appointed to evaluate Norwegian higher education. He was a member of the 'Norwegian Petroleum Price Board' from 1996–2006 and is currently the head of the Norwegian 'Conflict Resolution Body for the Upstream Gas Pipeline Network', from 2012, in both cases appointed by the King in Counsel. Professor Nordtveit has been engaged as an arbitrator in several Norwegian petroleum cases.

**John Paterson**

John Paterson is Professor of Law at the University of Aberdeen, Scotland, UK and Co-Director of the Centre for Energy Law. He is a co-founder of the LLM in Oil and Gas Law and also helps to run the North Sea Energy Law Programme, which brings together the universities of Aberdeen, Copenhagen, Groningen and Oslo to offer specialist education to energy lawyers in practice. He co-edits Gordon, Paterson and Usenmez's *Oil and Gas Law: Current Practice and Emerging Trends*, now in its second edition, and has researched and written on regulation in the energy industries for the past two decades. He has been involved in a number of international projects both in research and teaching. He has provided specialist training in oil and gas law and corporate governance for the corporate sector, expert advice in international arbitration, and consultancy to international organizations.

**Eduardo G. Pereira**

Dr Pereira has been active in the oil and gas industry for several years and is an international expert on joint operating agreements. His experience in this area – both academic and practical – is extensive. Dr Pereira is a managing director of STR Holdings. He played a key role in assisting Petra Energia to become Brazil's leading onshore oil and gas company, and he is now assisting the STR group to become a leading onshore player in Africa. He has practical experience in over 40 jurisdictions covering America, Europe, Africa and Asia. He concluded his doctoral thesis on oil and gas joint ventures at the University of Aberdeen.

Dr Pereira is the author and editor of several leading oil and gas textbooks: *Brazilian Upstream Oil and Gas: A Practical Guide to the Law and Regulation* (Globe Law & Business (GLB), 2012), *Joint Operating Agreements: Controlling Risks to the Non-Operator* (GLB, 2013), *Upstream Law and Regulation: A Global Guide* (GLB, 2013) and *Joint Operating Agreements: Mitigating Operational and Contractual Risks in Exclusive Operation* (GLB, 2013). He also has several books due for publication, including *Encyclopaedia of Oil and Gas Law: Volume 1* (GLB, due March 2014), *Brazilian Midstream and Downstream Oil and Gas: A Practical Guide to the Law and Regulation* (GLB, due August 2014), *Encyclopaedia of Oil and Gas Law: Volume 2* (GLB, due December 2014), *African Oil and Gas Law and Regulation: A Practical*

*Guide* (GLB, due April 2015) and *Joint Operating Agreements: Challenges and Dynamics between NOC and IOC* (GLB, due August 2015).

Dr Pereira is also an adjunct professor of energy law at the University of Eastern Finland Law School and an associate editor of *OGEL*.

**Kristoffer Svendsen**

Kris graduated from Bond University, Australia, with an LLB in 2006 and an LLM in 2007. He also holds a two-year LLM in Russian oil and gas law from Moscow State Institute of International Relations (MGIMO-University). He previously worked for a Russian investment bank in Moscow and a non-governmental organization in Brussels on the European Union's foreign policy towards Russia. Kris has been a visiting researcher at the Scandinavian Institute of Maritime Law, Tulane Maritime Law Center (2012–13) and Duke University School of Law. He regularly guest lectures in Russia and the US and publishes academic articles globally.

Kris is currently a PhD candidate at the K.G. Jebsen Centre for the Law of the Sea, Faculty of Law, University of Tromsø. His PhD topic addresses liability from oil platform spills in the Barents Sea. He also maintains an Of Counsel position at a Moscow law firm.

Kris is admitted as a lawyer in New South Wales, Australia, an attorney-at-law in New York State and a legal practitioner in Russia. He is also a member of the NORDEM (Norwegian Resource Bank for Democracy and Human Rights) roster.

**Alex Wawryk**

Alex Wawryk received First Class degrees in Economics and Law and a PhD in Law from the University of Adelaide, Australia. She is currently a senior lecturer at the University of Adelaide Law School in contract law, mining and energy law and international energy law, and undertakes guest lecturing in international energy law at the Adelaide Campus of the University College of London. Alex also taught environmental law for a number of years at Adelaide University.

As well as conducting research into petroleum law, Alex's areas of research have focused on renewable energy law, including measures to encourage renewable energy generation, and wind energy, and mining and environmental law. In 2014 she was awarded a Barbara Kidman Research Fellowship to pursue her research into petroleum law and regulation. Alex has published articles in a range of journals, including the Oxford *Journal of Environmental Law*, the *Journal of Energy & Natural Resources Law*, the *Australian and New Zealand Maritime Law Journal*, the *University of New South Wales Law Journal*, the *Melbourne University Law Review*, the *Australasian Journal of Natural Resources*

*Law and Policy* and the *Environmental and Planning Law Journal*. She is an associate editor for renewable energy for *OGEL*, a specialist on-line database for oil, gas and energy law.

Alex is a barrister and solicitor of the Supreme Court of South Australia, and she has been a member of the South Australian Law Society's Planning, Environment and Local Government Committee and a committee member of the Environmental Defenders Office (SA). She is currently a member of the World Conservation Union's Specialist Group on Energy Law and a member of the Resources and Energy Law Association (AMPLA). She is a member of, and the Law School representative to, the cross-disciplinary Institute for Mining, Energy and Resources at the University of Adelaide and a fellow of the university's Indo-Pacific Governance Research Centre.

# Table of cases

## Australia



## Canada



## Japan

**United States of America**

## International

# Table of legislation

## Australia

## Brazil



## Canada

## Japan

## Nigeria

## Norway



## Russian Federation

## United Kingdom

## United States of America

## European Union

## Treaties, Conventions and other International Instruments

# PART I

# Principles of petroleum regulation

# 1. Petroleum regulation in an international context: The universality of petroleum regulation and the concept of *lex petrolea*

**Alex Wawryk**

> While energy and resources law is on the one hand national law – sometimes supplemented in the case of regional economic integration as in the European Community – the organisational and contractual practices are greatly affected by the specific technical and economic logistics of the industry. An international 'lex mercatoria' of energy and resources law has emerged. Commercial, financing and other contractual practices are often alike, if not identical, all around the globe; with sometimes only scant influence from the idiosyncrasies of national law. The frequent recourse to arbitration tends to enhance the international customary law aspect of energy and resources law. Contractual innovations leapfrog from country to country and over the barriers of different legal systems. Comparative law therefore needs to understand how industry practices and imperatives on the one hand, [and] national law constraints on the other, shape legal instruments and concepts.[1]

## 1. INTRODUCTION

Petroleum exploration and production activities present complex legal, technical, economic, financial, political and environmental problems.[2] Upstream petroleum activities require considerable financial resources and infrastructure, and involve a high degree of geological and commercial risk. Most of the world's oil reserves are in developing countries with

---

[1] Thomas Wälde and Alan Page, editorial, (1993) 11 *Journal of Energy & Natural Resources Law* 1, 4.

[2] Claude Duval, Honoré Le Leuch, André Pertuzio and Jacqueline Lang Weaver, with contributions by Owen L Anderson, R Doak Bishop and John P Bowman, *International Petroleum Exploration and Exploitation Agreements: Legal, Economic and Policy Aspects* (2nd ed, Barrows, New York, 2009) 6.

limited financial resources, infrastructure and technical capabilities, whose governments are unable to bear the risks of exploration and production and thus contract with international oil companies (IOCs) to undertake operations on their behalf.[3] Because these contracts are for very long terms, IOCs are exposed to sovereign risk, that is, the risk of political events that affect petroleum developments. A classic example is the risk of a change of government, which might lead to a change in the regulatory regime, the renegotiation of contracts, a unilateral change in the terms of a contract by the host government, or the expropriation of property in oil reserves.

In order to spread risk and obtain access to capital, oil companies engage in co-operative business relationships. Rather than risking many millions of dollars alone in one project, a company may invest smaller amounts in several projects through joint ventures or other types of industry arrangements with other companies. Major international oil companies work around the world, as do independent companies specializing in operations such as offshore drilling and cementing, regularly moving personnel and equipment from one country to another. Integrated operations may cross national boundaries.[4] Business transactions also take place across nations, and capital flows globally.

Although oil companies operate globally, petroleum exploration and production activities worldwide are regulated by national legal regimes. These usually consist of a petroleum law, regulations and, in many countries, a contract between the government and international oil company which undertakes upstream activities. There is no multilateral treaty which governs the exploration and production of onshore or offshore oil, placing particular legal obligations upon nation states regarding industry regulation. This is because energy supply and consumption, onshore resource exploitation and environmental protection have historically been viewed as matters for internal control by nation states, and not matters of international concern. Also, as most offshore oil exploration and production takes place on the continental shelf, over which states have sovereign rights to explore and exploit resources as an extension of their land territory, the regulation of offshore oil and gas has also been regarded as a matter of domestic concern.

One aspect of the international oil industry that is heavily regulated by treaty is the international legal regime for the protection of the marine

---

[3] Duval et al (n 2) 6.

[4] International Regulators Forum, Global Offshore Safety, 'IRF Terms of Reference', accessed 12 November 2012 at www.irfoffshoresafety.com/about/.

environment from pollution caused by the maritime transport of oil. This is no coincidence, as the environmental impacts of marine pollution are often transboundary and require international co-operation for effective regulation.[5] Also, environmental and human rights treaties place obligations upon the states that are party to them, thereby influencing the content of national laws which affect the oil and gas industry as well as other industries. Thus, a range of environmental and social issues are dealt with through national petroleum and/or separate environmental laws.

Despite this, it is still very much the case that the regulatory aspects of upstream petroleum activities are not the subject of treaty law. And yet, to characterize regulation of the international petroleum industry as a set of unrelated, standalone national petroleum regimes is to misrepresent the reality of the governance of this global industry. Although there is no treaty giving rise to common principles of law, there has been a process of internationalization of domestic principles and rules and ways of conducting business, which has led to the development and spread of many legal principles now common to petroleum arrangements around the world. This process of internationalization reflects the international nature of world oil markets and the nature of onshore and offshore oil exploration and exploitation as truly global industries.

The global nature of the industry and the internationalization of principles of national law have given rise to an increasing sense that the industry is governed to some extent by 'international' or 'transnational' petroleum law. This international or transnational petroleum 'law' covers a multitude of legal issues, which may include laws relating to exploration, production/generation, transportation, investment and financing, business and contractual arrangements, access to pipelines, royalties and taxation, trade, dispute resolution and environmental and safety issues, to name but a few. International petroleum law transcends legal boundaries

---

[5] Legal initiatives aimed at dealing with oil tanker accidents, including those regarding safety at sea, the construction standards of ships and liability and compensation for oil pollution from ships, are addressed by multilateral and regional treaties, the common principles of which are applied internationally through national laws. Other matters regulated by treaties include the decommissioning and disposal of energy installations such as offshore oil platforms at sea, the routine operational discharge of oil from tankers into the ocean and the discharge of oil and wastes from land-based installations. However, existing conventions governing marine pollution caused by oil generally do not extend to liability for, and restoration of, damage caused by oil pollution arising from exploration and production activities conducted from offshore installations.

and may encompass, for example, aspects of contracts, property, administrative law, taxation law, environmental law and competition law.[6]

It is crucial to recognize that this is not an easily identifiable international petroleum 'law' established by traditional sources of international law. Rather, this 'law' stems from a number of broad and disparate sources of principles and norms that regulate the allocation of rights and duties concerning the exploitation of petroleum resources between individuals, between individuals and governments, and between states. It encompasses a concept of international governance and regulation that is broader than that of traditional international law, being a conglomeration of, and complex interaction between, the rules of custom, treaties, national and regional laws, the principles of intergovernmental organizations (IGOs) and non-government organizations (NGOs) and codes of conduct and guidelines of industry associations, which together regulate the various facets of oil exploration, production, transport, consumption and trade.[7]

It is within this context that this chapter aims to examine the concept of *lex petrolea*. The phrase is most often used to refer to an idea that there is a specialized 'transnational petroleum law' that derives from the internationalization of business practice, usages and customs of the members of the international petroleum industry or community.[8] This '*lex*

---

[6] Adrian Bradbrook, 'Energy Law as an Academic Discipline' (1996) 14 *Journal of Energy & Natural Resources Law* 193, 211.

[7] Global governance has been defined as 'a continuing process through which conflicting or diverse interests may be accommodated and co-operative action may be taken. It includes formal institutions and regimes empowered to enforce compliance, as well as informal arrangements … There is no single model or form of global governance, nor is there a single structure or set of structures. It is a broad, dynamic, complex, process of interactive decision-making': Commission on Global Governance, *Our Global Neighbourhood* (Oxford University Press, Oxford, 1995), cited in Patricia Birnie, Alan Boyle and Catherine Redgwell, *International Law and the Environment* (3rd ed, Oxford University Press, Oxford, 2009) 43.

[8] R Doak Bishop, 'International Arbitration of Petroleum Disputes: The Development of a Lex Petrolea' (1998) 1 XXIII YB COM ARB 1131; Thomas CC Childs, 'Update on *Lex Petrolea*: The Continuing Development of Customary Law Relating to Oil and Gas Exploration and Production' (2011) 4(3) *Journal of World Energy Law and Business* 214; Kim Talus, Scott Looper and Steven Otillar, '*Lex Petrolea* and the Internationalization of Petroleum Agreements: Focus on Host Government Contracts' (2012) 5 *Journal of World Energy Law and Business* 181; A Timothy Martin, '*Lex Petrolea* in International Law' in Ronnie King (ed), *Dispute Resolution in the Energy Sector: A Practitioner's Handbook* (Globe Law and Business, London, 2012); Alfredo De Jesús O, *The*

*petrolea*' is a type of *lex mercatoria*, or international 'law of merchants'; the principles and norms, and their legitimacy as law, are derived from their use and acceptance by the members of the international petroleum industry, to serve its own needs, separately and autonomously from principles of customary international law derived from state practice.[9] There is a focus on the internationalization of common contractual principles and practices in relation to the contracts between oil companies engaged in upstream activities, and the contracts between host governments (and/or their national oil companies) and international oil companies, reinforced by the decisions (where published) of international dispute resolution bodies in relation to contractual disputes.

There is, however, no generally accepted definition and understanding of the scope, jurisprudential basis for and sources and content of *lex petrolea*. *Lex petrolea* may also be sourced from the common norms and principles of petroleum law contained in domestic legal systems that have come to be 'internationalized', that is, applied worldwide by different countries, even though there is no treaty binding state parties to apply these principles of law. It also may be derived from the principles enumerated in traditional sources of international law, such as environmental and human rights norms contained in treaties, and customary international law; from so-called soft law, including treaties expressed in non-mandatory language, and the codes, guidelines, resolutions and standards of IGOs; and from the guidelines and standards of international petroleum industry associations, NGOs and international lending institutions, which regulate behaviour in the energy industries and are of key importance in the governance of the international petroleum industry.

The structure of this chapter is as follows. In order to provide the necessary background on the development of *lex petrolea*, section 2 briefly describes the internationalization of the petroleum industry over the course of the 20th century and demonstrates how common principles of contractual and commercial law of specific application to the petroleum industry, contained in contracts between companies and common clauses in host government contracts, have spread globally. It also explains the importance that international dispute resolution in the international petroleum industry plays as a source of *lex petrolea*, in particular international commercial arbitration and investment arbitration.

---

*Prodigious Story of the* Lex Petrolea *and the Rhinoceros: Philosophical Aspects of the Transnational Legal Order of the Petroleum Society*, TPLI Series on Transnational Petroleum Law, Vol I, no 1 (2012), accessed 1 November 2013 at www.lexpetrolea.org/#!publications/c1ger.

[9] De Jesús O (n 8).

Section 3 contains the main discussion of the scope and meaning of *lex petrolea* – the different conceptualizations and meanings attributed to *lex petrolea* by scholars and practitioners, the jurisprudential basis for the doctrine and the very real challenges posed in recognizing the existence of a *lex petrolea*. Section 4 concludes.

## 2. THE INTERNATIONAL PETROLEUM INDUSTRY AND THE EVOLUTION OF PETROLEUM AGREEMENTS

To understand the process of internationalization, and thus claims regarding the existence of a *lex petrolea*, which will be explored in some depth in section 3, it helps to have an understanding of the history and structure of the international petroleum industry. The structure of the industry, the role and conduct of international oil companies and host governments, and the interaction between the two, has affected the development of norms and principles contained in international petroleum agreements. While the nature, type and content of international petroleum agreements have changed from the infancy of the international petroleum industry in the early 20th century, there are common types of agreements and clauses that are used across the globe.

### Brief History of the International Petroleum Industry

The origin of the petroleum industry is generally traced to the United States in 1859.[10] The early discovery of oil in the US fostered the growth of the domestic oil industry in that country well before oil was discovered in other countries. Discoveries in Russia (1873), Iran (1908), Mexico (1910) and Venezuela (1922) created important oil-producing centres outside the US, with oil discovered in the Middle East later, in the 1930s.

---

[10] For an account of the early years of the American petroleum industry, see D Yergin, *The Prize: The Epic Quest for Oil, Money and Power* (Free Press, New York, 2008); R Anderson, *Fundamentals of the Petroleum Industry* (University of Oklahoma Press, Norman, 1984); J McLean and R Haigh, *The Growth of Integrated Oil Companies* (Harvard University, Boston, 1954); A Sampson, *The Seven Sisters: The Great Oil Companies and the World They Made* (Hodder and Stoughton, London, 1975).

Before World War I, the international petroleum industry was still in its infancy and primarily involved US, British and Dutch interests.[11] These countries, as part of their colonial empires, were involved in the search for oil through their national oil companies (NOCs). By 1938, the international oil industry was largely controlled by seven companies, the 'majors' or 'seven sisters'. These were large, vertically integrated oil companies that dominated the world's oil reserves, assuming near-complete control over oil exploration, production and transportation, and building complex networks of refining and marketing facilities to sell their products on a global scale.[12] The six majors were: the US companies Standard Oil of New Jersey (now merged in ExxonMobil), Standard Oil of California (Chevron), Gulf (now merged into Chevron) and Texaco (now merged into Chevron); the UK company British Petroleum (BP); and the Dutch company Royal Dutch Shell. CFP of France (now Total) could have been added as an eighth sister.[13]

Until World War II, the dominance of the seven sisters over non-US sources of oil was reflected in the petroleum arrangements the companies made with host countries. In the US, petroleum was first found on lands belonging to private landowners, and a body of contract law developed to expedite exploration and production on private land. When the US oil companies extended their activities across the globe over the course of the 20th century, they carried with them their customary forms of contracts, business organization and jurisprudential concepts.[14] Based on the US domestic oil and gas lease, the key type of petroleum arrangement for the exploitation of oil from non-US sources in the early 20th century was the concession contract.[15]

---

[11] K Hossain, *Law and Policy in Petroleum Development* (Frances Pinter (Publishers), Nichols Publishing Company, London, New York, 1979); J Hartshorn, *Oil Companies and Governments: An Account of the International Oil Industry in its Political Environment* (Faber & Faber, London, 1967).

[12] Duval et al (n 2) 43.

[13] For the history of the seven majors, see Anderson (n 10); McLean and Haigh (n 10); Yergin (n 10); G Barrows, *International Petroleum Industry,* Vols I and II (International Petroleum Institute Inc, New York, 1967); Hartshorn (n 11).

[14] Ibid 42.

[15] Duval et al (n 2). For a history of international petroleum agreements, see Hossain (n 11); Z Gao, *International Petroleum Contracts, Current Trends and New Directions* (Graham & Tortman, London, 1994); R Mikesel, *Petroleum Company Operations and Agreements in the Developing Countries* (John Hopkins University Press, Washington D.C., 1984).

The early concessions were extremely advantageous to the companies having the right to develop petroleum.[16] Common features were: exclusive access for the IOC over massive areas for very long terms; no obligations on the company to drill on the lands granted or to release the territory if exploration and drilling were not undertaken; and no right for the host country to participate in managerial decisions. The only financial benefits received by the host country were a fixed sum paid to the host country upon signing the contract and the right to royalty, usually calculated at a flat rate per tonne or barrel of oil produced. Many early concessions exempted the oil company from obligations to pay taxes such as income tax.[17]

As the major oil companies of the US and Europe explored and produced in the Middle East, Africa and Asia, the principles of the concession contract spread globally.[18] The first concession agreements were largely *ad hoc*, but over time, as the number of concessions increased and exploration became more widespread, a more systematic way of dealing with petroleum agreements developed, with common features. Changes in political power and the structure of the industry after World War II saw the renegotiation of concession contracts (into the 'modern concession contract') and the development of new types of petroleum arrangements (including the production sharing contract and service agreements/risk service agreements) on terms much more favourable to host countries. These agreements have now spread internationally and are common across the globe.

From World War II the Middle East became the centre of world oil production. Major discoveries in Africa occurred from the 1950s onwards, particularly in Libya, Nigeria and Algeria, and these countries have become large producers of oil, with the international petroleum industry viewed as truly global since this time. Crucial events that contributed to the renegotiation of concessions and the introduction of new types of agreement included: the entry of the independents into the international petroleum industry in the 1950s, these being companies that typically invested only in upstream operations into the international oil

---

[16] See, for example, the agreement between the government of His Majesty, the Shah of Persia and William Knox D'Arcy, Appendix to Annex 1419c, in the League of Nations Official Journal, 1932, pp. 2305–7, cited in Ernest E Smith, John S Dzienkowski, Owen L Anderson, John S Lowe, Bruce M Kramer and Jacqueline L Weaver, *International Petroleum Transactions* (3rd ed, Rocky Mountain Mineral Law Foundation, Westminster, Colorado, 2010) 429.

[17] Duval et al (n 2) 43, 59–62.

[18] Ibid 44.

industry; the nationalization of oil industries in the major exporting countries in the 1970s (along with the expropriation of host country oil reserves); and the formation of the Organization of the Petroleum Exporting Countries (OPEC) in the 1970s. In the 1980s, following the surge in prices as a result of the Iran–Iraq war in 1979, investment in oil exploration and new technology brought the previously uncompetitive offshore oil fields of the North Sea and the Gulf of Mexico into production. Currently, the 'shale revolution' is redefining economic and political analysis of the industry.

Today, when reference is made to the 'international petroleum industry' or 'international oil industry', this includes a number of actors and institutions operating on the world stage. First, the largest oil and gas producing companies in the world include many NOCs, which now hold approximately 90 per cent of the world's oil reserves.[19] These corporations were either created, or had their roles significantly expanded, by the nationalization of domestic foreign-owned assets in the major oil-producing and exporting countries in the 1970s.[20] In the country where the industry is nationalized, NOCs may carry out upstream operations or they may act as 'partners' of IOCs in joint projects, where the state decides to participate by acquiring an interest in the IOC ventures.

Many NOCs operate as an extension of the state, pursuing development and other objectives of the government. Others, such as Statoil (Norway) and Petrobras (Brazil), operate independently, with strategic and operational autonomy; they may support their country's goals but are primarily profit-driven. In the 1970s, most NOCs operated solely within the borders of their own country (distinguishing them from state-owned independent oil companies). However, in the last decade, many of the NOCs that once operated only domestically adopted a strategy to diversify internationally into upstream investments abroad.[21] They are now competing with the supermajors on extremely large projects outside

---

[19] Silvana Tordo with Brandon S Tracy and Noora Arfaa, 'National Oil Companies and Value Creation', World Bank Working Paper No. 218, The World Bank, Washington DC (2011) xi.

[20] Z Mikdashi, *Transnational Oil: Issues, Policies and Perspectives* (Frances Pinter (Publishers), London, 1986) 7. The USSR, following the Russian revolution, nationalized the industry after 1917; Bolivia in 1937; Mexico, following the Mexican revolution, expropriated domestic oil assets of the IOCs and created its own national oil company Pemex in 1938; and Iran nationalized its industry in 1951, following conflict with Britain.

[21] Duval et al (n 2) 52; Saud M Al-Fattah, 'The Role of National and International Oil Companies in the Petroleum Industry', USAEE Working Paper No. 13-137, January 2013, SSRN-id2299878.

their home country, investing in in-house research and development and developing in-house service companies, looking to buy smaller exploration and production (E&P) companies and also targeting specialized shale drilling companies in the US for investment, in order to obtain access to technology for shale oil and gas drilling which can be transferred to their home countries.[22]

Secondly, the international petroleum industry includes the large, vertically integrated privately owned international oil corporations such as BP, Exxon Mobil, Chevron and Shell. These IOCs became 'supermajors' after a wave of mergers and acquisitions conducted since the 1990s, by which the majors incorporated a number of the competing international oil companies of the 20th century. This group comprises former national oil companies that have been privatized since the 1990s, including not only BP and Shell, but also Total (France), ENI (Italy), Repsol YPF (Spain's Repsol and Argentina's YPF) and Petro-Canada. The new Russian corporations formed from the large oil and gas companies of the former Soviet Union, such as Lukoil, Rosneft and Gazprom, are often referred to as supermajors, despite varying degrees of state ownership.[23]

Thirdly, the actors in the international petroleum industry include the 'independents'; the smaller oil companies that entered the international petroleum industry and began to challenge the dominance of the seven sisters from the 1940s and 1950s.[24] These include companies such as Andarko and Apache, which engage in upstream exploration and production but have no downstream operations,[25] and highly specialized service companies and independent contractors offering support services such as

---

[22] 'Supermajordämmerng: The Day of the Huge Integrated Oil Company is Drawing to a Close', *The Economist*, 3 August 2013, accessed 3 August 2013 at www.economist.com/news/briefing/21582522-day-huge-integrated-international-oil-company-drawing.

[23] Duval et al (n 2) 52; Al-Fattah (n 21) 4.

[24] In 1945, only six US companies other than the five majors had any foreign exploration interests. Between 1953 and 1972, some 350 privately owned oil companies had entered the international industry or increased their activity: Anderson (n 10) 53; Hossain (n 11) 8.

[25] In the domestic industry, the term 'independents' refers to non-integrated companies whose activities are largely concentrated in one function of the petroleum industry, such as crude oil production, transport, refining or marketing. Independents comprise a wide and diverse spectrum of operators. Mikdashi (n 20) 9.

offshore rig operation, drilling and cementing, for example Schlumberger, Halliburton, BakerHughes and Weatherford.[26]

Furthermore, a myriad of industry associations can be seen as part of this international community, of which just a few are the International Association of Oil and Gas Producers (OGP), the International Association of Drilling Contractors (IADC) and the International Association of Petroleum Negotiators (IAPN). This is from a purely commercial, industry perspective. If we take an extended and holistic approach to the social/community 'licence to operate' we might include other stakeholders with interests such as local communities, indigenous peoples and environmental groups as actors within the international petroleum industry.

## Contracts between Oil Companies

Because of the degree of risk involved in exploration and protection, and the long timeframes required for investments, international oil companies frequently join together in co-operative business arrangements in order to spread risk and obtain access to capital. The industry has come to be characterized by the use of numerous standardized model contracts which govern the commercial relations between international oil companies worldwide.[27]

Modern model contracts originated from the contractual documents that companies have used independently for over 100 years.[28] When the international oil companies extended their activities globally over the course of the 20th century, they carried with them their customary forms and sets of contracts incorporating their own individually preferred precedents, which would be altered after one-on-one negotiations with another company. This, however, took up significant time and resources, because of the vast sums of money, substantial risk and massive liabilities, and complex and difficult issues involved. With some contracts taking years to draft, negotiate and sign, the whole process became very time-consuming and inefficient.[29]

Many model contracts or model agreements have now been developed by national and international petroleum industry organizations to improve

---

[26] Al-Fattah (n 21) 7–9.

[27] A Timothy Martin and J Jay Park, 'Global Petroleum Industry Model Contracts Revisited: Higher, Faster, Stronger' (2010) 3 *Journal of World Energy Law and Business* 4, 4.

[28] Ibid 5.

[29] Ibid 5–6.

efficiency and save time.[30] The continuing development, widespread acceptance and use of model petroleum contracts drafted by national and international organizations such as the Association of International Petroleum Negotiators (AIPN), which span a whole range of commercial issues and incorporate common clauses and principles, including (to name but a few) model farmout agreements, unitization agreements, joint international operating agreements, lifting agreements and confidentiality agreements, has contributed to a global standardization of upstream petroleum contracts.[31]

The existence of common clauses worldwide, in model industry contracts governing commercial transactions between oil companies, is a fundamental aspect of the internationalization of petroleum law. Furthermore, this standardization of the principles of law applicable at an international level to the global petroleum industry is integral to the argument that there is in existence a nascent *lex petrolea* – a specific legal regime, or body of international norms – which instructs or regulates the international petroleum industry.[32]

## Contracts between Host Governments and International Oil Companies

Unlike contracts between oil companies, there are no generally accepted model petroleum agreements between international oil companies and the governments of oil-producing countries, as these tend to be drafted to meet the specific ends and goals of the country involved. However, across the world today, we can observe the existence of common types of host government contracts, namely the production sharing contract, the modern concession contract and service agreements. Furthermore, these contracts contain similar provisions, clauses, structures and approaches, regardless of the identity of the host state.[33]

A range of terms will typically be included in a modern host government contract to reflect the interests of the parties. The aims of host governments now include, in addition to the historical and classic aim of seeking revenue: the greater involvement of their national oil

---

30 For a discussion of the advantages and disadvantages of model contracts, see ibid 5–6.

31 A Timothy Martin, 'Model Contracts; A Survey of the Global Petroleum Industry' (2004) 22 *Journal of Energy & Natural Resources Law* 281; Martin and Park (n 27) 5–6; Talus, Looper and Otillar (n 8) 184.

32 Bishop (n 8); Childs (n 8).

33 Talus, Looper and Otillar (n 8) 185, 192.

companies in managerial decisions and control; the development of their own technology and trained labour force; and the achievement of sustainable development. The interests of IOCs have remained reasonably constant, and include ensuring the following: access to petroleum resources; an acceptable level of profitability in upstream operations; and the most investment-friendly terms in contracts, such as lower taxation, and regulatory and legal certainty in the face of sovereign risk.[34] Modern host government contracts will generally contain clauses addressing the following matters, among others:[35]

- the ownership of petroleum produced;
- the area for, and term of, development;
- host country control over the rate and extent of exploration and production;
- an option for the host country to participate in upstream operations;
- the right of the host country to an equity percentage of the production;
- access to oil for the domestic or export market;
- the possibility of state participation in midstream or downstream operations;
- the generation of revenue through a variety of mechanisms, such as royalty, rentals and taxes;
- social benefits such as training and employment of nationals and a preference for purchasing local goods and services for use in petroleum operations;
- environmental protection;
- stabilization clauses, which provide certainty by limiting the right of a host country to unilaterally modify the petroleum agreement;
- adaptation clauses, to allow the host country and IOC to mutually renegotiate the agreement; and
- dispute resolution clauses.

The basic features of various types of host government contracts are shared to such an extent that it has been estimated that at least 80 per cent of the contents of these agreements consist of the same clauses, 'irrespective of their label'.[36]

---

[34] Duval et al (n 2) 31–2.
[35] Ibid.
[36] Ibid 54.

> [W]hile each contract category provides for different concepts and different levels of control, the modern contracts provide very similar end results and the choice between various contractual regimes has much to do with historical developments in the country. Furthermore, the content of various Host Government Contracts is often similar because the contract parties 'borrow terms and conditions' from other pre-existing contracts. Specific contract language tends to vary only when dependent on country- or time-specific external factors such as oil prices, global economic situation and the host government's internal political forces.[37]

The global spread of common national principles of petroleum law occurred as host country governments looked to other national legal regimes to find ways to better protect their interests. For example, the existence of one entity, OPEC, representing the interests of similarly situated countries, helped efforts to renegotiate the original concessions in the 1970s through the sharing of information among its members.[38] More recently, the use of common types of host government contracts has become increasingly globalized through: (a) their application by actors such as the World Bank, to foster development of petroleum resources in developing countries, and a relatively small group of private lawyers with expertise advising host governments internationally; (b) by the desire of governments to learn from previous experience, in order to both attract foreign investment and to obtain benefits and opportunities offered in other jurisdictions, while protecting their interests in matters such as environmental protection and fiscal terms; and (c) by growing international education and awareness in this area.[39]

Although the particular degree to which individual contracts will favour host governments or oil companies depends on prevailing market and political conditions, common principles have come to be incorporated in host government contracts across the globe, 'not so much a result of a conscious choice to include internationally accepted contract provisions', but as the result of a process of internationalization of principles of law, whereby, 'in order to stimulate foreign direct investment, host governments must establish terms that reasonably assure investors that their investments will earn the profits for which they bargain'.[40] Again, it is the standardization of the principles of law applicable at an international level – in this case to agreements between international oil companies and host governments – which is integral to the argument that

---

37 Talus, Looper and Otillar (n 8) 188.
38 Ibid 192.
39 Ibid 191.
40 Ibid 192.

there is in existence a nascent *lex petrolea*, or 'body of international norms', which instructs or regulates the international petroleum industry.[41]

**The Role of International Arbitration in Contractual and Investment Disputes**

Because the international petroleum industry invests in very large, complex, highly capital-intensive projects with long lifespans of 30 years or more, the need for legal and political stability is 'particularly acute'.[42] However, the long timeframes of oil and gas projects mean that 'circumstances, economics, governments and parties invariably change' over the lifetime of a project, leading to a significant risk of disputes.[43] International arbitration is the major type of dispute resolution procedure for the international petroleum industry. It allows companies to use a neutral forum,[44] and overcomes the concern that foreign court judgments may not be enforced at home or abroad, as countries that are parties to arbitration treaties are obliged to enact laws to ensure international arbitral awards are recognized and enforced domestically.[45] The positing by scholars and practitioners of the existence of a *lex petrolea* has gone hand in hand with the determination of disputes through international arbitration procedures.

There are two major types of dispute that arise in the international petroleum industry. The first of these are disputes between an international oil company and the host government, called investor-state or state investment disputes. Investor-state disputes arise when a government significantly changes the conditions or terms of a host government contract, or nationalizes or expropriates an investment.[46] The expropriation by the state of the rights of private oil companies to oil reserves is

---

41 Bishop (n 8); Childs (n 8).

42 Yulia Selinova, 'The Energy Charter and the International Energy Governance' (2012) 3 *European Yearbook of International Economic Law* 307, 315.

43 Paul Blyschak, 'Arbitrating Overseas Oil and Gas Disputes: Breaches of Contract versus Breaches of Treaty' (2010) 27 *Journal of International Arbitration* 579, 585; A Timothy Martin, 'Dispute Resolution in the International Energy Sector: An Overview' (2011) 4 *Journal of World Energy Law and Business* 332, 332. For the distinction between arbitrating breaches of petroleum contracts and breaches of investment treaties, see Blyschak.

44 Martin (n 43) 334.

45 The major treaty is the Convention on the Recognition and Enforcement of Foreign Arbitral Awards (New York), 330 UNTS 38 (entered into force 1959).

46 Martin (n 43) 334.

a major issue of sovereign risk. Historically, the process of nationalization, particularly where it involved the expropriation of the IOC's assets, has been very contentious. Expropriation without compensation has been a major topic of international arbitration proceedings, and indeed, some of the earliest cases adjudicated by international arbitral tribunals in the 20th century concerned the expropriation of international oil companies' property interests by host governments.[47]

Modern international petroleum contracts and international investment treaties (including numerous bilateral investment treaties, the International Convention for the Settlement of Investment Disputes, and the Energy Charter Treaty) contain various provisions that seek to achieve stability by reducing political and regulatory risk, in order to protect investments and facilitate investment flows and trade.[48] Disputes between investors in the international oil industry and host governments regarding breaches of international petroleum contracts and investment treaty provisions continue to form a key part of the case load of international arbitral panels. The international petroleum industry is a major global investor, and the energy sector forms the highest percentage of international commercial and state investment disputes in the world.[49] While many awards pre-1998 concern expropriation through nationalization of the industry, since 1998, most expropriation cases concern claims relating to the host state's fiscal regime, that is, through 'creeping expropriation', and a range of other contractual matters are now also determined by arbitration.[50]

The second major type of dispute to arise in the international oil industry is international commercial disputes between oil companies. These include, first, disputes between joint venture participants regarding contracts such as joint operating agreements, unitization agreements, farmout agreements, sale and purchase agreements and confidentiality

---

[47] Markus Krajewski, 'The Impact of International Investment Agreements on Energy Regulation' (2012) 3 *European Yearbook of International Economic Law* 343, 343–4.

[48] 1965 Convention on the Settlement of Investment Disputes between States and Nationals of Other States [1991] ATS 23, 17 UST 1270, TIAS 6090, 575 UNTS 159, 4 ILM 532 (1965) (in force 14 October 1966); Energy Charter Treaty [1994] ATSD 4217; 2080 UNTS 95; UKTS 2000/78 (Cm 4761), Cm 2952; 34 ILM 360 (1995) (in force 16 April 1998). Thomas Wälde, 'International Energy Investment' (1996) 17 *Energy Law Journal* 191–215.

[49] Martin (n 43) 332.

[50] Martin (n 8) 99.

agreements, and, secondly, disputes between operators and service contractors under agreements such as drilling and well service agreements, seismic contracts, construction contracts and equipment and facilities contracts.[51]

It has been stated that the determination of investor-state disputes and international commercial disputes in international arbitral tribunals is fundamental to claims that there is a *lex petrolea*.[52] When determining a contractual dispute, the arbitrator(s) must apply the substantive law identified in the parties' arbitral agreement as the law governing the contract. The agreement may specify that the national laws of a particular country are to govern the contract. However, some international contracts do not expressly provide for a single domestic legal regime to govern the transaction. Instead, they may require the application of principles common to both the international oil company and host government regimes, and/or the agreement may empower the arbitrator(s) to apply 'general principles of international law' to the dispute. Also, where a contract does not specify the substantive law to be applied (and sometimes even when it does), the rules of the arbitral institution may require the arbitrator(s) to use principles of international law and of trade usage as a supplementary source of law. The rules may make reference to the 'guidelines, customs and principles of international commercial law', 'generally accepted commercial usage and practices', 'generally accepted principles of international commercial law', 'principles of trade usage', 'usages of international commerce' or some such variation.

Arbitrators seek to rely on credible sources of law in determining their awards, including principles from traditional sources of international law, and national systems of law. By their application to disputes, international arbitral tribunals confirm certain commercial practices have authority as custom; while the international petroleum 'community' accepts the norms/principles articulated by arbitral tribunals. Thus, as arbitral institutions identify and apply principles of international law to make their decisions, including principles of commercial or trade usage, they also legitimize these principles as standardized principles of international law of particular application to the international petroleum industry. This then becomes part of the *lex petrolea* – the 'body of international norms' which instructs or regulates the international petroleum industry.[53]

---

51 Martin (n 43) 335.
52 Bishop (n 8); Childs (n 8).
53 Ibid.

## 3. THE CONCEPT OF *LEX PETROLEA*

Although there is broad agreement that there are common or internationalized norms, rules, standards and best practices that apply to or exist in the international petroleum industry, regardless of geographical location, there is no one definition and accepted understanding of the existence, scope, jurisprudential basis for and content of *lex petrolea*.[54] The phrase *lex petrolea* is used narrowly or widely to encompass some or all of a number of sources of law or regulation. While some international scholars will not recognize these principles as sources of international *law*, most scholars who accept the existence of a *lex petrolea* or 'transnational petroleum law' also accept the idea that international law and/or governance in the transnational or international petroleum industry is comprised of more than the traditional sources of international law, such as treaty and customary international law. In particular, *lex petrolea* is often claimed to be a particular branch of a general universal *lex mercatoria*, or 'law of merchants'.

### *Lex Mercatoria*

While there is no single accepted universal definition of *lex mercatoria*, one is that it is a 'set of general principles, and customary rules spontaneously referred to or elaborated in the framework of international trade, without reference to a particular national system of law'.[55] Some jurists have claimed *lex mercatoria* is a third legal order, which is neither national nor international law, but a mixture with the characteristics of both. Recognition of the existence of a *lex mercatoria* stems from the conviction of many jurists that neither municipal law nor international

---

[54] Ibid; Talus, Looper and Otillar (n 8); Martin (n 8); De Jesús O (n 8); Carmen Otero Garciá-Castrillón, 'Reflections on the Law Applicable to International Oil Contracts' (2013) 6 *Journal of World Energy Law and Business* 129; Mary B Ayad, 'The Vienna Convention as Authority for the Use of Precedent as Customary Practice in International Arbitrations of Oil Concessions and Investment Disputes in North Africa and the Gulf Arab States; or a *Lex Mercatoria* for a *Lex Petrolea*' (2013) *The Journal of World Investment and Trade* 918.

[55] Bernard Goldman, 'The Applicable Law: General Principles of Law: The Lex Mercatoria' in Julian Lew (ed), *Contemporary Problems in International Arbitration* (Martinus Nijhoff Publishers, Netherlands, 1987) 113, 116.

law are appropriate or suitable for dealing with international commercial disputes when parties from different countries are involved.[56]

Although the basic premise sounds simple, it is a difficult concept to grasp jurisprudentially. There is disagreement among scholars on the definition, sources, scope and parameters, and purposes of *lex mercatoria*, as well as to its existence as a body of law, and whether it exists or qualifies as a law, legal system or legal order. There is also a lack of consistency in the use of terminology. It is beyond the scope of this chapter to discuss the jurisprudence associated with the idea of *lex mercatoria*, but there are some key ideas that can be identified for the purposes of this chapter.

First, a fundamental characteristic of *lex mercatoria* is that it arises 'spontaneously' from the trade usage and business practices of 'merchants'. It is not deposited or set down by a sovereign. *Lex mercatoria* is designed to serve the international business community's aspirations, interests, needs and expectations, in particular the need for effectiveness of businesses that operate beyond or across national boundaries.[57] It is practical and transcends national borders and jurisdictions.[58] The customary norms of international commerce – that is, the body of law respected and followed by merchants – are the basis for the *lex mercatoria*.

However, there is no general agreement on the sources of *lex mercatoria*,[59] for example, whether it may be derived only from the general principles of law or trade usages, or whether it may also be derived from the rules of international organizations, standard form contracts and the reporting of arbitral awards. There is some disagreement as to whether *lex mercatoria* can be drawn from the law deposited by states, with a number of jurists accepting it may be drawn from principles of public international law and uniform laws, if incorporated into the *lex mercatoria* through frequency of use and/or adoption in contracts and arbitral awards, and provided the principles are accepted as custom, serving the needs of the international business community. A corollary of this disagreement as to the sources of *lex mercatoria* is that there is also disagreement as to the content of this law.[60]

---

56 Abul Maniruzzaman, 'The *Lex Mercatoria* and International Contracts: A Challenge for International Commercial Arbitration?' (1999) 14 *American University International Law Review* 657, 658.

57 Ibid 713–14.

58 Ibid 669; Ayad (n 54) 925; De Jesús O (n 8).

59 For a discussion on this point, see Maniruzzaman (n 56), 665–7, 672–3.

60 However, it has been recognized that the development of the UNIDROIT Principles of International Commercial Contracts, and the Principles of European

One of the great debates is whether *lex mercatoria* is a separate legal order or legal system, or whether it is limited to comprising a set of certain norms or principles or rules. This chapter cannot do justice to that debate, but some jurists assert that it is a separate legal order; others do not, rather recognizing its existence (if at all) as a set of norms.[61] Some assert that it is autonomous to and independent from national and international legal orders, although it seems difficult to accept this in a practical sense as, at the least, national legal systems are necessary for the enforcement of arbitral awards.[62] Without the national enforcement of international arbitral awards, international arbitration would lose its effectiveness as a form of binding dispute resolution mechanism, and a fundamental source of the *lex mercatoria* would be lost.

Indeed, despite the disagreements over the sources and content of *lex mercatoria*, the arbitral tribunal is universally considered an 'indispensable instrument' for the creation and development of *lex mercatoria*.[63] Where arbitral tribunals are empowered to apply the 'principles of international law', or principles of law 'common' to the home jurisdiction of each party, or 'the generally accepted customs and usages of international trade', either because the arbitral agreement or rules of an arbitral institution specifically requires it, the international arbitrator must identify and confirm the principles/usages that apply. If common rules are not available, the arbitrator selects and applies the rule that appears most appropriate and equitable. The international arbitral tribunals confirm certain commercial practices have authority as custom; while the international community then accepts the norms/principles articulated by arbitral tribunals as authoritative.

Those who do not accept that *lex mercatoria* is a separate legal order, or complete supranational body of rules, that exists autonomously and independently of national or international legal systems, may recognize it as an additional or supplementary option in the search for the law applicable to governing contractual disputes.[64] Under this approach, *lex mercatoria* is recognized as an additional or supplementary set of norms derived from international business practice and usage that can be translated into 'hard' law through incorporation in contracts, legislation or other posited national sources of law, or in traditional instruments of

---

Contract Law, have collated certain principles considered suitable for cross-border transactions: Maniruzzaman (n 56) 673–8.

[61] For a discussion on this point, see Maniruzzaman (n 56) 697–705.

[62] For a discussion on this point, see ibid 671.

[63] Ibid 693.

[64] Ibid 715; Ayad (n 54) 928; Garciá-Castrillón (n 54) 137.

international law such as treaties and resolutions, and be interpreted, developed and enforced through decisions of arbitral tribunals and courts.

### *Lex Petrolea*

As a part of, branch or type of *lex mercatoria*, these difficulties extend to the *lex petrolea*. What appears to be a simple idea – that there is a *lex petrolea*, or set of customary rules valid for the international petroleum industry based on the needs, practices and usages of its members – is a complex concept, that has not been rigorously defined, is used inconsistently and has uncertain scope and parameters. The possibilities suggested to date are that:

- *lex petrolea* is comprised of norms of traditional customary international law between nation states, of specific relevance to the international petroleum industry;
- *lex petrolea* is comprised of the principles of law applied by international arbitral tribunals, of specific relevance to the international petroleum industry;
- the common principles of law and/or clauses found in host government contracts and in model industry contracts between oil companies, as a part of business usage or custom, are a part of, comprise norms of, and are also a source of, *lex petrolea*;
- that the norms, principles and standards of industry associations, NGOs and IGOs are also a part of, comprise norms of, and are also a source of, *lex petrolea*, including principles concerning human rights, indigenous rights and environmental protection; and
- *lex petrolea* is an autonomous legal system, separate from national and international legal systems.

The idea that there may be a *lex petrolea*, comprised of norms of traditional customary international law of specific relevance to the international petroleum industry, arose in *Kuwait v AMINOIL*.[65] Part of this dispute concerned the assessment of compensation for the expropriation of Aminoil's property through the nationalization of the oil industry by the Kuwaiti government. The Kuwaiti government suggested that a 'series of transnational negotiations and agreements about compensation' for expropriations conducted in the Middle East and elsewhere in the

---

[65] *Government of the State of Kuwait v American Independent Oil Co (AMINOIL)*, Award of 24 May 1982, 21 ILM (1982) 976, Yearbook IX (1984) 71.

world during 1971–77 'had instituted a particular rule, of an international and customary character, specific to the oil industry'.[66] In these agreements, the assessment of compensation was 'not identical but had certain common features': it was 'very incomplete' and was determined by a certain valuation method, whereby compensation was limited to the net book value of the companies' redeemable assets. The Kuwaiti Government argued these precedents had 'generated a customary rule valid for the oil industry – a *lex petrolea* that was in some sort a particular branch of a general universal *lex mercatoria*'.[67] The Arbitral Tribunal rejected both the contention that such a rule existed, and even if it had, its application to the present case.

First, the Tribunal distinguished Aminoil's case from other negotiations. In Aminoil's case, all relations had been severed with the Kuwait government and compensation was to be determined by the Tribunal; in the cases forming the precedents, compensation for expropriation was determined through negotiations where there was an ongoing relationship between the international oil companies and governments. The overall results of the negotiations in other cases were complex. In addition to an indemnity, the companies often received other benefits through ongoing contracts of service, and long term supply. A crucial point of distinction was thus that in the other cases referred to by the Kuwaiti government, in addition to the compensation determined according to the net book value of assets, the international oil companies also either received or maintained a preferential relationship with the host government. While it would be 'difficult to express in figures the value in terms of money of these preferential arrangements', as they depended in each case on this structuring and policy of the former concessionaire companies, the advantages for major integrated oil companies were often considerable, unlike for Aminoil, an independent exploration company for whom all ties with the government had been severed.[68] Thus, even if there were such a rule, it did not apply in Aminoil's case on the facts.

Secondly, the Arbitral Tribunal queried the formulation of the suggested rule, stating:

> at the most it might be possible to go so far as to say that certain large transnational groups may have preferred compensation that had no relation to the value of their undertaking, if it was coupled with the preservation of good relations with the public authorities of the nationalizing State with, possibly,

---

66 Ibid 1035.
67 Ibid 1036.
68 Ibid 1036.

> resulting prospects for the future giving promise of greater worth than the compensation forgone.[69]

Thirdly, citing the *North Sea Continental Shelf Case*,[70] the tribunal stated that if reliance were to be placed on the precedents presented by the Kuwait government, then 'it would still be necessary for them to constitute expression of *opinio juris*' to be viewed as a customary rule of international law.[71] Here, the policy of OPEC, the 'crucial preoccupations of concessionaire companies to ensure the continued supply of petroleum products to consumers', and the 'passivity' of the oil importing states, led the transnational petroleum companies facing expropriations in the 1970s to 'accept de facto what the exporting countries demanded'.[72] While such acceptance was 'wise', it 'would be somewhat rash to suggest that it had been inspired by juridical consideration: the *opinio juris* seems a stranger to consents of that type'.[73] Therefore, even the practice as to net book value could not be viewed as a rule of international law because of the lack of 'mental element' – a belief that the practice is binding or obligatory as a matter of law.

Finally, the Tribunal found the consents of investors to the use of the net book value had been given 'under the pressure of very strong economic and political constraints'. While in the case of Aminoil there was no duress, 'the economic pressures that lay at the root of [the consents] had nothing to do with law, and do not enable them to be regarded as components of the formation of a general legal rule' that could be applicable in other cases.[74]

In rejecting the contention that a rule existed, the Arbitral Tribunal does not distinguish nor discuss the concept of *lex petrolea* as comprised of rules of traditional customary international law between nation states and *lex petrolea* as part of a *lex mercatoria*, which is the law of merchants. Rules of traditional customary international law between nation states require evidence of state practice and *opinio juris*, the belief that a practice is obligatory as a matter of law, as per the *North Sea Continental Shelf Case* cited by the Tribunal. In contrast, the concept

---

69 Ibid.

70 *North Sea Continental Shelf Case,* ICJ Rep (1969).

71 *Government of the State of Kuwait v American Independent Oil Co (AMINOIL),* Award of 24 May 1982, 21 ILM (1982) 976, Yearbook IX (1984) 71, 1036.

72 Ibid.

73 Ibid.

74 Ibid 1037.

behind *lex mercatoria* is that it is comprised of the practices and customs accepted by the international business community, and there is no strict requirement of *opinio juris*. When it looks for an expression of *opinio juris* and rejects its existence, the Tribunal clearly refers to the lack of belief of the *transnational petroleum companies* that the net book value is binding as a matter of law, and not to the intention or belief of the nation states that the net book value is binding as a matter of law. While its acceptance of transnational petroleum companies is necessary for a rule to be part of *lex mercatoria*, it is not necessary for it to be a rule of traditional customary international law, which is only concerned with nation states. The Tribunal's reasoning is thus not particularly helpful in delineating the source, scope and basis of *lex petrolea*.

Although not accepted in that case, the existence of *lex petrolea* has become a subject of subsequent academic research and writing, although the existence, definition, scope and jurisprudential foundation of *lex petrolea* is as yet unclear.[75] One approach is that *lex petrolea* is comprised of principles of law applied by international arbitral tribunals. The seminal article on this topic was published in 1998 by R Doak Bishop, an experienced international practitioner and arbitrator, who surveyed the key published arbitration awards relating to the international petroleum industry, to identify the rules or range of rules established by these published arbitral awards. Stating that the public awards provide the source material from which customary law may be drawn, and noting the argument of the Kuwaiti government in *AMINOIL*, the thesis of the paper was that the published awards 'have created the beginnings of a real *lex petrolea* that is instructive for the international petroleum industry'.[76]

Many of the awards surveyed by Bishop relate to procedural issues concerning international commercial arbitration or, substantively, to disputes concerning host government expropriation of the property of international oil companies, although a few other matters of dispute are addressed. Bishop concludes:

> While the published international arbitral awards involving petroleum issues are not so numerous yet and do not indicate such unity of opinion in the international community as to create anything like blackletter law rules, nevertheless, immense progress has been made in the past 25 years. The publication of awards has allowed later arbitrators to learn from the opinions of earlier tribunals and to build upon the foundations already constructed, rejecting unfounded reasoning and showing a growing sophistication and

---

75 See references above, (n 8).

76 Bishop (n 8) 1113.

> clarity. The result of this progress is that on some petroleum issues, clear legal rules have evolved, while on others at least the proper range of rules has been identified. This has not yet created a mature set of legal regulations, but it has developed the beginnings of a lex petrolea that serves to instruct, and in a certain sense even regulate – within broadly-defined boundaries – the international petroleum industry. As international arbitration continues to grow (provided that the publication of awards also continues), this *lex petrolea* may yet mature into a fully-developed subset of international law.[77]

Continuing in this tradition, in 2011, Thomas CC Childs surveyed the key substantive rulings in arbitral awards published since 1998 that relate to the international petroleum industry.[78] Childs's thesis was also that 'the published awards relating to the international exploration and production industry have created a "*lex petrolea*" or customary law comprising legal rules adapted to the industry's nature and specificities'.[79] Childs's survey reveals a definite shift in the proceedings brought before international arbitral tribunals. First, many of the post-1998 awards relate to proceedings by IOCs brought under bilateral investment treaties (BITs); prior to 1998, no BIT arbitrations had been brought by an IOC. Secondly, post 1998, only one award relates to the nationalization of an IOC's assets; whereas a large number of the awards surveyed by Bishop related to nationalizations. Several of the awards since 1998 deal with claims relating to changes to the host government's fiscal/taxation regime.

Some of the substantive matters brought before tribunals since 1998 address preliminary issues (pre-contractual liability, the validity of the E&P agreement, the nature of the agreement and state responsibility for the conduct of its NOC). Other matters the subject of dispute range from application and issue of the exploration licence, project management, the conduct of petroleum operations, taxation, in particular the imposition of windfall profits tax, the disposal of produced hydrocarbons, the transfer of interests in the project and *force majeure* to the frustration and repudiation of host government petroleum agreements and remedies for breach of contract. In conclusion, Childs claims:

> the rulings on substantive issues summarized in this article reflect the specific characteristics of international oil and gas exploration and production projects. While the published awards represent only a small fraction of the universe of awards relating to the exploration and production industry, they

---

[77] Ibid 1208–9.
[78] Childs (n 8).
[79] Ibid 214.

> address a sufficiently wide range of issues to create a '*lex petrolea*' or customary law comprising legal rules adapted to the industry's nature and specificities.[80]

The approach of Bishop and Childs is that a specific regime, *lex petrolea*, has developed through dispute settlement in the international petroleum industry, in particular through the published awards of international arbitrators and arbitral tribunals. To resolve the disputes, the arbitrators apply 'general principles of international law', which are derived from sources of international law such as United Nations General Assembly resolutions and declarations, treaties and customary international law, or from comparative studies of national laws. From these published arbitral awards, norms or principles of law are developed and discerned. These rules, which can be found from a survey of published awards, form the *lex petrolea* – the rules accepted by members of the international petroleum industry, through the arbitral process, as being developed and adapted to serve the industry's needs and specificities.

More recently, scholars and practitioners have moved beyond this approach, arguing that the rules/norms/standards of *lex petrolea* can be found not only in the published decisions of arbitral tribunals but in the common clauses and principles found in petroleum legislation and Host Government Contracts and the industry business practices found in model contracts.

Timothy Martin has argued that *lex petrolea* has developed from governments' petroleum legislation and contracts.[81] In his analysis, he notes that many of the awards concerning state-investor disputes are now publicly available, in particular claims made under investment treaties, according to the rules of arbitral institutions such as the International Centre for Settlement of Investment Disputes (ICSID). Talus, Looper and Otillar also discuss the idea that the internationalization of host government contracts may contribute not only to the internationalization of petroleum law but also to the existence of a *lex petrolea*.[82] As discussed earlier in this chapter, although there are no model host government contracts in the same way there are model industry agreements, the latter part of the 20th century has seen the global adoption of common types of Host Government Contracts, as well as common principles, clauses, terms and conditions in those agreements. If *lex petrolea* can be derived directly from common principles contained in petroleum agreements, this

---

80 Ibid 259.

81 Martin (n 8) 95.

82 Talus, Looper and Otillar, (n 8) 189.

is a far broader idea than that posited by *Kuwait v AMINOIL* and Bishop and Childs.

It has also been suggested that *lex petrolea* has developed from international industry business practices, which are found in its model contracts.[83] Martin notes that disputes arising from international commercial petroleum contracts are mainly decided in international arbitral tribunals and the awards are not usually publicly available because of confidentiality requirements.[84] International commercial awards are therefore often of little help in establishing a *lex petrolea*, given the scarcity of such awards.[85] Martin thus argues the best way of determining the *lex petrolea* of international oil agreements is by referencing the industry's business practices, captured in model contracts and the guidance notes, commentary and research arising from these model agreements, as the manner in which the industry develops the model contracts is the 'most thorough, documented and peer-reviewed process for international oil and gas agreements and thus the most credible source of *lex petrolea* for such agreements'.[86] These contracts, which serve the needs and specificities of the industry, are not drafted or set down by national or international law (although their enforcement relies on national laws). Because they are drafted by industry associations such as the AIPN and are widely used and accepted, they have legitimacy and are accepted as part of custom and usage.

Carmen Otero Garciá-Castrillón also recognizes the existence and relevance of *lex petrolea* as a branch or type of *lex mercatoria*, being the 'self-produced professional norms regulating commercial transactions'.[87] These may arise specifically from the petroleum industry to meet its own needs or be general principles of law incorporated as part of the *lex petrolea*. For Garciá-Castrillón, the sources of *lex petrolea* include: the model contracts prepared by professional associations; general contractual principles codified in international law, for example, the UNIDROIT

---

[83] Ibid 95.

[84] Ibid 102.

[85] Martin also points out that much domestic oil and gas law thus falls short of establishing a relevant *lex petrolea* for the international oil and gas business. Much of the publicly available case law interpreting oil and gas agreements comes from the US, Canadian and English domestic courts, which deal with disputes arising from their own jurisdiction. These national/domestic operations and agreements can vary significantly from the international industry's business practices and agreements, which makes domestic oil and gas cases a limited source for understanding international oil and gas agreements. Ibid 106.

[86] Ibid.

[87] Garciá-Castrillón (n 54) 140.

(International Institute for the Unification of Private Law) Principles of Contract Law; and the principles, standards, recommendations or guides elaborated by international organizations or associations, whether general or specific. She argues that *lex petrolea* is applied directly when it is incorporated or mentioned in contracts, treaties, and/or national legal systems and when it is viewed as a rule of customary international law; and that it can also be established through the arbitral resolution of disputes.[88]

García-Castrillón and Martin have also extended their claims regarding the content of *lex petrolea* to include principles concerning human rights and the environment and recognize that traditional law posited by states (such as treaties) can form part of the *lex petrolea*.[89] García-Castrillón argues that codes of conduct and principles/norms and rules regarding matters such as human rights, environmental protection, public transparency, etc, which may be required as conditions for obtaining financing from multilateral organizations such as the World Bank, can also be seen as part of self-regulation or a governance regime for the petroleum industry and part of a *lex petrolea*.[90] Martin has also extended the concept of *lex petrolea* beyond the published results of commercial and investment arbitration to include the application of norms of traditional international law, particularly those in relation to the environment and human rights. In this regard, Martin specifically embraces the sources of *lex petrolea* to include the application of traditional sources of international law, such as treaties.

> *Lex petrolea* covers a wide area of international law, given the size and significance of the industry. It can be viewed as either the application of

---

88 Ibid 142.

89 It is beyond the scope of this chapter to examine the codes of conduct, principles, norms and rules regarding matters such as human rights, indigenous rights and environmental protection and their relation to the international petroleum industry. For further reading, see Jay Wagner and Kit Armstrong, 'Managing Environmental and Social Risks in International Oil and Gas Projects: Perspectives on Compliance' (2010) 3 *Journal of World Energy Law and Business* 140; Jonathon Drimmer, 'Human Rights and the Extractive Industries: Litigation and Compliance Trends' (2010) 3 *Journal of World Energy Law and Business* 121; Duval et al (n 2) 31–8; A Wawryk, 'The Adoption of International Environmental Standards by Transnational Oil Companies: Reducing the Impact of Oil Operations in Emerging Economies' (2002) 20 *Journal of Energy & Natural Resources Law* 402.

90 García-Castrillón (n 54) 141–2.

international law to the petroleum sector or as a specific legal regime that has evolved to meet the needs of the international oil and gas sector – or as both.

> The growing development of *lex petrolea* in areas such as boundary disputes, human rights and environmental claims is more akin to the former – that is, the application of international law to the petroleum sector, whereas the areas of international commercial disputes and state investment disputes are more the latter – that is, a customary law of the international petroleum sector that has been adapted to the industry's nature and specificities.[91]

Finally, in what is perhaps the most extreme position, Alfredo de Jesús O has formulated an idea of *lex petrolea* existing as an autonomous, transnational legal order completely independent of both public international law and national law, comprising more than a set of norms developed by international arbitration or set out in model contracts. His conceptualization of *lex petrolea* is derived heavily from the idea of *lex mercatoria*. *Lex petrolea* arises from and attains legitimacy through acceptance and usage by the 'transnational petroleum society', independently of national and traditional public international legal systems. To him, the debate on transnational legal orders is not focused on whether they exist, but on their 'plurality and their content and methodology'.[92]

> [T]he *Lex Petrolea* exists and the idea roots date back to at the least the 1970s ... In [this] theoretical framework, the *Lex Petrolea* is not reduced to a bunch of rules designed to govern transnational petroleum contracts or the ones that can be extracted from the analysis of the international arbitral awards related to the oil and gas industry, nor the principles that can be inferred from the model contracts made by professional associations of the petroleum world etc. Those aspects are mere parts of a larger and more complex phenomenon. ... [T]he *Lex Petrolea*, as the *Lex Mercatoria*, is a system of law, more precisely a transnational legal order that is autonomous from both national and international legal orders as well as from other transnational legal orders ... The abovementioned transnational rules merely constitute the normative aspect of the whole construct.[93]

He argues that in the last decade, the transnational petroleum community has seen the emergence of new rules of a private origin specifically designed to govern petroleum contracts, which compete with state rules, are treated as rules of law and applied by arbitral tribunals, and which better satisfy the needs and interests of the transnational petroleum industry. These new rules are created spontaneously by the 'transnational

---

91 Martin (n 8) 108.

92 De Jesús O (n 8) 21.

93 Ibid 22.

petroleum society' itself, which is comprised mainly of oil producers, national and international oil companies and, to some extent, also some corporations engaging in 'parapetroleum activities'.[94] These rules are in addition to traditional sources of petroleum law, such as national petroleum codes or statutes, regulations and the 'overall legal, fiscal and regulatory texts' applicable to petroleum operations.[95]

As to the source of the normative rules, de Jesús argues that the 'ever evolving series of *best oilfield practice*' are the origin of much of the transnational rules; but also the rules from the standardization and internationalization of contracts and clauses in international petroleum contracts; the publication of 'doctrinal codifications of rules applicable to oil and gas contracts'; and the works and reflections of both contract and investment treaty international tribunals.[96] Arbitral jurisprudence is also a crucial source of rules; he states that although arbitral awards are meant to be confidential, even commercial awards meant to be confidential are easily accessible in the petroleum society.[97]

Thus, by the 'practices and progressive construction of specific types of contracts (the standardization and modelization of such contracts and clauses) and the arbitral jurisprudence, the members of the transnational oil and gas industry have created a number of rules that are specially designed and conceived to govern transnational petroleum contracts that simply transcend the concept of the Nation-State'.[98] For de Jesús, *only* the rules and codifications that satisfy the needs and interests of the transnational petroleum industry are considered as part of the transnational legal order, that is, of the *lex petrolea*. The rules are specialized. Thus, for example, even though general codification of contract law principles such as the UNIDROIT principles of International Commercial Contracts are useful, they are not part of the *lex petrolea* unless and until incorporated into the *lex petrolea* through, for example, their adoption in oil industry practices, standards and usages, or by application by an arbitral tribunal, if the general rule or principle satisfies the needs and interests of the transnational petroleum society.[99]

While the *contractual* legal order created by petroleum contracts is fundamental, de Jesús argues that this is only one aspect of the wider transnational legal order created by the 'transnational petroleum society'.

---

[94] Ibid 39.
[95] Ibid 23.
[96] Ibid 23.
[97] Ibid 25. Cf Martin (n 8).
[98] De Jesús O (n 8) 19.
[99] Ibid 45–7.

Nonetheless, his work is very much focused on the contractual/commercial practices of the transnational petroleum industry and the idea that various rules/norms/principles/standards have developed by the transnational petroleum society to meet the commercial needs of its members. He does, however, refer to good oilfield practice – the idea that members of the community are 'constantly seeking to determine the best oilfield practices that over time will become industry standards leading to the creation of new transnational rules'.[100] He seems to have a concept of *lex petrolea* beyond contractual and commercial practices and rules, referring in his analysis to the norms, rules and standards concerning the environment/health/safety/engineering/geological aspects of the industry. However, this aspect is very much underdone, in terms of the definition and idea of 'best oilfield practice' (a tricky concept at the best of times) and the sources that are relevant to determining 'best oilfield practice'.

A corollary of de Jesús's conceptualization of *lex petrolea* as a separate, independent and autonomous legal order is that parties to transnational petroleum contracts may chose the *lex petrolea* as the system of law that is the law governing their contract, and international arbitrators may apply the rules/norms/standards of this particular body of law to solve disputes.[101] *Lex petrolea* may thus serve an important function in international arbitrations – as a resource for legal principles common to domestic regimes and as a substantive law when commonality is not present.[102] However, whether practitioners and clients are prepared to do so is another question. Given the gaps and uncertainties in the *lex petrolea* – like *lex mercatoria*, it is not yet a complete body of laws, nor is there one easily identifiable and accessible source of law – practitioners and clients may prefer to nominate a national legal regime as the law governing the contract, as choosing a concept that is still so vague and uncertain may lead to appeals from the decision of the arbitral body.[103]

Despite the various positions taken on the source and content of *lex petrolea*, there is a very definite general consensus among those who accept its existence that it arises primarily from the decisions of international arbitration and court cases resolving disputes within the

---

100 Ibid 40.

101 See also Garciá-Castrillón on this point: (n 54) 136–40.

102 Kevin Jacobs and Matthew Paulson, 'The Convergence of Renewed Nationalization, Rising Commodities, and "Americanization" in International Arbitration and the Need for More Rigorous Legal and Procedural Defenses' (1999) 43 *Texas International Law Journal* 359, 387.

103 Ayad (n 54) 928.

international oil and gas sector, 'as this is where the contracts, legislation and treaties that affect the petroleum sector are tested and interpreted'.[104] While the decisions of arbitral tribunals are not binding on other tribunals, the global commercial and investment sectors seek consistency and predictability, and as more arbitral awards are published, arbitrators will look to previous decisions for consistency and excellence in legal reasoning.[105] Thus Martin, for example, has claimed that as arbitrators 'make their decisions in context and not in a vacuum' and as 'counsel use precedent in arguing their cases and arbitrators refer to precedent in writing their awards', the 'practical result is that precedent is relied on in international arbitration and a *lex petrolea* has developed accordingly'.[106]

Nonetheless, there are limitations regarding the development of law through international arbitral tribunals. At least as far as disputes between private international oil companies are concerned, the decisions and reasoning of international arbitration remain confidential, while in the absence of a court hierarchy, a doctrine of *stare decisis* and formal processes for the reporting of major cases or decisions, there are limitations to the extent to which decisions will bind future tribunals. Also, the fact that some oil-exporting countries have been unwilling to recognize international arbitral boards in relation to contractual disputes has created some uncertainty for the development of law through this means.[107]

## 4. CONCLUSION

The exploration for and production of petroleum is a global business, involving transnational flows of capital, labour and equipment. Although upstream activities are largely regulated by national petroleum regimes, as the industry has become truly globalized, we have witnessed a corresponding internationalization of law. Model contracts are evidence of international best practices being applied globally regardless of the location of individual commercial transactions, while the existence of similar types of host government contracts and similar clauses within

---

[104] Martin (n 8) 95–6.

[105] Talus, Looper and Otillar (n 8) 189. For a discussion of *lex petrolea* and precedent, see Ayad (n 54).

[106] Martin (n 8) 96.

[107] Talus, Looper and Otillar (n 8) 189.

those contracts are evidence of the internationalization of initially domestic principles of oil and gas law.[108]

In this context, the claim that there is a *lex petrolea* – norms or principles of law of particular relevance and application to the international petroleum industry, developed through the usage and practices of the industry to meet its needs and specificities and which attain their legitimacy as 'law' through their acceptance by the industry and the decisions of international arbitral tribunals – resonates within the industry.

However, there has been very little rigorous jurisprudential discussion, analysis and criticism of *lex petrolea* as a legal doctrine. Despite the posited existence of *lex petrolea*, there is no universal agreement regarding its sources, content and nature. The work of de Jesús is a major contribution to the literature in its attempts to define the jurisprudential basis for *lex petrolea*. While it has been well beyond the scope of this chapter to critically examine de Jesús's thesis in depth, jurists who disagree with the existence of *lex mercatoria* as a separate and autonomous system of law will also take exception to the posited existence of *lex petrolea* as an autonomous legal system. In addition to jurisprudential criticisms, there remain very practical difficulties with *lex petrolea*, in particular its identification as a set of norms or standards that can be applied as blackletter law. Until their legitimization by an arbitral tribunal, it remains extremely difficult to identify which rules, norms and practices are accepted as binding as forming part of 'law' – as opposed to something less than law – and their scope.

Despite the ambiguities and different approaches to the source and content of *lex petrolea,* the recognition that there has been and is a continuing internationalization of law and practice in the transnational petroleum industry, with an impact beyond systems of national law, is beyond doubt. The existence of international best practices and norms that are applied globally irrespective of national boundaries has had a direct impact on the way companies and governments conduct their petroleum activities today. The transnational petroleum industry is part of a global governance regime that is far more complex, broad, dynamic and interactive than can be conceived of by examining national regimes alone.

---

108 Kim Talus, editorial, 'OGEL Ten Years Special Issue: Internationalisation of Energy Law' (2013) 10(3) *Oil, Gas & Energy Law Intelligence* 1, 2, accessed 12 November 2012 at www.ogel.org.

# 2. Access to petroleum under the licensing and concession system

**Tina Hunter**

## 1. INTRODUCTION

Sound petroleum regulatory frameworks should perform a number of functions. They should encourage the responsible, metered development of petroleum resources for the benefit of both the state and all members of the host community. They should also encourage the participation of international oil companies (IOCs) to develop the resources, but not allow the IOCs to control the development.[1]

A regulatory framework needs to be flexible enough to respond to perceived issues relating to global energy supply. Long-term global security of petroleum supply depends upon whether the investment needed to expand supply capacity is forthcoming. For example, an estimated $4 trillion in investment was required from 2004 to maintain and expand petroleum capacity.[2] If a petroleum regulatory framework is not flexible to respond to these needs, the host state is likely to suffer, especially if the framework is rigid and incapable of responding to changes in global conditions such as supply, price and security.

Considering the finite nature of petroleum resources, and the necessity for petroleum in modern society, regulatory systems have been developed and implemented in petroleum producing nations. There are two classes

[1] Norwegian Petroleum Directorate, *Plan for Development and Operation* (2014), accessed 14 December 2014 at www.npd.no/en/Topics/Resource-accounts-and–analysis/Temaartikler/Norwegian-shelf-in-numbers-maps-and-figures/Plans-for-development-and-operation-PDO/.

[2] International Energy Agency, *Energy Policies of IEA Countries* (IEA, Paris, 2004), 34.

of petroleum regulatory systems, generally divisible by the granting of title and by taxation arrangements.[3]

The first is contractual systems, dominated by production sharing contracts (PSCs), and used almost exclusively in developing and emerging nations.[4] The second is the licensing and concession system (LCS) which is primarily utilized by developed nations with strong legal structures and institutions. It is the LCS that is the subject of this book and therefore the focus of this chapter. However, in order to understand the LCS, it is essential to understand the basis of contractual systems.

## 2. CONTRACTUAL SYSTEMS FOR ACCESS TO PETROLEUM

There are three main types of contractual system between the contractor and the state in the petroleum industry: risk-bearing service contracts; non risk-bearing service contracts; and production sharing contracts.

### Risk-Bearing Service Contracts

Under risk-bearing service contracts the contractor[5] receives compensation for work performed in convertible currency only (usually USD), rather than a share of the petroleum. The amount of remuneration can be converted to petroleum, which is purchased at market value price.[6]

### Non Risk-Bearing Service Contracts

In non risk-bearing service contracts the contract is between an IOC and a state agency. The IOC renders technical and financial services to the state for the operation and development of existing oil fields. The services are reimbursed, and the IOC will often also receive an additional reward linked to the extra production generated by the IOC's special

---

[3] Daniel Johnson, *International Petroleum Fiscal Systems and Production Sharing Contracts* (PenWell, Tulsa, 1994), 22.

[4] Ibid, 21.

[5] A 'contractor' is the company that undertakes the exploration work, and is usually an IOC.

[6] Bernard Taverne, *An Introduction to the Regulation of the Petroleum Industry: Laws, Contracts and Conventions* (Kluwers, Netherlands, 1994), 20.

development work. In this type of work contract there is no exploration risk, as the petroleum has already been found.[7]

## Production Sharing Contracts

Production sharing contracts (PSCs) are the most common form of contractual systems. Under a PSC an IOC and host state enter into a contract for the exploration and production of the host state's petroleum resources. The IOC assumes all of the risk in the exploration and development of a state's petroleum resources, in exchange for a share of the petroleum produced. It is based on the philosophy that the ownership of the petroleum shall always stay with the host country, and that the international oil companies are just acting as contractor to explore and exploit it.[8] PSCs are extensively used since they isolate a project from general risk by providing a stable contractual framework to underpin investments, thereby encouraging project financing.[9] Under this system, the host government owns all the mineral resources and retains ownership of the resources, but offers IOCs areas to explore for and exploit petroleum.[10] The host state gives the oil company the right to receive a share of the production (or revenue), in accordance with the PSC. Generally, the terms of the contract are determined through legislation, but many aspects are usually negotiated.[11]

It is unanimously agreed that the PSC was born in Indonesia in the early 1970s, and had been defined essentially as a contract based on the concept of the owner of the resources (the state) engaging a third party (an IOC) as contractors to explore for and exploit the oil. The proceeds of the contractor's work or activity (the petroleum) are shared between the state and the contractor on the basis of a previously agreed formula, after costs are subtracted.[12]

---

[7] Ibid, 21.

[8] Johnson, above n 3, 21–2.

[9] Andrew Seck, 'Production Sharing Arrangements as a Means of Drawing Large Scale Investment into the Energy, Mining and Other Sectors (an Oil and Gas Industry Perspective from Russia)' (2002) 12 *CEPMLP Internet Journal* 12–18.

[10] US Department of the Interior, Coastal Marine Institute, *Fiscal System Analysis: Concessionary and Contractual Systems Used in Offshore Petroleum Arrangements* (2004), 32.

[11] Ibid, 32.

[12] As defined by the former Indonesian Foreign Minister Dr Mochtar Kusumaatmadja in 'Indonesia's National Policy on Offshore Mineral Resources:

In PSCs the oil company contracts with the host state to undertake petroleum exploration, production and investment in that country's oil resources at its own risk.[13] The host state is not required to invest any capital into the exploration and production. Rather all investment and risk is assumed by the IOC. If the IOC does not find oil, it loses the money it has invested in the exploration operations. Conversely, if a large-scale discovery is made, the IOC recovers its original investment costs and gains an additional large amount of profit. It must also be remembered that profitability of a field will also be dependent upon the current market price for oil – the higher the market price the larger the profitability for the IOC.

As compensation for rendering these services, the contracted IOC will receive a certain part of the production that might result from the work.[14] Conversely, should the exploration and investment yield no production, the IOC will receive no compensation, monetary or in petroleum, for its services.[15] These risks are normally accepted as the price paid for the prospect of making considerable profit if the prospectivity of the field is as believed, or better.[16] The percentage of petroleum received as compensation is partly related to the capital and operating costs incurred by the IOC, and partly represents a profit element.[17]

At present the PSC is used in many developing countries. There is no one global uniform or model PSC.[18] However, many states have model contracts which are frequently used as a basis for bidding and the start of negotiations between the IOC and the host government.[19] The terms of the model contract are also frequently subject to negotiation as political and economic conditions change.[20] Usually, PSCs are individually adapted for the needs of different countries, but will generally specify a fixed exploration period[21] and divide the oil and gas discovered into 'cost' oil and 'profit' oil. 'Cost oil' is oil provided to remunerate the IOC

---

Some Legal Issues' in E Borgesse, N Ginsberg and J Morgan (eds), *Ocean Yearbook 9* (Brill, Halifax, 1991), 92.

13 Ian Rutledge, *The Sakhalin II PSA – A Production 'Non-Sharing' Agreement: Analysis of Revenue Distribution* (2004), 3 accessed 14 December 2014 at www.foe.co.uk/sites/default/files/downloads/sakhalin_psa.pdf.

14 Taverne, above n 6, 20.

15 Ibid, 20.

16 Rutledge, above n 13, 3.

17 Taverne, above n 6, 20.

18 Rutledge, above n 13, 13–14.

19 US Department of the Interior, Coastal Marine Institute, above n 10, 32.

20 Ibid, 32.

21 Rutledge, above n 13, 13.

for costs associated with drilling, infrastructure, operating costs, etc, and is calculated at the market value. The IOC's capital costs are either expensed in the current year or depreciated over a number of years. Generally the IOC receives the cost oil which it sells either on the open market or back to the state oil company (SOC).[22] There is usually a cap on the annual amount of cost oil that can be recovered from the gross annual revenue (usually a percentage), ensuring that the state receives some profit from the commencement of production, even if there is no royalty payment element. After the costs have been recovered, the remaining oil is known as 'profit oil'.

## 3. LICENSING AND CONCESSION SYSTEMS FOR ACCESS TO PETROLEUM

Licensing may be defined as:

> the identification by government of potential (upstream) petroleum investment opportunities in the national territory, their subdivision into discrete contract areas of prospective size, their offering to the international oil companies by a suitable tendering process and the establishment and negotiation of technical, financial and contractual terms and conditions (for award) consistent with their petroleum prospectivity and with the national interest.[23]

### Overview of Licensing and Concession Systems

A petroleum licensing system is one where a license is granted for a specific type of petroleum operation, usually for exploration or production. It is granted for a specific area (the license area), for a specific period.[24] Depending on the regulatory framework governing the LCS, the license may confer exclusive rights, in the sense that for as long as the license is valid, and subject to certain conditions, the licensee is authorized to exercise the rights conferred in the license against third parties.[25] Often, non-exclusive licenses are conferred for geological and geophysical prospecting, while an exclusive license is usually conferred

---

[22] Ibid, 13.

[23] Michael Bunter, *The Promotion and Licensing of Petroleum Prospective Acreage* (Kluwer, Norwell, 2002), xxii.

[24] Bernard Taverne, *Petroleum Industry and Governments* (Kluwer, Netherlands, 1999), 147.

[25] Ibid, 146.

for exploration work that involves drilling, as well as for production operations.[26]

The LCS is often utilized in countries with relatively low reserves and high costs associated with offshore oil production, and the royalty/tax system dominates.[27] A survey of countries that utilize the modern licensing system indicates this to be an accurate reflection, with developed countries such as the United States, the United Kingdom, Norway, Australia and the Netherlands all implementing the LCS for the regulation of oil and gas.[28]

One of the main characteristics of the LCS is that the state retains the prerogative sovereign right to modify at any time those terms and conditions that are not negotiated but fixed by legislation.[29] This is especially applicable to the rate of taxation that the state imposes on the income from petroleum operations. The possibility of alteration to the taxation regime is one of the most significant threats to the predictability of oil operations' economic result, and, if abused by the state, can seriously prejudice the feasibility of a project by reducing the profitability of a project.[30] This may represent a risk which is too high compared to the return from the project, in some instances it can induce investors to refrain from the investment itself, since oil projects assume considerable investments and commitment of financial, human and technical resources for several years before the investor starts seeing a payback. Some investors may avoid such a significant commitment if the predictability of the economic conditions cannot be ensured.

Oil companies nonetheless invest considerable time, money and resources in projects governed by an LCS, even though there is some element of uncertainty connected with fiscal conditions. This is because the remaining aspects of that country's legal, social, economic and political system indicate that the projects can rely on a stable investment climate.[31]

---

[26] This is typefied by the grant of a non-exclusive exploration license and an exclusive production license in Norway under the Petroleum Activities Act 1996. See ibid, 146.

[27] Abdulaziz Al-Attar and Osamah Alomair, 'Evaluation of Upstream Petroleum Agreements and Production Costs' (2005) 29(4) *OPEC Review* 242, 245–6.

[28] Ibid, 245–6.

[29] G C Moss, 'Contract or License? Regulation of Petroleum Investment in Russia and the Role of Foreign Legal Advice' (1998) 3 *CEPMLP Internet Journal* 3–11.

[30] Ibid, 3–11.

[31] Ibid, 3–11.

The process for the allocation of a petroleum license is similar in all petroleum jurisdictions that utilize the LCS. It initially involves the state identifying areas for petroleum exploration, and the subdivision of these areas into discrete contract areas of a predetermined size.[32] These areas (also known as acreage) are offered to national and international oil companies by a suitable tendering process (the licensing round), and the license awarded. This process may include negotiations relating to the technical, financial and contractual terms and conditions for the award of the license, consistent with their petroleum prospectivity and with the national interest.[33]

The allocation of a petroleum license encapsulates not only the process of the award of a petroleum license, but also marries the method of allocation with the national petroleum interests. State petroleum interests are generally articulated in a state's petroleum policy and regulated by the legislative framework. The LCS allows for as little or as much state participation as the state desires or requires at any given time, and also allows changes in the level of participation over time.[34] Norway, for example, has changed its system for, and level of, direct state participation several times.[35]

The licensing of petroleum acreage occurs between the petroleum rights holder (the state, usually represented by the relevant ministry) and international oil companies. The process of allocating a license for the development of petroleum reserves can broaden the number of applicants for an area, ensuring that the largest number of oil companies is able to apply for access to the acreage on offer.[36] Furthermore, the process is regarded as politically legitimate by the community, since it is viewed as open and transparent.[37] This ensures that petroleum extraction can occur with little negative electoral reaction.[38]

---

32 Bunter, above n 23, xxii.

33 Ibid, xxii.

34 Jerome Davis, 'Does One Size Fit All: Reflecting on Governance and North Sea Licensing Systems' (2004) Background Paper: BC Offshore: Potential and Problems, a MASC Workshop for Lawyers, Dunsmuir Lodge, Sidney BC, 18–21 March 2004, 6.

35 For an outline of the history of the Norwegian licensing system, see Brent F Nelsen, *The State Offshore: Petroleum, Politics and State Intervention on the British and Norwegian Continental Shelves* (Praeger, Westport, 1991).

36 Michael Bunter, 'A New Approach to Petroleum Licencing' (2003) 1(1) *Oil, Gas and Energy Law Intelligence*, accessed 14 July 2014 at www.gasandoil.com/ogel/samples/freearticles/roundup_15.htm.

37 Davis, above n 34, 7.

38 Ibid, 7.

The award of a petroleum license is essential for the commencement of petroleum extraction.[39] Upon the allocation of an exploration or production license, exclusivity over the license area is awarded to the licensee, who gains a proprietary right to explore and produce petroleum.[40] Property rights regarding petroleum resources for both the state and the licensee are recognized and enforced under the LCS.[41] Whilst the petroleum remains in situ, even if part of the grant of a production license, the petroleum resources are owned by the state.[42]

A production license confers ownership of petroleum when it is produced in accordance with the terms of the license. 'Produced' is generally defined to be when the petroleum enters the well that has been legitimately drilled by the licensee,[43] although this may vary in an individual country's legislation. This system of ownership is related to the ancient 'rule of capture' and title of the petroleum is transferred to the oil company upon the petroleum reaching the wellhead, as the petroleum is 'caught' by the licensee in the wellhead.[44] The license gives a right to explore for and produce petroleum, and thereby acquire ownership to the produced petroleum.[45] This exclusivity protects the licensee against exploration from other parties in the license area for the period that the license is valid.

The objective of a state in the exploitation of the petroleum resources, and the benefits it wishes to reap, varies depending on the petroleum policy formulated by the state. The allocation of petroleum licenses should reflect the petroleum objectives of the state, to ensure that the exploitation of petroleum resources is capable of meeting these policy objectives.[46] The allocation of a petroleum license is crucial in optimizing petroleum development. As John Strongman of the World Bank

---

[39] Ibid, 7.

[40] Kjell J Sunnevåg, 'Designing Auctions for Offshore Petroleum Lease Allocation' (2000) 26 *Resources* Policy 3, 3.

[41] Davis, above n 34, 7.

[42] US Department of the Interior, Coastal Marine Institute, above n 10, 5.

[43] Bernard Taverne, *Petroleum Industry and Governments: An Introduction to Petroleum Regulation, Economics and Government Policies* (Kluwers, Netherlands, 1999), 147.

[44] Colin Roberts, 'Resource Regulatory Systems' (2006) *Australia's Paydirt* 93, 93.

[45] For example, the grant of exclusive rights for a Norwegian production license and ownership of the oil produced is guaranteed under s 3-3 of the Petroleum Activities Act 1996 (Norway).

[46] Department of Resources, Energy and Tourism, *Resources* (2009), accessed 14 July 2014 at www.ret.gov.au/resources/Pages/Resources.aspx.

noted, in order to translate natural capital[47] into societal capital,[48] you need the right policies and good implementation capacity from the allocation of exploration rights to how governments spend their revenue.[49] Hence, the optimal development of petroleum resources starts with the allocation of the petroleum license. The difficulty lies in determining which method of allocation is most likely to enable optimal development.

## Process of Awarding Licenses

The licensing of petroleum acreage is a transaction between the petroleum rights holder (usually the government, represented by the relevant ministry) and an oil company, especially IOCs.

In most instances the award of petroleum licenses occurs within formal licensing rounds.[50] Formal licensing rounds are discrete acts of licensing that take place on a regular basis, often annually or biannually, at which

---

[47] 'Natural capital' is defined as assets in their role of providing natural resource inputs and environmental services for economic production. It is generally considered to comprise three principal categories: natural resource stocks, land and ecosystems. All are considered essential to the long-term sustainability of development for their provision of 'functions' to the economy, as well as to mankind outside the economy and other living beings. See Organisation for Economic Cooperation and Development, 'Glossary of Statistical Terms' (2007), accessed 15 August 2014 at http://stats.oecd.org/glossary/detail.asp?ID=1730.

[48] 'Societal capital' may be seen as a form of social capital, which is defined as the norms and social relations embedded in the social structures of societies that enable people to coordinate action to achieve desired goals. The term social capital has found its way into economic analysis only recently, and there is no consensus on which aspects of interaction and organization merit the label of social capital, nor on how to measure it and how to determine empirically its contribution to economic growth and development. See Organisation for Economic Cooperation and Development, 'Glossary of Statistical Terms' (2007), accessed 15 August 2014 at http://stats.oecd.org/glossary/detail.asp?ID=3560.

[49] John Strongman, 'Sustaining the Social License: Enduring Challenges, Evolving Solutions' (2009), paper presented at the Sustainable Development 2009 Conference (SD09), Mineral Council of Australia, Adelaide, South Australia, 26–28 October 2009, slide 28.

[50] Michael Bunter, *The Promotion and Licencing of Offshore Acreage* (Kluwer, Norway, 2002), xxii. In Australia there are annual licensing rounds, whereas in Norway the licensing occurs on a less frequent basis, generally every two years.

predefined acreage is publicly offered for licensing by the state.[51] Formal licensing rounds can assist the state to identify the most suitable party to develop a petroleum field since the formal process encourages companies to apply for a license, and be assessed according to the stipulated criteria.[52] They create competition between companies wishing to be active in the exploitation of petroleum within that province.[53] Licensing rounds are used in some jurisdictions (for example Norway) to ensure that the parties that are selected for a license area are the most capable for the particular geology and conditions of that field.[54] They can also assist the state to lessen its risk by ensuring that companies assume the exploration and production risk.[55]

This formal licensing round process is often supplemented with 'out of round', or informal, licensing rounds, in order to fulfil policy goals. Oil companies may approach a state informally for access to petroleum acreage (which may or may not be specified), and a bargain is struck, without competition from other companies.[56]

Another licensing method is the special licensing round (for example in Norway under the 'award in predefined area' (APA) rounds).[57] Such rounds award acreage mostly in mature provinces to encourage exploration, since there is an expectation of smaller oil discoveries.[58] They are made as part of the implementation of the policy of prudent production, with the aim of discovering and developing these petroleum resources in mature areas using new technologies, before existing infrastructure in the

---

51 Bernard Taverne, *Petroleum Industry and Governments: An Introduction to Petroleum Regulation, Economics and Government Policies* (1999), 137. An example of this is the announcement requirement found in s 3-5 of the Petroleum Activities Act 1996 (Norway).

52 Bunter, above n 36, 2–3.

53 Ibid, 2–3.

54 Petroleum Regulations 1997 (Norway), s 10.

55 Bunter, above n 36, 2–3. See also Kenneth Dam, *Oil Resources: Who Gets What How?* (University of Chicago Press, Chicago, 1976), 4–5, and Bernard Taverne, *Petroleum Industry and Governments: An Introduction to Petroleum Regulation, Economics and Government Policies* (1999), Chapter 5 on conditions regarding the award of petroleum licenses.

56 Ibid, 137. This often occurs in developing countries that utilize the LCS, for example in Nigeria.

57 Ministry of Petroleum and Energy, 'APA 2009: Continued Broad Interest for Mature Areas on the Norwegian Continental Shelf' (2009) Press Release 16 September 2009, accessed 14 December 2014 at www.regjeringen.no/en/dep/oed/press-center/Press-releases/2009/apa-2009-continued-broad-interest-for-ma.html?id=577167.

58 Ibid.

fields are decommissioned.[59] In addition, these rounds encourage sustainable development by ensuring petroleum resources are discovered and developed before existing infrastructure is closed down.[60]

There are two principal systems of awarding licenses within the LCS.[61] The first is the bid or auction system, where licenses are 'sold' to the highest bidder. This bid or auction may include either cash bidding, where the license is sold to the highest bidder, or work program bidding (WPB), where the license is awarded to the applicant that bids to spend the highest amount of work, in dollar terms, on exploration for petroleum in the license area. The second is the discretionary system of allocation. The system of allocation is outlined in Figure 2.1.

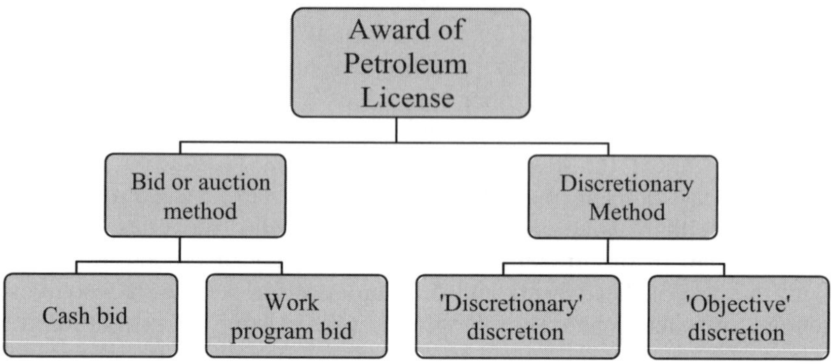


*Source:* Compiled by author

*Figure 2.1 Method of allocation of petroleum license: Bid/auction method or discretionary method*

## Award of License by Bidding

Under the bid system, whether the auction is made by cash auction or a work program bid system, the bidder with the highest value (either cash or work program) is awarded the license.

[59] See Ministry of Petroleum and Energy, Norway, 'APA 2008 – Broad Interest for Mature Norwegian Shelf' (2008), accessed 14 December 2014 at www.regjeringen.no/en/dep/oed/press-center/Press-releases/2008/apa-2008–broad-interest-for-mature-area.html?id=528899.

[60] Ministry of Petroleum and Energy, above n 57.

[61] Dam, above n 55, 4–5.

**Cash bid system**

The bid system of allocation enables an oil company to assert control over the licensing process by stating what they are prepared to pay for the property rights over the acreage being offered for license by the government.[62] The value of those property rights is expressed in terms of how much a company will pay as a cash bonus or special royalty rate.

Under the cash bid (or auction) system, a license is awarded to the applicant that is the highest bidder for a defined acreage.[63] The award is made following strict economic criteria, where the true market value of the acreage is the price it fetches in a competitive marketplace.[64] It is based on the economic theory that the company willing to pay the most for the acreage is the one most suited for the job.[65]

Cash bidding is theoretically the most efficient approach to the allocation of a petroleum license since the state allocating the rights receives resource rent payment and benefits from an efficient allocation system (as the most efficient company can afford to bid the highest).[66] Allocation by cash bid is seen as an incorruptible method of maximizing the price paid for the grant of a petroleum license.[67] In addition, it is seen as a system that avoids the shortcomings that may affect the discretionary allocation of licenses, particularly corruption.[68]

However, the inherent uncertainty regarding the petroleum quality and quantity of a particular field means that bidders will discount their cash

---

[62] Knut Sinding, 'Auctions and Discretion in Oil and Natural Gas Licensing' (1999) *CEPMLP Publication 1/99*, 2.

[63] Dam, above n 55, 5.

[64] Bunter, above n 36.

[65] Espen Myhra, 'Introduction to Different Licencing Systems' (2006) 5th Workshop of Philippines Petroleum Policy and Management Case Study, 14–18 March 2006, Cebu City, accessed 14 December 2014 at www.ccop.or.th/ppm/document/PHWS5/PHWS5DOC07_myhra.pdf www.ccop.or.th/ppm/document/PHWS5/PHWS5DOC07_myhra.pdf, 3–5.

[66] Trudi Rodgers and Stewart Webster, 'Resource Rent Mechanisms in Australian Primary Industries: Some Observations and Issues' (2007), paper presented at the 51st Annual Conference of the Australian Agricultural and Resource Economics Society, 13–16 February 2007, Queenstown, New Zealand, 12.

[67] Tim Warman and Lauren Goldblatt, 'The Work Program Bidding System for Exploration Permits under the Petroleum (Submerged Land) Act 1967 (Cth)' (2008) 27 *Australian Resources and Energy Law Journal* 178, 179.

[68] Walter Mead, 'Towards an Optimal Oil and Gas Leasing System' (1994) 15(4) *Energy Journal* 1, 10–11.

bonus bids accordingly.[69] A company bidding for a license in an area where they do not know for sure whether or not petroleum is present will have to calculate a considerable risk in the bid. This means that sometimes, particularly in a province of unproven geology, the cash bid may not be an accurate reflection of the real economic worth of the province since there is no comprehensive data pertaining to the province.[70]

The objective of the cash bid system is twofold. It serves as a method for the allocation of licenses, as well as a resource rent taxation tool.[71] The cash bid system in its pure form 'approximates an optimal system'[72] where the state seeks to maximize and collect the net present value of the economic rent. This is optimal since the amount bid reveals the bidders' valuation of the acreage offered.[73] Therefore, not only does the cash bid system collect the maximum amount of resource rent, but it also reveals to the government how valuable the bidding company believes the acreage to be, and which company values that acreage the most.[74] That information is crucial to a government since it furnishes the state with the capacity to assess the value of the acreage to the state itself and the companies that wish to exploit it.

The cash bid system attempts to capture the economic rent by pitting bidders for the resource against each other.[75] In this competitive situation, bidders give up some of the prospective economic rent to the licensing authority in return for a petroleum license that confers the exclusive right to petroleum exploration and production.[76] Since theoretically each bidder gains an advantage, the expectation is that the auction system will result in bidders attempting to outbid each other, until the state has captured all of the economic rent.[77]

A concern of the auction system is the low number of bidders. Where this occurs, it has been demonstrated that sealed bids are preferred over oral bids, since they ensure that uncertainty is captured.[78]

---

69 Rodgers and Webster, above n 66, 12.
70 Ibid, 12.
71 Sunnevåg, above n 40, 5.
72 Mead, above n 68, 5.
73 Sunnevåg, above n 40, 5.
74 Ibid, 5.
75 Dam, above n 55, 5.
76 Ibid, 5.
77 Ibid, 5.
78 Mead, above n 68, 7–8.

The cash bid system was initially used almost exclusively by the United States,[79] which has a long history of the use of cash or 'bonus' bidding for oil and gas licenses.[80] The system has been used for the allocation of both onshore tracts of publicly owned land[81] and for the allocation of offshore licenses in the Gulf of Mexico.[82] This has altered in recent times with several other states, including Nigeria and Australia, using cash bidding as a system of allocation of licenses. Chapter 12 provides an analysis of jurisdictions utilizing the cash bid system for the allocation of licenses.

When implementing the Outer Continental Shelf Lands Act 1953 (OCSLA),[83] the US federal government incorporated a simple means of cash bid allocation of leases.[84] Today, the OCSLA requires the allocation of licenses on the outer continental shelf to ensure the protection of the environment and returns revenues to the federal government in the form of bonus bids, rents and royalties.[85] It also incorporates a number of other alternative bidding systems,[86] mixing the capacity for broad discretion

---

79 Dam, above n 55, 6.

80 Bunter, above n 36.

81 Sunnevåg, above n 40, 4. Petroleum licensing in the USA is a combination of lease contracts on public lands, designed according to the patterns developed by the private sector, plus general taxation. See Bernard Mommer, *Private Landlord–Tenant Relationship in British Coal and American Oil: A Theory of Mineral Leases*,(1997), 44, accessed 14 December 2014 http://daxabek.files.wordpress.com/2014/09/private-landlord-tenant-relationship-in-british-coal-and-american-oil-a-theory-of-mineral-l.pdf. Under the US legal system, mineral resources are the property of the public or private owner of the land on or beneath which the mineral resource is located. See Taverne, above n 24, 174. Ownership of petroleum in the USA is governed by the 'rule of capture', where the landholder can extract as much oil and gas they desire from their land, so long as they conduct their operations without trespassing or interfering with the rights of neighbours to drill into the same geological formation under their own lands. For a discussion of ownership of mineral resources in the USA, see John S Lowe, *Oil and Gas Law* (West Publishing, Racine, 1988), 9–10.

82 Mead, above n 68, 1.

83 The Outer Continental Shelf Lands Act 1953 (US).

84 Terence Daintith, 'Administering the Petroleum (Submerged Land) Act: Too Much Discretion or Too Little?' (2004) *AMPLA Yearbook 2004* 1, 34.

85 Marc Humphries, *Outer Continental Shelf: Debate over Oil and Gas Leasing and Revenue Sharing* (2008) CRC Report for Congress, accessed 8 June 2014 at http://ncseonline.org/NLE/CRSreports/07Dec/RL33493.pdf, 4.

86 Daintith, above n 84, 34.

with a number of highly prescriptive planning and program arrangements. Only the auction system has been used for the allocation of licenses to date.[87]

**Work program bidding**

Similar to the cash bidding process, WPB utilizes a system of bids in order to allocate petroleum licenses to companies. It is a competitive bidding system based on the concept that the winning company is the company able to perform the most exploration.[88] Under WPB, petroleum licenses for exploration are awarded to companies that bid to perform specific activities that are desired by the state licensing the petroleum resources.[89] The rationale behind the WPB system is to encourage and provide an incentive for companies to effectively explore for petroleum.[90]

The number and type of areas that are bid for by potential licensees, as well as the quality of those bids, determine the success of the WPB.[91] The quality of the bids refers to the amount and timing of the proposed work programs, as well as the technical and financial resources.[92] Generally, the optimal work program for an area (which is analogous to the investment outlay) is likely to be positively related to the expected profitability of exploring (and the subsequent development) of the area, and influenced by factors such as future oil prices and prospectivity of the acreage.[93]

The decision by a company to participate in the bidding process is influenced by an assessment of the business environment, and whether the oil company is prepared to engage in activities with the perceived return-risk profile.[94] If embarking on a bid, the company assesses the acreage and formulates an optimal (profit maximizing) work program for the area. Similar to the cash bid, this assessment is based on such factors as prospectivity of the area, technology available, expected price for the oil that will be sold and whether there are potential buyers if petroleum is

---

87 Ibid, 34.

88 Mead, above n 68, 10–11.

89 Ibid, 2.

90 Warman and Goldblatt, above n 67, 180.

91 Athol Maritz, 'Work Program Bidding in Australia's Upstream Oil and Gas Industry, 1985–1999' (2003) ABARE Report 03.14, Prepared for the Department of Industry, Tourism and Resources, accessed 12 June 2014 at www.abare.gov.au/publications_html/energy/energy_03/er03_work_program.pdf, 2.

92 Ibid, 2.

93 Ibid, 3–4.

94 Ibid, 3.

discovered.[95] Other factors that may also be taken into account include the type of crude the acreage is likely to contain, water depth and costs associated with the development of the field.[96]

The optimal work program forms the basis of the competitiveness of the bid, although the submitted bid will also be a function of the competitiveness of the tender, where competitiveness is related to the number of companies submitting competitive bids for the area.[97] It is also related to the number of companies actively participating in the industry and the likely bids from competing companies. Similar to cash bidding, the more companies that bid for acreage, the greater the competition between the companies, and therefore the greater the likelihood that the WPB will be optimal.[98]

Ideally, the selection of the winning bid should be based on terms and criteria that are final and binding (enforcing the concept that it is a contract between the state and the oil company).[99] The successful work bid should bind the licensee to their commitments in a manner that bonus or royalty systems do not, therefore limiting operator discretion.[100] This is accomplished through the specified work program, which generally requires the licensee to drill a certain number of wells at particular locations and to run a certain amount of seismic surveys.[101] Unfortunately, often the bidder is required to speculate as to what criteria the state considers important when awarding petroleum licenses thus introducing an element of discretion into the allocation of petroleum licenses.[102]

### Assessment of the bidding system

A difficulty with the bidding system (both cash and WPB) is a lack of information. Sometimes the companies may not know whether a field contains petroleum, as there may be limited data upon which a company can base its bid, although this is not always the case.[103] This means that

---

[95] Ibid, 3.

[96] Ibid, 3.

[97] Ibid, 4.

[98] The bid can be greater than optimal where there is a presumption that the field will contain more oil and gas than it actually does. For example, in the Cornea Field in north-western Western Australia. Mead, above n 68, 2–4.

[99] Ibid, 2–4.

[100] Davis, above n 34, 5–6.

[101] Ibid, 6.

[102] Mead, above n 68, 2.

[103] Often, in developed regimes such as the UK and Australia, there may be seismic data available that identified possible petroleum resources. An example of this is pre-bid seismic data from Geoscience Australia. For an example of this,

companies decide the potential of a field, and the value of a field is left to market forces prevailing upon the parties that bid, with the most optimistic bid winning. The winning company might be the company most able to maximize the potential of the field, but it may also be an over-optimistic company that believes they can find oil in the field. However, if the field is dry, no company, optimistic, wealthy or competent, can find oil when it does not exist, although at least a competent company has the greatest chance of finding and developing petroleum resources.[104]

The value of the WPB system is that it is able to establish geological information regarding prospectivity of the field, since the selection of a bidder based on the exploration work program means that the award of a license under a WPB will at least realize information regarding the field.

## Discretionary Allocation of Petroleum Licenses

Under the second method of allocation of petroleum licenses, the discretionary system, the state allocates licenses according to administratively or politically created criteria defined by the state,[105] enabling the government to define and assert a set of legal conditions to which the participating oil company is expected to conform.[106] Often, the criteria used and the transparency of the process is a source of concern,[107] although this has been somewhat mitigated by the requirement for objective, transparent criteria in the award of licenses, particularly in the EU and EEA countries.[108] The discretionary system is a feature of the North Sea system of licensing, developed by Norway since it began awarding licenses in 1963.[109]

### Outline and objective of the discretionary system

The discretionary system for the allocation of petroleum licenses has been used in many jurisdictions, including Norway, the United Kingdom

---

see Geoscience Australia '2014 Release Areas and Geology' (2014), accessed 13 August 2014 at www.petroleum-acreage.gov.au/2014/release-areas-and-geology.

[104] An example of this is the Norwegian giant field Ekofisk.

[105] Taverne, above n 24.

[106] Davis, above n 34, 5.

[107] Dam, above n 55, 4–5.

[108] Refer, for instance, to the EC Hydrocarbon Directive: Directive 94/22/EC of the European Parliament and of the Council of 30 May 1994 on the Conditions for Granting and Using Authorizations for the Prospection, Exploration and Production of Hydrocarbons (1994).

[109] Nelsen, above n 35, 23.

and Denmark.[110] It is a flexible system of awarding exploration and production acreage on the basis of a committed work program set either by the oil company upon application for acreage, or by the state developing the resources.[111] It awards acreage to applicants on the basis of predetermined criteria. These criteria may include technical and financial competence and management capacity of the applicant.[112] However, typically the primary criterion for the allocation of the acreage is the level and type of work the applicant is prepared to undertake in an exploration.[113] It is important to note that there is much variation in the criteria stipulated by different states, and the criteria for award of license will differ according to the goals of the state in exploiting the petroleum resources.

The objectives of the discretionary system are to provide states with the capacity to select the most appropriate companies for the grant of exploration and production rights, to enable the state to define the criteria for the award of licenses according to the national petroleum objectives and field maturity, and to develop frontier areas where required.[114] Discretionary allocation enables a state to determine which factors will be used to select the licensee, rather than being confined to purely economic criteria. Under discretionary allocation, other factors can be used in selecting the winning applicant, and the selection criteria can be aligned with national petroleum objectives. Where optimizing petroleum extraction is a goal for the state, the discretionary method may be a valuable tool.

Discretionary allocation can provide a state with a plethora of options in developing its petroleum resources. It allows a government to rapidly explore their fields if so desired[115] and promote timely and effective development of a field in the event of a commercial discovery.[116] Furthermore, it can be used as an effective tool to slow down exploration as a method of controlling production by limiting the number of licenses

---

[110] Bunter, above n 36.

[111] Ibid.

[112] Bunter, above n 23, 87.

[113] Geoff Frewer, 'Auctions vs. Discretion in the Licencing of Oil and Gas Acreage' in G MacKerron and P Pearson (eds), *The International Energy Experience: Markets, Regulation and the Environment* (Imperial College Press, London, 2000), 166.

[114] Bunter, above n 23, 87; Frewer, above n 113, 166; Nelsen, above n 35 and Davis, above n 34.

[115] Nelsen, above n 35, 19.

[116] Peter Cameron, *Property Rights and Sovereign Rights: The Case of North Sea Oil* (Academic Press, London, 1983), 15–16.

that are granted to petroleum explorers.[117] It also enables the state to control or limit the number of foreign oil companies that exploit oil within that state[118] and can be used to obtain geological and geophysical information about a country's oil reserves and potential.[119] It is able to respond to environmental and political change with greater sensitivity than auction systems.[120] This means that there may be differing terms for differing licensing rounds in any jurisdiction.[121]

Discretionary allocation also gives a state some control over the investment in its petroleum industry, since the government is able to determine which of the proposed programs for exploration are carried out.[122] With this control comes a need for the state to have sufficient human, financial and knowledge resources to be able to evaluate the proposals for exploration made by oil companies.[123] Thus, this system vests the discretionary power to award a license in the petroleum management body and tender evaluators (usually a licensing regulator), who consider the formal applications against the state-defined criteria which encompasses financial, technical and managerial aspects of the applicant.[124] This flexible system for the allocation of licenses for exploration and production awards licenses according to administratively or politically derived criteria, in order for the state to maximize the benefits to that state from the exploitation of the petroleum resources that the state owns.[125]

The use of discretion in the award of a petroleum license is not an 'all or nothing' system. Rather, discretion is a continuum, ranging from total government control at one extreme, through various degrees of government discretion, to award through objective publically available criteria at

---

117 Dam, above n 55, 4.

118 Cameron, above n 116, 15–16.

119 Ibid, 15–16.

120 Davis, above n 34, 5.

121 An excellent discussion of the differing conditions attached to early petroleum licensing rounds is found in Nelsen, above n 35. See also a description of the licensing rounds in Norwegian Petroleum Directorate, *Annual Report, 1998* (1998) accessed 14 December 2014 www.npd.no/Global/Engelsk/3-Publications/NPD-annual-reports/Annual-report-1998.pdf.

122 Frewer, above n 113, 166.

123 Ibid, 166.

124 Bunter, above n 36.

125 Dam, above n 55, 4.

the other end.[126] This spectrum can range from the 'discretionary' discretionary allocation of petroleum licenses, such as those allocated by the Norwegian government prior to 1994,[127] to the 'objective' discretionary allocation of licenses, where licenses are awarded under the discretionary method of allocation utilizing objective and transparent criteria. The difficulty faced by governments therefore is not whether or not they should allocate acreage using the discretion system of allocation, but rather how far along the discretion continuum they should move in allocating licenses and how much control should remain with the state.[128]

The exercise of discretion along this spectrum enables the state to award licenses to the applicant whose proposed exploration program is most likely to fit with national petroleum objectives.[129] These objectives may include growth of industry sectors, management of other natural resources, or political or environmental objectives.[130] By allocating licenses using discretion it is possible to determine, define and publicize the criteria for the award of licenses based on those national objectives. Such criteria may encompass experience, technical knowledge, financial capacity or nationality. In addition, where the selection criteria are determined and advertised by the state during the licensing round, this enables a state to attract companies that satisfy the criteria upon which the award of license will be based, rather than only attracting companies with the financial capacity to pay the highest bid up front.

Often discretion enables the state to exert control over the activities of the licensee in petroleum activities. Frewer notes that the greater the level of control exhibited by the state in the award of licenses, the greater the level of influence a state is able to exert over the activities of the oil company in the exploration and production of petroleum.[131] However, it is important to note that this level of control may also be attributable to the amount of control that a state exerts in the development of a field, rather than the conditions of award alone.

---

[126] Frewer, above n 113, 165. This spectrum can range from the award of a license using advertised, objective criteria, to a government decision to award the license based on apparent nonsensical criteria that are neither open, nor transparent.

[127] For example, the allocation of licenses in Norway that specified that any licensees awarded licenses on the Norwegian Continental Shelf were required to use Norwegian labour and companies wherever possible.

[128] Frewer, above n 113, 165.

[129] Ibid, 168.

[130] Ibid, 168.

[131] Ibid, 168.

**Assessment of the discretionary system**

The UK government initially debated the merits of using a discretionary versus an auction system in the 1960s, ultimately rejecting the auction system since it 'did not permit the same range of detailed control that discretionary licencing systems do'.[132] This choice of discretionary allocation enabled the UK Ministry of Power to allocate licenses to companies that were willing to carry out the wishes of the UK government for the rapid development of the resources on the UK Continental Shelf (UKCS).

In the early licensing rounds the UK government encouraged, through its discretionary award of licenses, the rapid and thorough exploitation of the UKCS, in order to exploit the petroleum resources for their economic worth.[133] The UK argued that the allocation of licenses using discretion with weighting on the work program was required because such emphasis promoted rapid and thorough exploration and exploitation of its North Sea oil resources.[134] Oil companies proposed a work program and entered into negotiations with the UK government regarding the required work program. When agreement was reached on the terms of the license, the companies committed to a minimum program of work on the UKCS in a specified period of time. Hence, there was not only an element of work program bidding, but also negotiation with the UK government to determine the most suitable applicant for that license area.

Discretion is at the core of the Norwegian petroleum licensing allocation system, as noted by the Norwegian Petroleum Directorate:

> [The Norwegian licensing] system differs from the approach taken in many other countries in that licenses go to the companies which are best qualified for the job, rather than those making the highest bid in an auction.[135]

Unlike the UK government, the Norwegian government has exclusively used the discretionary system for the allocation of petroleum licenses since the commencement of petroleum exploration and production.[136]

---

[132] Nelsen, above n 35, 19.

[133] Ibid, 19.

[134] Select Committee in Energy, Third Report Session 1981–2, *North Sea Oil Depletion Policy* (House of Commons, London, 1982) [55], in Ursula M H Kretzer, 'Overcapitalisation in Licencing Systems Based on Size of Work Programme' (1993) 19(4) *Resources Policy* 299, 299.

[135] Johannes Kjøde, in Norwegian Petroleum Directorate, 'A Long Road to Travel: How Norway Hands out Petroleum Acreage' (2007) 2007(1) *Norwegian Continental Shelf* 24, 24.

[136] Nelsen, above n 35, 23.

The Norwegian state used discretion in the allocation of licenses as a method to stabilize offshore exploration rather than control production.[137] It was attracted to the use of discretion since it afforded the state a greater degree of control over offshore activities,[138] allowing the Norwegian government to implement its petroleum policy objectives. These policy objectives include slower rates of exploration, development of Norwegian oil companies and encouraging Norwegian industries to develop sufficient capacity to play a major role in the development of Norwegian petroleum resources.[139]

An important reason for the use of discretion in the award of licenses in Norway is the high regard for other natural resource sectors, particularly fishing. Petroleum production is seen as one of the many natural resources that are available for exploitation by the Norwegian state,[140] although there have been rather strong political conflicts, resulting in legal and political discussions on this question, especially in the 1980s, that led to the current regulation. The question of whether the rich fisheries in Lofoten should be opened for petroleum activity is currently causing a strong political struggle within the coalition government. Before any decision occurs, an analysis of the economic and environmental impact of petroleum activity in the area is carried out. When the ministry selects acreage for production licenses, resource and exploration considerations are of particular importance, and special emphasis is placed on environmental concerns and issues related to fisheries.[141] As such, if there is a clash in the exploitation of petroleum and other natural resources, the state has the right to delay or suspend petroleum activities.[142] This is easier to accomplish in a system with discretionary selection of licensees than in a system where the company has paid for the license. The importance of other natural resources is also reflected in the liability requirements for petroleum operations, where a licensee is liable, regardless of fault, for financial losses incurred as a result of pollution and waste from petroleum activities.[143]

---

[137] Ibid, 95.

[138] Ibid, 23.

[139] Ibid, 62.

[140] Petroleum Activities Act 1996 (Norway), ss 3-13.

[141] Norwegian Petroleum Directorate, *Newsletter* No. 15, October 1998 (1998), accessed 21 June 2014 at www.npd.no/engelsk/npetrres/newsletter/no1598/default.htm#Ekofisk%20IOR.

[142] Petroleum Activities Act 1996 (Norway), ss 3-13.

[143] Ibid, s 8-3.

Some academics perceive the use of administrative procedures for the allocation of petroleum licenses as questionable, since it is non-transparent, relies on discretion and decision-making authorities have limited information on relative cost efficiency.[144] However, the voluntary participation by a company in the discretionary allocation of petroleum licenses indicates their acceptance of the conditions of the award of license and confidence in the method of allocation. Further, a company's participation in petroleum activities under discretionary allocation indicates their acquiescence to the level of control imposed by the state over petroleum exploration, development and production.

## 4. CONCLUSION

Access to petroleum can be granted either using contracts, such as production sharing contracts, or through petroleum licenses. Generally, these licenses are granted through bidding, whether it be cash or work bidding, or a discretionary system that assigns licenses using selection criteria. Although the award of petroleum licenses varies between jurisdictions, there are two general systems of grant of license – the 'North Sea' system, arising in the UK and Norway in the 1960s and focusing on the discretionary allocation, or the North American system, with its emphasis on the bid system. This chapter has outlined the types of systems and the theoretical basis of each system. Chapters 3 to 11 will focus on the award of licenses in individual countries, both developed and developing. An analysis of the grant of licenses will be undertaken in Chapter 12, demonstrating the similarities and differences in the grant of access to petroleum in developed and developing petroleum jurisdictions.

---

144 Sunnevåg, above n 40, 3.

# PART II

# Comparative petroleum regulation in mature petroleum provinces

# 3. Licensing and concession system for developing Australia's conventional petroleum resources

**Tina Hunter**

## 1. HISTORY AND OVERVIEW

The history of Australia's offshore petroleum resource development parallels that of other offshore provinces, particularly that of the North Sea. Whereas the history of petroleum in the US is dominated by the recovery of petroleum from onshore fields, and then the gradual movement into calm shallow waters offshore, offshore petroleum development in Australia centred on the offshore zone. Similar to the North Sea, petroleum was discovered in the mid 1960s in the torrid offshore area of Bass Strait. In 1960 Standard Oil's geologist Lewis Weeks recommended that BHP (now BHP Billiton) explore Bass Strait, predicting that it would be an oil rich province. His predictions proved correct, with the discovery of oil in Marlin field (part of the Gippsland Basin) in Bass Strait by BHP and its farm-in partner, Esso, in 1966. The nearby Halibut and Kingfisher oil fields were discovered by 1968. All three fields came online by the early 1970s, and Bass Strait became the dominant petroleum-producing province in Australia.[1] From the oil boom of the late 1960s, through to the early 1990s, Bass Strait has been Australia's dominant oil-producing area. This was further enhanced by the discovery and development of oil fields in the offshore Gippsland Basin. By 1985 the Gippsland Basin reached its peak, producing an average of 450,000 bbl/d. Since this peak, basin production has steadily declined, with production levels hovering at

[1] For an excellent outline of the history of petroleum in Australia to 1988, see Australian Academy of Technological Sciences and Engineering, *Technology in Australia, 1788–1988* (Australian Academy of Technological Sciences and Engineering, Melbourne, 1988), Chapter 11.

or below 80,000 bbl/d since 2006. Even with this decline, the Gippsland Basin still contributes around 19 per cent of Australia's crude production.

Despite the production from the fields in Bass Strait, petroleum exploration activity continued in other offshore areas. Woodside, Shell and Burmah began drilling on the North West Shelf (NWS) off the north-western coast of Western Australia (WA) in 1968, followed by drilling in the waters off Barrow Island in 1969. Exploration activity continued on the NWS during the 1970s. However, water depth and geology limited exploration activities until the development of anchorless drill ships in the late 1970s enabled drilling in deeper provinces.

Throughout the 1980s discoveries and developments on the NWS continued. Rather than an abundance of oil, discoveries indicated that the NWS contained huge gas fields, including the Gorgon, Pluto, Wheatstone and Ichthys fields. However, distance to markets, lack of interest in gas as a petroleum commodity and cost of developing the fields meant that the development of these abundant gas reserves was delayed under a system of retention titles. The development of onshore pipeline infrastructure, technological development in the transport of LNG, domestic gas requirements and subsequent DomGas reservation,[2] and the increased energy requirements of Australia's Asian neighbours has enabled the development of NWS gas resources since the late 1990s.

Whereas the southern fields, particularly the Gippsland and Otway Basins in Bass Strait, dominated the first 30 years of Australia's offshore petroleum history, today the NWS, particularly the Browse, Carnarvon and Bonaparte Basins, contains an abundance of gas (almost 160 trillion cubic feet, or tcf), as well as oil reserves, and is Australia's dominant petroleum hub. Together these NWS basins are home to almost 80 per cent of Australia's offshore petroleum reserves and hold approximately 85 per cent of Australia's gas reserves, enabling Australia to provide approximately 10 per cent of the world's liquefied natural gas (LNG), with the majority produced in WA from the NWS LNG trains. The demand for WA resources is evidenced by the placement of WA in the top 25 global jurisdictions for investment in petroleum exploration in 2009.[3] The attractiveness of WA as a global petroleum heavyweight is

[2] DomGas reservation refers to the policy of the Western Australian government that requires all gas producers to reserve a percentage of their production for domestic use in Western Australia. The policy was implemented to ensure a secure and affordable domestic gas supply to meet Western Australia's long term energy needs and sustain economic growth in the state.

[3] Fraser Institute, *Global Petroleum Survey, 2010* (Fraser Institute, Vancouver, 2010), 17.

also demonstrated by WA's ranking as the sixth largest global producer of LNG, with the capacity to become the second largest global producer when planned new LNG facilities come online. While the majority of Australia's remaining offshore petroleum reserves are located in the north-western geological provinces, other basins, such as the Perth and Mentelle Basins, are newly explored or awaiting development. As such, Australia claims both mature and frontier offshore petroleum resources.

Onshore petroleum exploration and production dates back significantly further than offshore activities, with onshore resources meeting Australia's gas requirements. Initial exploration activities primarily took place in South Australia, extending back as early as 1886, with wells sunk in the area near the Otway Basin. The Cooper and Eromanga Basins in Central Australia have been the most abundant and important onshore petroleum basins, with exploration activities culminating in the Moomba gas field in the 1960s, which was developed in the 1970s. This field has been extensively developed and connected to domestic markets since the 1970s, with the construction of a vast pipeline network to Sydney, Brisbane and Melbourne. Other conventional petroleum discoveries include the discovery of the Mereenie oil and gas field in the Amadeus Basin (Northern Territory) in the 1960s and the nearby Palm Valley gas field in the 1960s. The development of these fields was delayed until the 1980s, primarily due to the inaccessibility of markets for these remote petroleum resources. These fields have now been largely depleted, with the Northern Territory (NT) relying on gas from the NT Blacktip field to fulfil domestic energy requirements.

While conventional gas reserves onshore (particularly from the Cooper-Eromanga Basin) have been developed since the 1970s, the development of the unconventional coal seam gas (CSG) reserves in Queensland's Surat Basin has increased exponentially since the early 2000s. Initial CSG development in Australia began in the Bowen Basin (Queensland (Qld)) in 1976 when Houston Oil and Minerals of Australia Incorporated drilled two unsuccessful wells. In February 1996 the first commercial methane gas project commenced at the Moura mine in Central Queensland with a methane drainage project from the coal mine. The first stand-alone commercial production of CSG commenced in December 1996 at the Dawson Valley project adjoining the Moura coal mine in the Surat Basin. With the development of technologies and techniques to recover gas resources from unconventional geologies, the Bowen Basin and adjoining Surat Basin coals have become a vital source of unconventional gas resources, with large-scale exploration and development undertaken since the early 2000s.

Australian petroleum activities today are dominated by offshore petroleum, primarily on the NWS. There has been the development of unconventional fields, particularly CSG in Queensland, and emerging exploration of shale gas in Western Australia. However, given that this book addresses the recovery of conventional petroleum, the regulation of shale and coal seam gas shall not be considered.

## 2. LEGAL FRAMEWORK FOR PETROLEUM ACTIVITIES

### Overview of Regulation of Offshore Petroleum Activities in Australia

Jurisdiction over Australia's offshore petroleum resources is conferred under the United Nations Convention on the Law of the Sea (UNCLOS) 1982. The maritime zone is divided into three zones, as defined under UNCLOS:

- the Territorial Sea: from the baseline (meaning the low-water mark) seaward 12 nautical miles;
- the Exclusive Economic Zone (EEZ): from 12 nautical miles seaward to 200 nautical miles;
- the Continental Shelf: from 200 nautical miles to the edge of the Continental Shelf, as decided by the United Nations Commission on the Limits of the Continental Shelf, to a minimum distance of 200 nautical miles from the baseline.

Relevant Articles of UNCLOS regulate these maritime zones. Article 2 of UNCLOS confers sovereignty of Australia's Territorial Sea upon the Commonwealth. Similarly, sovereignty over Australia's EEZ is conferred under Article 56 of UNCLOS. Australia made a formal claim to its Continental Shelf in 1953 and that claim was recognized by the Convention on the Continental Shelf 1958 and then by UNCLOS. In April 2008 Australia's offshore jurisdiction was conferred an area of the Continental Shelf as a result of a successful submission to the United Nations Commission on the Limits of the Continental Shelf.[4] With this extension to Australia's maritime zone, Australia attained sovereignty over the

---

[4] Geoscience Australia, 'Australia's Maritime Jurisdiction Extended' (2008), available 12 December 2013 at www.ga.gov.au/ausgeonews/ausgeonews 200806/inbrief.jsp.

mineral and other non-living resources of the seabed and subsoil of the extended Continental Shelf under Article 77 of UNCLOS.

Since the Australian government is based on federalism, there are particular constitutional, legislative and administrative challenges that govern the exploitation of offshore petroleum in Australia, and these have influenced Australia's regulatory regime for its offshore petroleum resources. Australia's six colonies became states upon Federation, and two territories (the Northern Territory and the Australian Capital Territory, which has no petroleum resources) were subsequently established. If you include the Commonwealth, in all there are nine jurisdictions, all with the constitutional capacity to make laws with respect to Australia's maritime area. As such, this makes the regulation of the offshore jurisdiction complex and at times unwieldy.[5]

In order to understand the offshore petroleum regulatory regime that exists today it is important to understand the historical framework. Under colonial law, a colony's land finished at low-water mark (LWM) as an inherited legal regime from the English legal system.[6] Upon Federation the question of jurisdiction over the seas remained unimportant, with the 1901 constitution largely silent on offshore jurisdiction except for s 51(x) (fisheries in Australian waters beyond territorial limits). Colonial jurisdiction over the seas finished at 3 nautical miles (nm), which was the jurisdiction that had been established for criminal and fishery jurisdictions within the colonies.[7] Consequently, former colonies (now states) continued to assert jurisdiction over the offshore areas they had previously asserted control over, especially in the area of fisheries.

The issue of the offshore jurisdiction of the states and Commonwealth arose with the discovery of petroleum in Bass Strait in the 1960s. Initial arrangements between the Commonwealth and states for the exploration and production of offshore petroleum were created under the 1967 Agreement Relating to the Exploration for, and Exploitation of, the Petroleum Resources and Certain Other Resources of the Continental Shelf of Australia and of Certain Territories of the Commonwealth and of Certain Other Submerged Land (the Agreement),[8] that was forged

---

[5] As noted by Terence Daintith in T Daintith, 'A Critical Evaluation of the Petroleum (Submerged Lands) Act as a Regulatory Regime' (2000) *AMPLA Yearbook 2000* 91, 93.

[6] M White, *Australian Offshore Laws* (Federation Press, Sydney, 2009), Chapter 2.

[7] Ibid.

[8] C Hunt, *The Offshore Petroleum Regimes of Canada and Australia* (Calgary Institute of Resources Law, Calgary, 1989), 63.

between the Commonwealth, states and affected territories. The Agreement did not intend to create legal relationships enforceable in a court of law.[9] Rather, it noted that petroleum activities would be encouraged by uniform legislative measures on the Continental Shelf beyond territorial limits, and that the state and national governments would cooperate to ensure effectiveness of authorities over petroleum resources.[10] The legal status of the Agreement was made clear in clause 26:

> the Governments acknowledge that this Agreement is not intended to create legal relationships justiciable in a Court of Law but declare that the Agreement shall be construed and given effect to by the parties in all respects according to the true meaning and spirit thereof.[11]

To achieve constitutional legitimacy and maintain practicality, each state and territory government legislated with respect to offshore petroleum operations in identical terms to the Commonwealth (Cth) petroleum legislation, the Petroleum (Submerged Lands) Act 1967 (Cth) (PSLA), known as 'mirror' legislation.[12] In addition, all governments agreed not to make, amend or repeal regulations under the legislation except under a prior agreement to do so.[13]

This regulatory relationship between the states and the Commonwealth altered in 1973 when the Commonwealth claimed the offshore maritime zones from baseline in the Sea and Submerged Lands Act 1973 (Cth) (SSLA). This raised the ire of the states, who argued that the colonies had historical jurisdiction over the territorial sea area at Federation, and therefore continued to exert jurisdiction post-Federation. The New South Wales government challenged the constitutional validity of this legislation in the High Court.[14] New South Wales contended that they held rights in the territorial waters from the baseline seaward 3 nautical miles, the same as those rights for fishing as addressed in *Bonser v La Macchia*.[15] The full High Court in *New South Wales v Commonwealth*[16] held that sovereign rights in relation to the Continental Shelf outside the

---

[9] Ibid, 64.

[10] Ibid, 63.

[11] See M Crommelin, 'The Legal Character of Petroleum Production Licenses in Australia' in T Daintith, *The Legal Character of Petroleum Licences: A Comparative Study* (University of Dundee, Dundee, 1981), 61.

[12] Ibid, 62; and *Bonser v La Macchia* (1969) 122 CLR 177.

[13] Crommelin, above n 11, 62.

[14] *New South Wales v Commonwealth* (1975) 8 ALR 1.

[15] *Bonser v La Macchia* (1969) 122 CLR 177.

[16] *New South Wales v Commonwealth* (1975) 8 ALR 1.

territorial waters were vested in the Commonwealth. Furthermore, with the exception of Gibbs and Stephen JJ in dissent, the Justices also concurred that the sovereignty in territorial waters was vested in the Commonwealth.[17] As a consequence of this decision, the states were denied property rights in the seabed and subsea resources of the territorial waters, since their territory ended at low-water mark.[18] This decision had a major impact on the states' jurisdiction over, and income from, offshore petroleum, and prompted negotiations between the states/territories and the Commonwealth.

A permanent solution to the constitutional issue was reached in the Offshore Constitutional Settlement (OCS) in 1979 with the states granted property and regulatory rights from the baseline seaward to 3 nautical miles (known as Coastal Waters). The OCS was legitimized through the mirroring of the Petroleum (Submerged Lands) Acts 1967 by the states and the Northern Territory (states/NT) for Coastal Waters.[19] In addition, a plethora of other necessary legislation was enacted to enable the implementation of the OCS.[20] The terms of the OCS, which remain today, are outlined in s 5(2) of the Offshore Petroleum and Greenhouse Gas Storage Act 2008 (Cth) (OPGGSA):

(a) Commonwealth offshore petroleum legislation should be limited to the area that is outside the coastal waters of the States and the Northern Territory [Commonwealth Waters]; and
(b) for this purpose, the outer limits of State and Northern Territory Coastal Waters are three nautical miles from the baseline of the territorial sea; and

---

17 Ibid.

18 P Brazil, *Offshore Constitutional Settlement 1980: A Case Study in Federalism* (Centre for International and Public Law, Faculty of Law, Australian National University, Canberra, 2001), 2.

19 Petroleum (Submerged Lands) Act 1967 (Cth); Petroleum (Submerged Lands) Act 1981 (NT); Petroleum (Submerged Lands) Act 1982 (NSW); Petroleum (Submerged Lands) Act 1982 (Qld); Petroleum (Submerged Lands) Act 1982 (SA); Petroleum (Submerged Lands) Act 1982 (Tas); Petroleum (Submerged Lands) Act 1982 (Vic); and Petroleum (Submerged Lands) Act 1983 (WA).

20 Required Acts include Coastal Waters (State Powers) Act 1980 (Cth); Coastal Waters (Northern Territory Powers) Act 1980 (Cth); Coastal Waters (State Title) Act 1980 (Cth); and Coastal Waters (Northern Territory Title) Act 1980 (Cth).

(c) the States and the Northern Territory should share, in the manner provided by [the OPGGSA], in the administration of the Commonwealth offshore petroleum legislation; and
(d) State and Northern Territory offshore petroleum legislation should apply to State and Northern Territory Coastal Waters; and
(e) the Commonwealth, the States and the Northern Territory should try to maintain, as far as practicable, common principles, rules and practices in regulating and controlling the exploration for, and exploitation of, offshore petroleum beyond the baseline of Australia's territorial sea.[21]

The regulatory framework for offshore petroleum activities in Australia is complex. There are three regulatory zones for petroleum activities and therefore three different laws. Under s 5(2)(c) of OPGGSA, the states/NT are required to share in the administration of the Commonwealth offshore petroleum legislation. As part of the shared regulation of petroleum resources, a joint authority (JA) for each state or territory was initially established under the PSLA, and maintained under the OPGGSA.[22] Under this shared arrangement, the states/NT regulate activities in State Waters, which are those areas wholly within the states' jurisdiction. The states/NT also regulate petroleum activities in Coastal Waters (as outlined in s 52(b) of the OPGGSA). Petroleum activities in Commonwealth Waters are regulated under the OPGGSA and its subordinate legislation.

The principal legislation for the regulation of offshore petroleum activities in Coastal Waters is the 'mirror' PSLA legislation that was implemented by each state/NT in the early 1980s. The legislation enacted by each state/NT 'mirrors' the principal Commonwealth legislation (now the OPGGSA). The principle of mirror legislation is that the law of the states reflects, or mirrors, that of the Commonwealth law in regulating petroleum activities in order to achieve and maintain consistency across the jurisdictions.[23] Since the implementation of the OPGGSA in 2008, most jurisdictions have made alterations to the existing petroleum legislation, although only Victoria has rewritten and renamed its legislation to

---

21 Offshore Petroleum and Greenhouse Gas Storage Act 2006 (Cth) s 5(2) (OPGGSA).

22 Ibid, Pt 1.3.

23 See Australasian Parliamentary Council Committee, 'National Uniform Legislation: Acts of Jurisdictions Implementing Uniform Legislation [as at June 2012]' (2012), available 1 June 2013 at www.pcc.gov.au/uniform/National%20 Uniform%20Legislation%20table.pdf.

mirror OPGGSA. However, the key underlying principles for the regulation of petroleum activities remain the same despite the differences that exist in the legislation in some states. The relevant legislation for each state for offshore petroleum activities in State Waters and Coastal Waters is outlined in Table 3.1.

Given that these arrangements are mirror arrangements and that most offshore petroleum activities in Australia occur in Commonwealth Waters, this chapter will focus on the grant of licenses and regulation of petroleum activities in Commonwealth Waters, noting that the system used in state/NT waters generally emulates the Commonwealth arrangements.

## Regulatory Arrangements for Commonwealth Waters 1980–2012

Australia's legislative regime for the regulation of petroleum activities can be characterized as a rule-based legislative system. The original petroleum legislation, the Petroleum (Submerged Lands) Act 1967 (Cth) (PSLA),[24] was conceived as an ingenious legal mechanism to give effect to the Petroleum Agreement,[25] securing offshore petroleum development without having to resolve the jurisdictional issues between the Commonwealth and the states.[26] It addressed the constitutional demarcation of jurisdictions by enacting a comprehensive legislative 'code' in 1967, creating joint Commonwealth–state administration of petroleum titles.[27] Much of the detail ordinarily contained in regulations was incorporated into the PSLA, since if administrative delegation from the Commonwealth to the states/NT occurred, there was a risk of variation or conflict between the Commonwealth and states/NT.[28] To reduce the capacity for the states/NT to go their own way, the PSLA was necessarily detailed, outlining each state or territory's legislative capacity to grant dual titles to

---

[24] Note that the original petroleum legislation was the PSLA. It was rewritten and entered into force in 2006 as the Offshore Petroleum Act 2006 (Cth) (OPA). In 2008 the OPA was amended to contain greenhouse gas provisions, and the title was amended to become the Offshore Petroleum and Greenhouse Gas Storage Act 2006 (Cth) (OPGGSA).

[25] Hunt, above n 8, 64.

[26] T Daintith, *Discretion in the Administration of Offshore Oil and Gas* (AMPLA, Melbourne, 2005), 13.

[27] Ibid. This joint management required the establishment of two authorities that regulate petroleum activities: the joint authority, which comprises the relevant Commonwealth minister and the responsible state minister, and the designated authority, comprising the responsible state or territory minister.

[28] Ibid, 13.

*Table 3.1 Relevant Acts and Regulations for State and Coastal Waters*

| Jurisdiction | Relevant legislation | |
|---|---|---|
| | *Onshore and internal waters* | *Offshore – Coastal Waters* |
| New South Wales | – Petroleum (Onshore) Act 1991<br>– Petroleum (Onshore) Regulation 2007 | – Petroleum (Offshore) Act 1982<br>– Petroleum (Offshore) Regulation 2006 |
| Northern Territory | – Petroleum Act 1984<br>– Schedule of Onshore Petroleum Exploration and Production Requirements 1993 | – Petroleum (Submerged Lands) Act 1981<br>– Petroleum (Submerged Lands) Regulations 2000 |
| Queensland | – Petroleum Act 1923<br>– Petroleum and Gas (Production and Safety) Act 2004<br>– Petroleum and Gas (Production and Safety) Regulation 2004<br>– Petroleum Regulation 2004 | – Petroleum (Submerged Lands) Act 1982 |
| South Australia | – Petroleum and Geothermal Energy Act 2000<br>– Petroleum and Geothermal Energy Regulations 2000 | – Petroleum (Submerged Lands) Act 1982 |
| Tasmania | | – Petroleum (Submerged Lands) Act 1982 (Tas) Petroleum (Submerged Lands) (Management of Environment) Regulations 2002<br>– Petroleum (Submerged Lands) (Diving Safety) Regulations 2008<br>– Petroleum (Submerged Lands) (Management of Safety on Offshore Facilities) Regulations 2008<br>– Petroleum (Submerged Lands) (Occupational Health and Safety) Regulations 2008<br>– Petroleum (Submerged Lands) (Pipelines) Regulations 2008 |
| Victoria | – Petroleum Act 1998<br>– Petroleum Regulations 2000<br>– Pipelines Act 2005<br>– Pipelines Regulations 2007 | – Petroleum (Submerged Lands) Act 1982<br>– Petroleum (Submerged Lands) Regulations 2004 |
| Western Australia | – Petroleum Act 1936<br>– Petroleum and Geothermal Energy Resources Act 1967<br>– Petroleum Pipelines Act 1969<br>– Petroleum Pipelines Regulations 1970<br>– Petroleum and Geothermal Energy Resources Regulations 1987 | – Petroleum (Submerged Lands) Act 1982<br>– Petroleum (Submerged Lands) Regulations 1990<br>– Petroleum (Submerged Lands) (Management of Well Operations) Regulations 2006 |

| Jurisdiction | Relevant legislation | |
|---|---|---|
| | *Onshore and internal waters* | *Offshore – Coastal Waters* |
| | – Schedule of Onshore Petroleum Exploration and Production Requirements 1991<br>– Petroleum and Geothermal Energy Resources (Management of Safety) Regulations 2010<br>– Petroleum and Geothermal Energy Resources (Occupational Safety and Health) Regulations 2010<br>– Petroleum Pipelines (Management of Safety of Pipeline Operations) Regulations 2010<br>– Petroleum Pipelines (Occupational Safety and Health) Regulations 2010 | – Petroleum (Submerged Lands) (Diving Safety) Regulations 2007<br>– Petroleum (Submerged Lands) (Management of Safety on Offshore Facilities) Regulations 2007<br>– Petroleum (Submerged Lands) (Occupational Safety and Health) Regulations 2007<br>– Petroleum (Submerged Lands) (Pipelines) Regulations 2007 |

oil companies under state authority and delegated authority from the Commonwealth.[29]

The PSLA was implemented as 'a combination of painstaking detail and grand-scale delegation' in 1967,[30] and remains today (in its new incarnation of the OPGGSA) as a coherent, but highly unusual, system of offshore petroleum regulation, articulated by administrative rules and powers within the principal Act.[31] It operates by prohibiting the activities it covers (for example, extracting petroleum), and then granting companies an administrative authorization to conduct the activity (this is known as a 'command and control' scheme of authorization).[32] This is demonstrated by s 97(1) of the OPGGSA, where 'a person commits an offence if that person explores for petroleum and the exploration occurs in an offshore area'.[33] It then excuses the offence of taking of petroleum under s 97(2) where the taking of petroleum is 'authorised by a petroleum exploration permit, or otherwise required by or under this Act'.[34]

As a result of 40 years of changes to petroleum licensing and activities, the detailed, rule-based PSLA required over 1000 amendments from 1968–2005, resulting in over 30 separate compilations of the Act.[35] A

29 Ibid, 13.

30 Ibid, 13.

31 Daintith, above n 5.

32 Ibid, 13–14. The prescriptive, rule-based, legislative structure created under the PSLA continues today under the OPGGSA.

33 Offshore Petroleum and Greenhouse Gas Storage Act 2006 (Cth), s 97(1).

34 Ibid, s 97(2).

35 The full legislative history of the Petroleum (Submerged Lands) Act 1967 (Cth) can be found at www.comlaw.gov.au, available 14 April 2014.

replacement, the Offshore Petroleum Act 2006 (Cth) (OPA), was enacted in 2006, as a 'plain English rewrite' of the PSLA.[36] By the new millennium, industry and governments alike had identified the PSLA as cumbersome, unwieldy and complex, as the result of numerous amendments and updates.[37] The OPA contained only alterations to the structure and style of the legislation, and implemented only a few minor policy changes from the regime set out in the PSLA.[38]

Whereas the PSLA had been 391 pages, the rewritten OPA was over 630 pages. However, it would appear that the OPA was no better than its predecessor. The PSLA was described by Professor Daintith as 'old, fat and ugly, and not likely to score highly in a legislative beauty contest'. Furthermore, Daintith expressed concern over its replacement (the OPA), noting in 2004 that 'replacing an Act … by one which is even fatter would be a profoundly disappointing result'.[39] Unfortunately, Daintith's concern was realized, with the rewrite of the PSLA being a fatter and uglier 630 page prescriptive tome that required over 100 amendments prior to its commencement in July 2008.[40]

The OPA became even 'fatter and uglier' in its new incarnation as the OPGGSA in 2008. This new incarnation (and new name) arose as a result of the addition of offshore greenhouse gas storage provisions into the OPA. It continues the trend set by its predecessors (the PSLA and OPA) in that it is rule-based and highly prescriptive, demonstrated by the length and level of detail of the legislation.[41] At over 830 pages, the legislation addresses in minutiae the award and management of licenses and titles, safety arrangements for offshore petroleum activities and the jurisdiction of the regulatory authorities.[42]

The OPGGSA regulates all aspects of offshore petroleum exploration, production and geosequestration of greenhouse gas in Commonwealth Waters (beyond 3 nautical miles) in nine chapters and seven schedules:

---

36 Explanatory Memorandum, Offshore Petroleum Bill 2005 (Cth), 2.

37 Ibid, 2.

38 Ibid, 2.

39 Daintith, above n 13, 43.

40 In its first incarnation as the Offshore Petroleum Act 2006 (Cth) (OPA).

41 The OPGGSA found its first incarnation as the OPA in 2006. Although enacted in 2006, the OPA did not enter into force until 1 July 2008 after all relevant jurisdictions had enacted 'mirror' legislation. In November 2008 the OPA was renamed the OPGGSA, and the greenhouse gas storage provisions were incorporated by the Offshore Petroleum Amendment (Greenhouse Gas Storage) Act 2008.

42 Offshore Petroleum and Greenhouse Gas Storage Act 2006 (Cth).

- Chapter 1: Introduction;
- Chapter 2: Regulation of activities relating to petroleum, including licenses;
- Chapter 3: Regulation of activities relating to greenhouse gas substances;
- Chapter 4: Registration of transfers and dealings in petroleum, including farm-ins;
- Chapter 5: Registration of transfers and dealings in greenhouse gas titles;
- Chapter 6: Administration, including the functions and powers of the regulatory bodies;
- Chapter 7: Information relating to petroleum (including data management and data submission requirements);
- Chapter 8: Information relating to greenhouse gas (including data management);
- Chapter 9: Miscellaneous;
- Schedule 1: Scheduled areas for the states and territories;
- Schedule 2: Area that includes the area to be avoided;
- Schedule 3: Occupational health and safety (the primary safety provisions);
- Schedule 4: Petroleum production licenses to which the Petroleum Royalty Act applies;
- Schedule 5: Release of technical information given to the Designated Authority before March 2000;
- Schedule 6: Transitional provisions; and
- Schedule 7: Greater sunrise areas.

This detailed legislative regime is accompanied by general regulations, the Offshore Petroleum and Greenhouse Gas Storage Regulations 1985 (OPGGSR),[43] and is also supported by three operational regulations: Offshore Petroleum and Greenhouse Gas Storage (Safety) Regulations 2009, Offshore Petroleum and Greenhouse Gas Storage (Environment) Regulations 2009, and the Offshore Petroleum and Greenhouse Gas Storage (Resource Management and Administration) Regulations 2009.

The initial regulatory framework for all offshore petroleum activities under the OCS was that of the Joint Authority/ Designated Authority arrangement. The Joint Authority (JA) comprised the responsible state

[43] Offshore Petroleum and Greenhouse Gas Storage Regulations 1985 (Cth). These regulations began life as the Petroleum (Submerged Land) Regulations 1985 (Cth), with a name change on 1 July 2008.

minister and the responsible Commonwealth minister,[44] with the Commonwealth minister having the power to decide a matter where the Commonwealth and state ministers disagree.[45] Similarly, there was a Designated Authority (DA) for each state/NT offshore Commonwealth area, with the responsible minister of the state or territory being the DA for that offshore area.[46]

The responsibility for the administration of the exploration and production of offshore petroleum from 1982 was therefore divided between:

- the JA for the state or territory (the JA is constituted by the responsible state/territory minister and the responsible Commonwealth minister); and
- the DA of the state or territory (responsible to the state/territory minister).[47]

In the event of a disagreement in the JA, the relevant petroleum legislation dictated that the Commonwealth view prevailed.

Initially health and safety arrangements for offshore petroleum activities were shared between the Commonwealth and states/NT under the PSLA. However, the Commonwealth recognized the need for a single offshore petroleum safety regulator to cover both Commonwealth and states/NT offshore areas after the Piper Alpha platform disaster in 1988. Legislative reforms in the early 2000s to the then principal petroleum Act, the PSLA, created a new national offshore petroleum regulator, the National Offshore Petroleum Safety Authority (NOPSA), and also defined its role, structure and governance. Amendments were also made to the occupational health and safety provisions of the PSLA. Furthermore, major legislative reform to all state/NT mirror legislation was undertaken to implement the new national regulator in State and Coastal Waters where those states conferred upon the Commonwealth the right to regulate offshore petroleum safety.

As a consequence of the establishment of NOPSA in 2005 and the OCS, multiple agencies were responsible for the regulation of petroleum activities under the OPGGSA, as illustrated in Figure 3.1:

- the day-to-day petroleum operations, pipelines and subsea facilities in Commonwealth Waters are regulated by the JA (comprising the

[44] Offshore Petroleum and Greenhouse Gas Storage Act 2006 (Cth), s 56.
[45] Ibid, s 59.
[46] Ibid, s 70.
[47] Ibid, s 4.

relevant Commonwealth and state ministers), with the JA delegating the day-to-day regulation of petroleum operations, including pipelines and subsea facilities, to the DA but retaining decision-making authority;

- the day-to-day petroleum operations, pipelines and subsea facilities in Coastal and State (internal) Waters are regulated by the relevant state agency;
- environmental aspects of offshore petroleum activities in all waters are regulated by the relevant state agency under the JA/DA arrangement; and
- safety of all offshore facilities in all waters, except WA State and Coastal Waters, are regulated by NOPSA.

This regulatory arrangement is reflected in Figure 3.1.

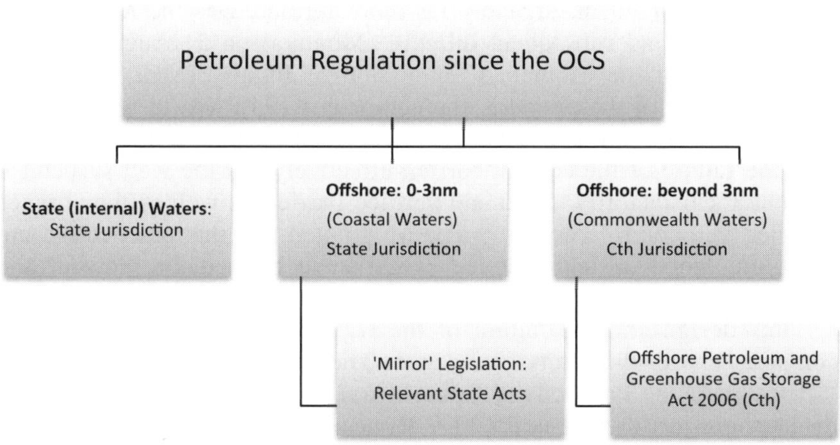


*Figure 3.1 Petroleum regulatory framework for offshore jurisdictions under the Joint Authority/Designated Authority arrangement 1980–2012*

**The offshore regulatory regime since 2012**

In 2008 and 2009, two potentially lethal petroleum facility incidents occurred in Australia's offshore petroleum jurisdictions. These incidents led the petroleum industry and governments alike to reassess the existing petroleum regulatory regime.

The first event was the Varanus Island gas pipeline explosion, which occurred at the Varanus Island offshore gas facility off the north-western WA coastline. It resulted from the rupture of a gas pipeline, cutting the supply of gas to WA by 30 per cent. Investigation into the incident indicated that the pipe rupture and explosion could have been avoided if facility integrity had been managed by a single agency rather than jointly by the NOPSA and the WA Department of Mines and Petroleum (WA DMP), since a single regulator would have been responsible for the facility and the pipelines carrying the gas from the production platform to markets onshore.[48] Essentially, the horizontally disjunctive regulatory system in force at the time of the Varanus explosion, which split regulation of the pipeline between state and Commonwealth jurisdictions, was a major contributor to the explosion. Furthermore, an assessment of the effectiveness of Australian offshore petroleum safety regulation by Bills and Agostini in 2009,[49] conducted after the Varanus Island explosion, noted that the failure to vertically integrate regulation of a facility to include well integrity in the regulation of offshore petroleum safety was detrimental to offshore petroleum safety.

The second incident, arguably far more serious, was the Montara H1 Well Blowout and subsequent oil spill (Montara) on 21 August 2009.[50] The findings of the Montara Commission of Inquiry (MCI) concluded that the failure of the operator, the company PTTEPAA, to use pressure control caps (PCCs) as a secondary well control barrier in conjunction with the failure of the cement barrier, effectively left the well suspended without a satisfactorily tested and verified barrier, as well as the absence of other well control barriers.[51] The MCI noted that this practice did not constitute 'good oilfield practice' (GOP), and blamed the blowout and consequential oil spill on the failure of the operator to adhere to the systems design and the failure of the regulator in its role of ensuring compliance with the safety case regime and maintaining GOP.[52]

The Australian Productivity Commission also considered regulation of the offshore jurisdiction in its 2009 *Review of Regulatory Burden on the*

---

48 K Bills and D Agostini, *Offshore Petroleum Safety Investigation: Varanus Island Incident Investigation* (Government of Western Australia, Perth, 2009).

49 K Bills and D Agostini, *Offshore Petroleum Safety Regulation: Marine Issues* (Australian Government, Canberra, 2009).

50 All other major oil spills in Australia have been the result of ship-sourced pollution. For details of all major oil spills in Australia's waters in the last 30 years, refer to Australian Maritime Safety Authority, 'Major Oil Spills in Australia' (2009) available 12 January 2014 at www.amsa.gov.au/Marine_Environment_Protection/Major_Oil_Spills_in_Australia/.

51 D Borthwick, 'Report of the Montara Commission of Inquiry' (2010), Department of Resources, Energy and Tourism, available 12 December 2014 at www.ret.gov.au/Department/Documents/MIR/Montara-Report.pdf.

52 Ibid, 16–17.

*Upstream Petroleum (Oil and Gas) Sector* (the PC Report).[53] It criticized the regulatory arrangements of the offshore petroleum sector, describing the regime as 'complex'.[54] This complexity was illustrated by highlighting that 22 petroleum and pipeline laws applied at both the Commonwealth and state/NT levels with more than 150 statutes governing upstream petroleum activities in such areas as occupational health and safety, native title and environmental protection.[55] Most importantly, the report identified that well over 50 Commonwealth, state and territory government agencies regulate upstream petroleum activities, incorporating the regulation of health and safety, facility integrity, resource management, well operations and environmental issues.[56]

A major area of regulatory inconsistency has been well regulation. In its submission to the Productivity Commission in 2009, the Australian Petroleum Production and Exploration Association (APPEA) concluded that the areas requiring major reform were the administration of Well Operations Management Plans (WOMPs), subsea equipment and pipelines.[57] This reform was required because regulatory responsibilities for these activities were divided between the DAs and NOPSA and these activities carry the most risks to the integrity of a facility. APPEA also noted that regulatory interaction and uniformity are critical to the safety performance of operations and should be regulated by a single body.[58]

The PC Report concluded that the sum effect of the large number of statutes regulating offshore petroleum, split over a large number of regulatory bodies, is a regulatory burden for stakeholders and regulatory inconsistencies. Such inconsistencies may result in either regulatory duplication or, worse still, regulatory gaps that result in incidents at offshore facilities.

The inquiries into the Varanus and Montara incidents attributed causation of both events to the existing regulatory regime, thus demonstrating that the structure and function of the petroleum regulatory regime

---

[53] Productivity Commission, *Review of Regulatory Burden on the Upstream Petroleum (Oil and Gas) Sector* (Productivity Commission, Melbourne, 2009), 34.

[54] Ibid, p XXIII.

[55] Ibid, p VXIII.

[56] Ibid, p XXIII.

[57] APPEA, 'Submission No 16 to Productivity Commission', *Review of the Regulatory Burden on the Upstream Petroleum (Oil and Gas) Sector*, 19 September 2008, available 12 June 2014 at www.pc.gov.au/__data/assets/pdf_file/0019/83422/sub016.pdf, 20.

[58] Ibid.

significantly contributed to facility incidents in Australia's offshore petroleum sector. In particular, the Montara Blowout and subsequent MCI demonstrated to the public that the robust regulatory system that should have been inherent was, in fact, absent. The MCI also concurred with the 2008 recommendations from the 2008 NOPSA Report, the 2009 recommendations of the Productivity Commission, and the recommendations of Bills and Agostini in 2009 that the legislated coverage of NOPSA should be extended to cover the whole of the petroleum production chain, similar to the Norwegian approach to facility integrity regulation.[59]

In 2009, the Commonwealth Government embarked on a program of regulatory reform. It commenced with the consolidation of numerous existing offshore petroleum regulations and guidelines, replacing the prior system of Directions of General Application. The Offshore Petroleum and Greenhouse Gas (Safety) Regulations (2009) (Cth) brought together previous safety regulations into a single set of regulations.[60] Also in 2009, significant amendments were made to the Offshore Petroleum and Greenhouse Gas Storage Act (Environment) Regulations 1999 (Cth). The final part of the reform to offshore petroleum regulations was completed in April 2011, where the remaining regulations were combined and amended, becoming the Offshore Petroleum and Greenhouse Gas Storage (Resource Management and Administration) Regulations 2011 (Cth) (RMA Regulations). In addition, reform was made to the OPGGSA, conferring upon NOPSA functions and powers in relation to the non-occupational health and safety structural integrity of facilities, wells and well-related equipment.

These reforms established a vertically integrated approach to petroleum regulation, with NOPSA responsible for the regulation of facility activities and well activities. This is a significant improvement on the vertically disconnected regulatory regime that existed previously where platform and facilities were regulated by NOPSA and the DA managed WOMPs/well operations.

As a consequence of the PC Report, NOPSA report, Bills and Agostini's report and the findings from the Varanus and Montara inquiries, the Commonwealth Government also announced in August 2009 an intention to create a national offshore petroleum regulator (NOPR), to

---

[59] Bills and Agostini, above n 49, 19–20.

[60] Offshore Petroleum and Greenhouse Gas (Safety) Regulations (2009) (Cth), s 1.4.

commence operations by 1 January 2012.[61] The Commonwealth noted that fundamental institutional reform for offshore petroleum regulation was required as a result of the JA/DA arrangement, which led to significant regulatory duplication, slow decision-making, unclear regulatory accountability, inefficient use of regulatory resources, inconsistent decision-making across Commonwealth offshore areas, and resistance to reform.[62] Furthermore, the Commonwealth noted that the creation of a national offshore petroleum regulator had the potential to significantly reduce the time for approvals processes by reducing administrative duplication, streamlining regulatory processes, providing greater transparency in decision-making and consolidating resources.[63] Demonstrating its commitment to institutional reform, the Commonwealth declared in February 2011 that:

> in dealing with recent approvals under the Act [Offshore Petroleum and Greenhouse Gas Storage Act 2006] and the uncontrolled release of oil and gas from the Montara Wellhead Platform in the Timor Sea, it has become more clearly apparent that decisions relating to safety, environmental management and resource management have significant overlapping implications.[64]

After much contemplation regarding the most suitable structure for a national petroleum regulator, the Ministerial Council on Mineral and Petroleum Resources (MCMPR) considered the Commonwealth Government's proposed regulatory reform in February 2011. The recommended reforms tabled at the meeting sought to expand NOPSA's powers to incorporate environmental management (with the enlarged regulator to be called the National Offshore Petroleum Safety and Environmental Management Authority, shortened to NOPSEMA) and a new National Offshore Petroleum Titles Authority (NOPTA).[65] The MCMPR adopted the proposed regulatory structure, rejecting the industry call for a single safety and title regulator. By accepting the Commonwealth reform proposals, the MCMPR endorsed the splitting of regulatory functions between NOPSEMA and the new NOPTA (similar to other petroleum jurisdictions, including the UK and Norway), as well as retaining the existing JA arrangement to ensure the relevant state/NT ministers

---

61 Department of Resources, Energy and Tourism, *Commonwealth Government Response to the Productivity Commission Review* (DRET, Canberra, 2011), 8–9.

62 Ibid.

63 Offshore Petroleum and Greenhouse Gas Storage Act 2006 (Cth), s 642.

64 Department of Resources, Energy and Tourism, above n 61, 8–9.

65 Ibid, 9.

retained a voice in offshore petroleum activities adjacent to their State and Coastal Waters.

Under the NOPSEMA/NOPTA reform proposed by the Commonwealth Government:

- The JA would be retained for key title decisions.
- NOPTA, an authority within the Department of Resources, Energy and Tourism, advises the JA on resource titles, registration of titles and collection of data.
- The regulatory functions of environmental plans and compliance under the OPGGSA were moved from the JA/DA arrangement to NOPSEMA.
- The expanded NOPSEMA regulates well and pipeline integrity, environment plans and day-to-day operations of petroleum activities within Commonwealth Waters.[66]

The new regulatory regime was implemented through the amendment of OPGGSA as well as other relevant Acts and regulations. In addition, the National Regulator Amendment Act 2011 (Cth) expanded NOPSA's functions to create the new NOPSEMA and established the new NOPTA (a separate organization responsible for titles administration). The new regulators and regulatory structure commenced 1 January 2012. The new regulatory structure is illustrated in Figure 3.2.

**Regulation of onshore petroleum resources**

Jurisdiction over onshore minerals and petroleum is prima facie the responsibility of each of the states and territories, with each jurisdiction legislating to regulate those activities within its boundaries. This arises since the scheme of the Commonwealth constitution is to allocate specific powers to the Commonwealth and residual powers to the states. The Commonwealth does not have a specific power to legislate for minerals and petroleum resources. As a general matter, jurisdiction over onshore mineral and petroleum resources does not fall within the ambit of the enumerated powers in s 51 of the Commonwealth constitution. Having said that, certain of the Commonwealth's powers (such as under s 51(xx) (Corporations), or s 51(i) (Trade and Commerce))[67] are sufficiently broad to allow control over major aspects of petroleum operations. Each state's constitution lays down a right for the state to make

[66] Ibid, 9.

[67] See, for example, *Murphyores Incorporated Pty Limited v Commonwealth* (1976) 136 CLR 1.

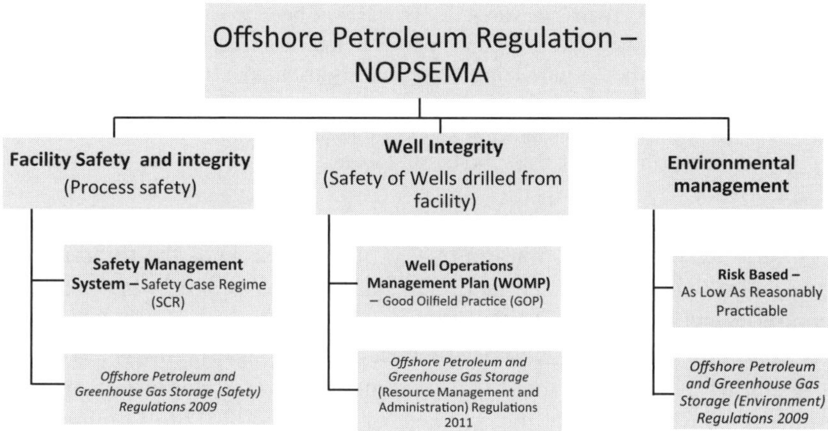


*Figure 3.2 Regulatory framework for petroleum in Commonwealth Waters since 2012*

laws with respect to the peace, welfare and good government. This enables a state to make laws with respect to the regulation of mining and petroleum activities. These powers may extend offshore. The Acts and regulations governing offshore petroleum exploration and recovery activities are outlined in Table 3.1.

## 3. OWNERSHIP AND ACCESS TO PETROLEUM RESOURCES

### Ownership of Petroleum

Australia claims its sovereign right to own and develop its petroleum resources onshore under the United Nations General Assembly Resolution 1803 (XVII) of 14 December 1962, Permanent Sovereignty over Natural Resources. Sovereignty is also protected by United Nations Resolution No 3281, Charter of Economic Rights and Duties of States, which states:

1. Every State has, and shall freely exercise, full permanent sovereignty, including possession, use and disposal, over its wealth, natural resources and economic activities …
2. Each State has the right …
   a. To nationalize, expropriate or transfer ownership of foreign property in which case appropriate compensation shall be paid by the State

> adopting such measures … In cases where the question of compensation gives rise to a controversy, it shall be settled under the domestic law of the nationalizing State and its tribunals, unless it is freely and mutually agreed by all States concerned that other peaceful means be sought on the basis of sovereign equality of States in accordance with the principle of free choice of means.[68]

Although the state claims sovereignty over the resources, there is a common misconception that the owner of the land owns the minerals and petroleum that lie under the land. This assumption is somewhat correct at common law. According to the maxim *cuius est solum eius est usque ad coelum et ad inferos* (to whomsoever the soil belongs, he owns also to the sky and to the depths), there is a presumption that a landowner owns all minerals on or beneath the surface of the land.[69] This common law presumption was subject to the Royal Prerogative on Royal Metals (gold and silver). During the period from the settlement of Australia until the enactment of the first mining laws in 1851, control over minerals and other natural resources vested with the British government.[70] When the governors of the colonies granted title to the land prior to 1851, the ownership of resources under the ground (including the minerals) passed to the landowner (the grantee) in accordance with common law concepts unless those rights were reserved from the grant. This excluded the right to the precious metals of silver and gold, which were vested in the Crown by virtue of the Royal Prerogative.[71]

This common law position has been significantly altered, and in some jurisdictions completely reversed, as a result of statutory enactments. The constitutions of New South Wales (NSW) and Victoria vested control over Crown Lands in 1855, thereby vesting ownership of minerals to the Crown in right of the state (that is, owned by the state). Since that period, ownership of minerals in all Australian jurisdictions has altered, with

---

[68] United Nations General Assembly Resolution No 3281, 12 December 1974, Charter of Economic Rights and Duties of States.

[69] For a discussion of this, see A. Bradbrook, 'The Relevance of the Cuius est Solum Doctrine to the Surface Landowner's Claims to Natural Resources Located Above and Beneath his Land' (1988) 11 *Adelaide Law Review* 462.

[70] Michael Hunt, *Minerals and Petroleum Law* (Lexis Nexis, Melbourne, 1996), 8.

[71] Ibid, 8. Royal Prerogative refers to the customs, privileges and immunities at common law conferred upon the Sovereign.

respective Mineral Acts in Australian states and territories vesting ownership of mineral resources in the Crown.[72]

The ownership of onshore petroleum resources in Australia is today vested in the Crown of each state in right of that state. Ownership of petroleum (except oil shale) in Tasmania is vested in the Crown, unless it was held prior to the declaration of Crown ownership.[73] Similarly, ownership of petroleum is vested in the Crown in Queensland under s 9 of the Petroleum Act 1923 (Qld) and s 26 of the Petroleum and Gas (Production and Safety) Act 2004 (Qld). In NSW ownership in petroleum existing in a natural state on or below the surface of any land in the state is the property of the Crown.[74] Petroleum that is found either on the surface or below the land is also owned by the state in WA, whether or not alienated in fee simple.[75] Ownership of onshore petroleum resources on Commonwealth and territories land is reserved for the Crown in right of the Commonwealth.[76]

Each state is generally recognized as having proprietary rights in the seabed of the Territorial Sea adjacent to its coast, and this is supported by Article 1 of UNCLOS. However, outside the Territorial Sea a state has sovereign rights over the seabed forming part of its Continental Shelf, rather than proprietary rights. Sovereignty over offshore resources within Australia's EEZ is clearly stated in Article 56 of UNCLOS, where:

> 1. In the exclusive economic zone, the coastal State has:
>    (a) sovereign rights for the purpose of exploring and exploiting, conserving and managing the natural resources, whether living or non-living, of the waters superjacent to the seabed and of the seabed and its subsoil, and with regard to other activities for the economic exploitation and exploration of the zone, such as the production of energy from the water, currents and winds;

Australia implemented its sovereign right over its maritime zones under the Sea and Submerged Lands Act 1973 (Cth) (SSLA), including sovereignty over internal waters under s 10 of the SSLA, the Territorial Sea (s 6 of the SSLA), the EEZ (s 10A of the SSLA) and the Continental

---

72 See, for example, s 5 of the Petroleum and Geothermal Energy Act 2000 (SA), where 'The property in petroleum and other regulated resources is vested (or continues to be vested) in the Crown'.

73 Mineral Resources Development Act 1995 (Tas), s 6.

74 Petroleum (Onshore) Act 1991 (NSW), s 6(1).

75 Petroleum and Geothermal Energy Resources Act 1967 (WA), s 9.

76 Atomic Energy Act 1953 (Cth); Minerals Acquisition Act 1981 (NT), s 3.

Shelf under s 11 of the SSLA. Consequently, sovereignty over and ownership of offshore petroleum is vested in the Crown.[77]

## Grant of Access to Petroleum

As the owner of *in situ* petroleum resources,[78] the Australian government has the unfettered right to award rights in its petroleum resources in order to exploit those resources.[79] Thus, while ownership of mineral and petroleum resources is vested in the Crown in respect of the relevant state or the Commonwealth, the Crown, pursuant to the relevant statute, grants various titles (also known as a lease, license, tenement or permit) to a lessee (also known as a licensee or permittee). This confers in a legal sense an authorization, or in some cases a right, to the licensee, lessee or permittee to enter onto land and take minerals.

In Australia, the Commonwealth Government is responsible for the granting of access to offshore petroleum resources outside the 3-mile Territorial Sea limit, while the states grant access to petroleum resources in Coastal and State Waters.

In Australia, the allocation of exploration rights occurs primarily through the use of the work program bidding (WPB) system.[80] From 1993 until 2014, Australia was one of the only petroleum-producing countries to allocate petroleum licenses under the licensing and concession system solely under the WPB.[81] Early awards of petroleum licenses (prior to 1993) in Australia were made using cash bonus bidding, particularly for highly prospective petroleum areas, with the values of winning cash bids ranging from A$1 million to A$20 million.[82] However, it should be noted that the inherent uncertainty regarding the quality of a

---

[77] Art 56 of UNCLOS.

[78] The Crown retains *in situ* rights over natural resources for two primary reasons. Firstly, the resources provide the state with high economic value. Secondly, ownership of petroleum resources enables the government to have control over the development of those resources.

[79] Art 1 of United Nations General Assembly Resolution 1803 (XVII) of 14 December 1962, Permanent Sovereignty over Natural Resources.

[80] Offshore Petroleum and Greenhouse Gas Storage Act 2006 (Cth), ss 104–9.

[81] T Daintith, *Research Report: Discretion and Prescription in the Administration of Petroleum Policy* (2005), available 6 January 2015 at www.esrc.ac.uk/my-esrc/grants/RES-000-22-0329/outputs/Download/8c52e4da-4130-46b5-a25b-3e00901126ee, 26.

[82] Trudi Rodgers and Stewart Webster, 'Resource Rent Mechanisms in Australian Primary Industries: Some Observations and Issues' (2007), paper

particular field meant that bidders discounted their cash bonus bids accordingly.[83] This form of allocation of petroleum licenses had not been used in Australia since 1993,[84] with all Australian acreage from 1993 allocated by WPB.[85] However, the Commonwealth retained legislative provisions for the award of licenses through cash bidding in Australia under OPPAGSA.[86] In 2013 the Commonwealth Government, on the recommendation of the 2010 *Australia's Future Tax System Review* implemented a system of cash bidding for licensing rounds from 2014.

The WPB system was utilized since it operates within a free market philosophy that welcomes international companies,[87] which have the necessary resources, knowledge and competence to develop the petroleum resources without state participation. This is important for Australia, since one of its primary national petroleum objectives is to encourage international investment in the offshore petroleum sector. To support the free market ethos, Australia has no mandatory local equity requirements and has no government-owned oil companies.[88] This also supports another goal of the Australian government, which is to not participate in petroleum activities.[89] By encouraging experienced international companies to bid for acreage in Australia, there are fewer requirements for state intervention or participation.

The process for the allocation of a petroleum license is outlined in Figure 3.3. It commences when applications for exploration licenses for offshore acreage areas are invited in the annual release under the work

---

presented at the 51st Annual Conference of the Australian Agricultural and Resource Economics Society, 13–16 February 2007, Queenstown, New Zealand, 12.

[83] Kenneth Dam, *Oil Resources: Who Gets What How?* (University of Chicago Press, Chicago, 1976), 6.

[84] Department of Industry, Tourism and Resources, *Cash Bidding System: A Guide in Relation to the Australian Government's Petroleum (Submerged Lands) Act 1967* (RET, Canberra 2004), 1.

[85] Rodgers and Webster, above n 82, 12.

[86] Offshore Petroleum and Greenhouse Gas Storage Act 2006 (Cth), ss 110–14.

[87] Department of Resources, Energy and Tourism, 'Offshore Acreage Release 2008: An Overview for Applicants' (2008), available 12 March 2014 at www.ret.gov.au/resources/Documents/Acreage%20Releases/2008/site/page2.htm, 3.

[88] Ibid.

[89] Ibid, 3–4.

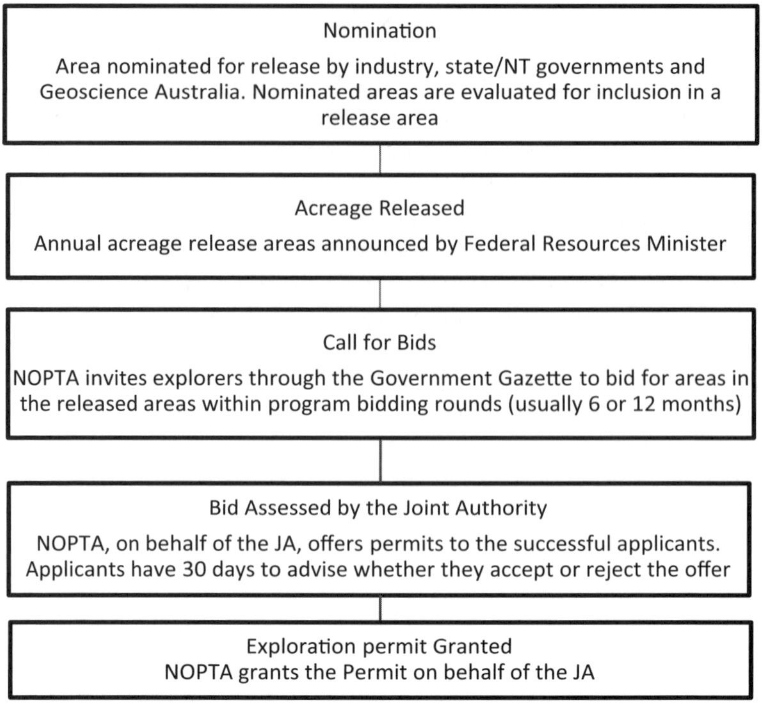


*Source:* Hunter and Chandler (2013)

*Figure 3.3 Process for the allocation of an offshore petroleum exploration permit in Australia*

program bidding system in accordance with OPGGSA.[90] These releases of acreage will usually comprise a mix of mature and frontier fields to ensure that exploration activities are equally split between known and lesser known or unknown provinces. Companies interested in a particular acreage form a commercial Joint Venture (JV) and appoint an operator. The JV then submits a work program bid for the acreage to the Joint Authority (JA). If that work program bid is accepted, an exploration license is awarded to the JV by the JA,[91] and exploration activities to fulfil the work program commence.[92]

---

[90] Offshore Petroleum and Greenhouse Gas Storage Act 2006 (Cth), s 126. See also s 104(1).

[91] Ibid, s 99.

[92] Ibid, s 99(5).

In deciding the winning bid, the JA is required to have regard to the predetermined, publicly outlined criteria that have been established by the state.[93] Current criteria for the award of WPB in Australia includes the number and timing of exploration wells to be drilled; the amount, type and timing of seismic surveying to be carried out; the amount, type and timing of any purchasing or licensing of existing data; and significant appraisal work over any previous petroleum discoveries within the area.[94] If a winning applicant cannot be chosen on the basis of the minimum guaranteed work program, the amount and timing of work proposed under the secondary work program will be assessed against the published criteria to determine the winning applicant.[95]

When awarding a petroleum license in Australia, the JA is required to take into consideration the past performance of the applicant, even where the applicant's proposed work program is the highest submitted.[96] Of particular importance is whether the applicants have been participants in previous licenses that have been cancelled as a result of a default in work program commitment and there was no agreement to maintain good standing.[97] In the first instance, an applicant must satisfy the JA of its capacity to undertake its proposed work program. In particular, the applicant company is required to prove the adequacy of financial resources and technical expertise available to the applicant.[98] In addition, the applicant must demonstrate the likelihood that it will continue to have access to sufficient resources to meet the requirements of the proposed work program and other commitments previously entered into in other permit areas. The future viability of any consortium that is lodging the application must be demonstrated, and include evidence that a satisfactory commercial Joint Venture Agreement (JVA) has been or can be reached and of the capacity of each party to work with the rest of the consortium.

---

93 As stipulated by ibid, s 106(3).

94 Department of Industry, *Bid Assessment Criteria* (2014), 1.

95 Ibid, 1.

96 Ibid, 2.

97 Ibid, 2.

98 The *Guidance Notes for Applicants* are published annually by the relevant Department concerned with the administration of petroleum titles, in conjunction with the release of petroleum acreage available for bidding. In 2014 the responsible body was the Department of Industry. See website available 12 October 2014 at www.petroleum-acreage.gov.au/2014/exploring-investing-and-bidding.

The exploration license is awarded for an initial term of six years.[99] The work program proposed will comprise two parts: the mandatory *Primary Work Program* (years 1–3), and the *Secondary Work Program* (years 4–6).[100] When bidding for a petroleum license under the WPB system in Australia, the applicant is required to propose a minimum guaranteed work program (including expenditure), for each year of the first three years. At a minimum, this should include exploration wells to be drilled, seismic and survey activities, and data evaluation.[101]

The granting of an *exploration license* authorizes the licensee to explore for petroleum in the license area,[102] as well as conferring an exclusive right to explore, including seismic survey and the drilling of wells.[103] The amount of seismic activity and the number of exploration wells to be drilled is stipulated by the JV group when applying for the license in the work program bid. Therefore, the level of activity will differ in each license area, depending upon the amount of work that was bid by the applicant.

While the exploration licenses (or permits) are awarded for an initial term of six years,[104] there is the capacity to extend the permits twice.[105] At each renewal of the license, 50 per cent of the permit area must be relinquished, with a maximum of two renewals permitted.[106] Effectively, this provides the licensee with 25 per cent of the original area awarded at the end of the renewal periods. If the exploration has yielded commercially exploitable reserves of petroleum, a production license may be applied for under OPGGSA.[107] Should the discovery be non-commercial at the time of discovery, but likely to be in the next 15 years, the licensee

---

99 Department of Industry, above n 94, 1.

100 Ibid, 2.

101 Ibid, 2.

102 Offshore Petroleum and Greenhouse Gas Storage Act 2006 (Cth), s 98.

103 Ibid, s 98.

104 Department of Industry, above n 94, 2. See also Offshore Petroleum and Greenhouse Gas Storage Act 2006 (Cth), s 102.

105 Ibid, s 102, s 119.

106 Athol Maritz, 'Work Program Bidding in Australia's Upstream Oil and Gas Industry, 1985–1999' (2003), ABARE Report 03.14, prepared for the Department of Industry, Tourism and Resources, available 12 January 2014 at www.abare.gov.au/publications_html/energy/energy_03/er03_work_program.pdf, 3.

107 Offshore Petroleum and Greenhouse Gas Storage Act 2006 (Cth), s 170, with conditions of the license stipulated in s 161.

is able to apply for a retention license under OPGGSA,[108] which permits the licensee to retain title to the acreage for five years[109] and the capacity to renew the retention lease.[110]

Upon making a potential commercial discovery of petroleum, the licensee is required to notify NOPTA in writing, nominating the 'block(s) for declaration of a location over the area' of a potentially commercial discovery. Following the declaration of the location, the licensee has two years (unless an extension is granted) to apply for a production license or a retention license.[111] Exploration permit holders have the right to develop economic discoveries by applying for a production license. Where the petroleum discoveries are not yet economic, the exploration titleholder can retain title through a retention lease.

*Production licenses* are required in order for petroleum to be removed from the acreage for commercial sale.[112] Chapter 2, part 2.4 of OPGGSA provides for the grant of production licenses over blocks in an offshore area,[113] and the conditions of petroleum licenses.[114] If an applicant wishes to proceed with the commercial development of a field, the grant of a production license occurs after the submission of a written assessment of the discovery. A consultation process with the JA NOPTA follows. Initially, the licensee produces a preliminary field development plan as part of the consultative process, to secure approval of the production license and associated infrastructure requirements. After approval of the preliminary field development plan, and in consultation with the government, the licensee submits a finalized field development plan (FDP) to the JA to facilitate formal field approval requirements. The granting of the production license confers production rights on the licensee.[115] If a licensee wishes to change an FDP, it is required to seek the approval of the relevant bodies in accordance with OPGGSA.[116]

While OPGGSA establishes the procedure for the award of an exploration license,[117] and the rights conferred by an exploration

---

108 Ibid, Chapter 2, Pt 2.3, Divisions 1 and 2, ss 134–46, with conditions of the retention lease stipulated in s 141(2).
109 Ibid, s 139(1).
110 Ibid, s 153.
111 Ibid, s 141(3).
112 Ibid, s 161.
113 Ibid, ss 159–67.
114 Especially under ibid, s 162.
115 Ibid, s 170.
116 See ibid, s 168(4) and s 168(5).
117 Ibid, s 104.

permit,[118] the Act does not outline the criteria or process for the assessment of applications for acreage under WPB. Rather, the criteria for the assessment of applications are published in the annual *Guidance Notes for Applicants.*

For the first time in over 20 years, the Commonwealth Government in 2014 released competitive cash bidding for mature provinces and in provinces where there are known petroleum reserves. The Australian government has returned to the use of cash bidding under the LCS as it is seen as more equitable, economically efficient and administratively simple, while preventing over-exploration in known or mature areas by not requiring an exploration work program.[119] Under the cash-bidding system of allocation of licenses, bidders are required to prequalify. Bidders prequalify by submitting a permit application to NOPTA. If the applicant is successfully determined to have prequalified, they are invited to make a bid for a license area.[120] In the 2014 licensing round, four areas (acreages) were released for cash bidding. These acreages were all located near the Rankin, Pluto, Wheatstone and Pluto gas accumulations in the Northern Carnarvon Basin. The vast majority of licenses in the 2014 licensing round will be allocated under the work program bidding system, with 26 licenses released for allocation by WPB.

Upon the award of a license in Australia, the licensee is granted certain rights (exclusive or non-exclusive rights) over the license area, depending on the type of license awarded.[121] These rights are proprietary in the sense that they are transferable, and can be sold, as is the case with other proprietary rights in property.[122] A transfer of a title to a licensee is authorized under ss 472–474 of the Offshore Petroleum and Greenhouse Gas Storage Act 2006 (Cth) (OPGGSA), and executed under ss 3–4 of the Offshore Petroleum and Greenhouse Gas Storage Regulations 1985 (Cth) (OPGGSR) using prescribed forms set out in Schedule 4 of the Regulations.

---

[118] Ibid, s 98.

[119] Department of Resources, Energy and Tourism, 'Competitive Cash Bidding' (2013), available 14 December 2014 at www.industry.gov.au/resource/Documents/upstream-petroleum/Cash-Bidding-Fact-Sheet.pdf, 1.

[120] Department of Industry, 'Cash Bid Exploration Permit Guidelines' (2014), see website available 12 October 2014 at www.petroleum-acreage.gov.au/2014/exploring-investing-and-bidding, 1.

[121] The grant of an exploration license in Australia confers the right to explore for petroleum in the Commonwealth's offshore zone under s 98 of the Offshore Petroleum and Greenhouse Gas Storage Act 2006 (Cth).

[122] But see the case of *Commonwealth of Australia v WMC Resources Ltd* [1998] HCA 8.

Section 161(1)(a) of OPGGSA authorizes the holder of a petroleum production license to recover petroleum from the license area. The unauthorized recovery of petroleum from offshore areas (without a production license) is prohibited,[123] implying that ownership of any recovered petroleum is conferred only upon the licensee authorized to carry out petroleum production and recovery operations. The transfer of property in extracted petroleum to a licensee or lessee is outlined in s 285(2) of OPGGSA.

## 4. REGULATING OFFSHORE PETROLEUM ACTIVITIES

### Operational Licensing

When petroleum activities are undertaken there is regulation of the activity. This regulation comprises two parts: regulatory approvals for operations (well integrity, safety and environment) and operational licenses required for petroleum production, especially infrastructure and petroleum pipeline licenses. There may also be specific conditions attached to the granting of a production license, as deemed necessary by the Joint Authority.[124]

**Infrastructure licenses**
These licenses are granted under Part 2.5 of the OPA, granting permission to construct or operate a facility in an offshore area. The grant of an infrastructure license continues indefinitely,[125] and confers rights to the licensee to construct and operate infrastructure facilities in the license area.[126] An infrastructure license is granted upon a written application through an offer document,[127] with the infrastructure license terminated if there has been no operation for five years.[128]

---

[123] Prohibited under s 160(1) of the Offshore Petroleum and Greenhouse Gas Storage Act 2006 (Cth).
[124] Ibid, s 195.
[125] Ibid, s 196.
[126] Ibid, s 194.
[127] Ibid, s 199.
[128] Ibid, s 197.

### Pipeline licenses

A pipeline is often constructed to transport petroleum from the offshore platform to a suitable storage facility or for loading onto ships. Pipeline licenses are granted under Part 2.6 of the OPA, granting permission to construct or operate a pipeline in an offshore area. The grant of a pipeline license to the licensee continues indefinitely,[129] and confers to the licensee the right to construct and operate pipeline facilities in the license area.[130]

An application for a pipeline license must be made to the Joint Authority,[131] with the application containing detailed design and construction plans, pipeline size and capacity, financing of the construction and operation, and accompanied by appropriate plans.[132] If there has been no construction or use of the pipeline in five years, the pipeline license will be terminated.[133]

### Safety Regulation Framework

Offshore petroleum safety is regulated by NOPSEMA who is responsible for both the safety of the platform that is located above the sea and from which petroleum operations occur, and the integrity of the well that is being drilled subsea. NOPSEMA also regulates the environmental management of the petroleum activity. This means that one body is responsible for the regulation of all three areas, as identified as best practice in the PC Report, the Bills and Agostini Report and the MCI. However, because no states/territory has yet conferred regulation of their Coastal or State Waters on NOPSEMA, the horizontal integration of regulation remains elusive, with the responsibility of NOPSEMA limited to the area seaward from 3 nautical miles.

The regulation of safety under NOPSEMA thus comprises two aspects: facility safety and integrity, and well integrity. The regulation of these activities is accomplished through the use of a combined framework, which is illustrated in Figure 3.4, comprising:

1. The use of legally enforceable laws (the Act and subordinate regulations); and

---

129 Ibid, s 214.
130 Ibid, s 211.
131 Ibid, s 217.
132 Ibid, s 217.
133 Ibid, s 215.

2. non-enforceable guidance, including policy, guidelines, guidance notes and advice.

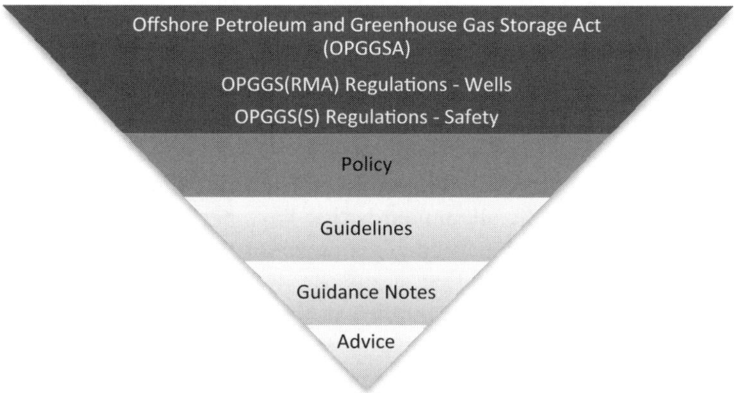


*Source:* Hunter and Chandler (2013)

*Figure 3.4 The regulatory triangle: Comprising legal and non-legal regulation*

NOPSEMA undertakes the regulation of facility integrity (also known as process safety)[134] and well integrity through the use of management plans. Given that there are separate regulations for facility integrity/safety and well integrity, these two aspects will be considered separately.

**Regulation of facility safety and integrity**

For the regulation of safety, the management plan is the *Safety Management System* (SMS), based on the safety case. For wells, the plan is the *Well Operations Management Plan* (WOMP). For both safety and well operations, the management plan is a permissioning document the operator (the duty holder) is required to produce to manage the safety of the facility (Safety Management Plan) or well (Well Operations Management Plan). The management plan is submitted to NOPSEMA who assesses and ultimately accepts the plan. Once accepted, the management plan dictates how the facility of well integrity will be preserved.

[134] 'Process safety' is defined by the American Institute of Chemical Engineers as 'the protection of people and property from episodic and catastrophic incidents that may result from unplanned or unexpected deviations in process conditions', available 21 August 2014 at www.aiche.org/ccps/about/process-safety-faqs#What%20is%20Process%20Safety.

The safety management system is underpinned by the safety case regime. NOPSEMA defines a Safety Case as a document produced by the operator of a facility that plays three roles: it identifies the hazards and risks associated with that facility, describes how the identified risks are controlled, and describes the safety management system (SMS) that is in place to ensure that the controls are consistently and effectively applied. The Safety Case must cover the design, construction, installation, modification and maintenance of the facility, while the SMS provides for all activities and operations that will occur at the facility. As part of the Safety Case, all standards that have been applied or will apply at the facility must be specified. This does not mean that the Safety Case needs to comply with a set of standards. Rather, it means that where a standard is specified as being applied to the facility in an accepted Safety Case, then that particular standard must be complied with.[135]

The Safety Case is the permissioning document that is owned by the operator of the facility. It is assessed by NOPSEMA, who accepts the Safety Case if satisfied that the control arrangements set out in the document demonstrate that the risks on the facility will be reduced to as low as reasonably practicable (ALARP). Following the acceptance of the Safety Case, NOPSEMA undertakes visits to the facility to monitor the application of the Safety Case in practice to ensure compliance.[136]

The principle of the Safety Case is that those who control the risk must manage the risk. Therefore, the Safety Case is prepared by the operator of the facility, rather than the regulator. It is the operator's job to assess the processes, procedures and systems on the facility to identify and evaluate the risks, and then implement appropriate controls. This is because it is the operator, rather than the regulator, that has the greatest in-depth knowledge of the facility. Given that the workers of the facility are those that are exposed to and create/control the risk, the involvement of the workforce in the preparation of the Safety Case is essential. Furthermore, worker involvement is necessary so that they understand what happens on the facility and why. This involvement makes it more likely that workers will do the right thing, since they are undertaking the behaviour to reduce risk, rather than complying with a 'rules-based' culture.[137]

---

135 NOPSEMA, 'Safety Case' (2013), available 12 April 2014 at www.nopsema.gov.au/assets/Uploads/document/QandA%20-%20Safety%20Case%20in%20Context.pdf.

136 Ibid.

137 Ibid.

The legal requirement for the Safety Case is set out in Schedule 3 of the Offshore Petroleum and Greenhouse Gas Storage Act 2006 (Cth). Clause 9(1) outlines the duty of the operator to 'take all reasonably practicable steps to ensure that the facility is safe and without risk to the health of any person at or near the facility', and that 'all work and other activities carried out at the facility are carried out in a manner that is safe and without risk to the health of any person at or near the facility'.

The fundamental underpinning of the Safety Case and the regime which it sits within is the reduction of the risk to as low as reasonably practicable (ALARP). Although a definition of ALARP does not appear in the Act or subordinate legislation, the concept of reasonably practicable was considered in *Edwards v National Coal Board*:

> Reasonably Practicable is a narrower term than 'physically possible' and seems to me to imply that a computation must be made by the owner in which the quantum of risk is placed in one scale and the sacrifice involved in the measures necessary for averting the risk (whether in money, time or trouble) is placed in the other; and that if it is to be shown that there is a gross disproportion between them – the risk being insignificant in relation to the sacrifice – the defendants discharge the onus on them.[138]

Therefore, ALARP is the point where the sacrifice required to reduce the impact and risk of the activity any further would be grossly disproportionate to the benefit gained.

The regulation of safety is a general duties regime that is performance-based, with some prescriptive elements. There is generally a tension between prescriptive and objective-based regimes. Prescriptive regimes tend to be less flexible, with less risk for the operator since the regulator stipulates the requirements for operations. This typically engenders a 'tick and flick' mentality in the operator, and creates greater risk for the regulator. Conversely, the objective-based regulatory regime provides greater flexibility, but also a greater need for guidance as to accepted practices. While an objective-based framework has been implemented in Australia post Piper Alpha, there is also a recognition that such a regime relies upon experienced operators and knowledgeable regulators. As the MCI so clearly demonstrated, the combination of an inexperienced operator and a poor regulator in such a regime can have devastating and far-reaching consequences.

---

[138] *Edwards v National Coal Board* [1949] All ER 743 (CA).

**Regulation of well integrity**

The RMA Regulations regulate well integrity through the *Well Operations Management Plan* (WOMP), which requires the operator to submit a plan for the construction and management of a well under part 5 of the RMA Regulations. A WOMP should demonstrate that there is a system in place to manage well integrity and well activities for the life of the well and include a description of the design, construction and management of well activities and a plan for managing the risks associated with the activities in accordance with sound engineering principles and good oil field practice. If the WOMP is accepted by NOPSEMA, the titleholder is obliged to implement the methodology as specified in the WOMP to conduct the well activities.[139] A WOMP may address a single well or cover a series of wells in a wider drilling campaign, or relate to numerous production wells for a facility. It is required to detail various programs and describe the equipment and procedures to be used in the well activity (that is, testing program, drilling program). Examples of well activities to be included in a WOMP are:

- Drilling, completion, re-completion, operation, testing, abandonment/suspension of a well;
- Exposing new formation(s) to the completed well bore (that is, perforating, sidetracking, etc);
- Altering or repairing an existing well bore or completion; and
- Altering the physical production capability of a well through down-hole intervention.[140]

The WOMP is underpinned by the concept of good oilfield practice (GOP). Good oilfield practice is a rather elusive term. It is defined in s 7 of the OPGGSA as 'all those things that are generally accepted as good and safe in the carrying on of exploration for petroleum or petroleum recovery operations'. Given this 'definition', the Department of Resources, Energy and Tourism has issued a guideline, which considers good oilfield practice. The guideline states that good oilfield practice is 'generally understood to reflect and incorporate all those practices which are generally accepted as safe, economical and efficient in exploring for

[139] NOPSEMA, 'Management of Offshore Well Operations: Interim Guideline' (2011) available 14 April 2014 at www.nopsema.gov.au/assets/Guidelines/N-04600-GL0936-Management-of-Offshore-Well-Operations-Plans-Interim-Guide line.pdf.

[140] Ibid.

and exploiting petroleum'.[141] It recognizes that Australian and international standards provide some guidance as to accepted practices.

However, the guidance notes recognize that further definition is required, and refers to the Explanatory Statement for the RMA Regulations, which states that:

- The term good oilfield practice is very similar to 'good production practice', which has been described as the production of crude oil or raw gas at a rate that can be sustained without adversely and significantly affecting conservation and without avoidable waste.
- The term 'good production practice' has at times been used interchangeably with 'maximum efficiency rate'. Maximum efficiency rate is the maximum rate at which oil and gas can be produced without excessive decline or loss of reservoir energy.
- Optimum long-term recovery of petroleum has been defined as the best possible, most advantageous recovery, which maximizes the economic recovery of hydrocarbon resources. In a decision based on cost-benefit analysis, the optimum approach would be one that is most economical in that it maximizes the benefits, and is associated with the lowest cost, outlay or risk.

Further, the guideline refers to s 569 of the OPGGSA, which requires petroleum titleholders to carry out operations in a proper and workmanlike manner in accordance with good oilfield practice. This requirement to *carry out operations in a proper and workmanlike manner* implies a management system. Existing legislation requires a safety management system, and it is expected that titleholders will apply similar principles to achieve the objectives of GOP.

**Environmental regulation**

The regulation of the environmental effects of offshore petroleum activities is shared between the states/NT and the Commonwealth, with three distinct environmental protection regulatory zones:

1. State (or internal) Waters, where environmental regulation lies wholly within the jurisdictions of the state/Northern Territory;

---

[141] Department of Resources, Energy and Tourism, 'Good Oilfield Practice: A Guideline in relation to the Offshore Petroleum and Greenhouse Gas Storage Act 2006' (2013). available 14 April 2014 at www.nopta.gov.au/_documents/guidelines/Good-Oilfield-Practice-Guideline.pdf.

2. Coastal Waters, from baseline seaward 3 nm, where environmental regulation rests with the states/Northern Territory; and
3. Commonwealth Waters, seaward from 3 nm, with jurisdiction over petroleum activities in these waters resting with NOPSEMA.

There are no petroleum titles in State Waters in Queensland or New South Wales. There are some titles in South Australian and Victorian waters and a large numbers of titles in WA waters. The WA State Waters include Barrow Island and the petroleum titles in the vast State Waters from Exmouth to Dampier. The majority of petroleum activities in Coastal Waters occur in WA and Victoria, with minor activities occurring in South Australia and the Northern Territory. At present no petroleum leases are active in the Coastal Waters of New South Wales or Queensland.

The regulation of the environment in Coastal Waters is undertaken by each state (where required) utilizing the environment plan system, unless a jurisdiction confers its regulatory powers on the Commonwealth. As at April 2013, none of the states/NT has conferred regulation of the environment on the Commonwealth. It is only through a combination of mirror legislation and memoranda of understanding that there is some degree of regulatory harmony regarding the environmental effects of offshore petroleum activities.

An overview of the relevant legislation that applies to the regulation of the environmental effects of petroleum activities in Coastal Waters is outlined in Table 3.2.

**Commonwealth Waters**

Similar to safety, the regulation of the environment by NOPSEMA is undertaken by utilizing a hierarchy of legally enforceable (Acts and regulations) and non-enforceable (policies, guidelines, guidance notes and practice notes) tools.

The regulation of the environment in Commonwealth Waters is based on a system of environmental assessment and approvals. Since 1 January 2012 the Australian Government has had jurisdiction for the regulation of petroleum activities in Commonwealth Waters. Environmental matters are regulated under the OPGGSA and the Offshore Petroleum and Greenhouse Gas Storage (Environment) Regulations 2009 (Cth). This framework regulates all aspects of the environment in Commonwealth Waters. An overview of the environmental management regulatory process is outlined in Figure 3.5.

*Table 3.2 Legislation regulating the environmental impacts of offshore petroleum activities in Coastal Waters*

| Jurisdiction | Acts | Regulations |
|---|---|---|
| Queensland | Petroleum (Submerged Lands) Act 1982 (Qld) | |
| Northern Territory | Petroleum (Submerged Lands) Act 1982 (NT)<br>Offshore Waters (Application of Territory Laws) Act 2001 (NT) | Petroleum (Submerged Lands) Regulations 2000 (NT)<br>Petroleum (Submerged Lands) (Application of Commonwealth Laws) Regulations 2012 (NT) |
| Western Australia | Petroleum (Submerged Lands) Act 1982 (WA) | Petroleum (Submerged Lands) (Pipelines) Regulations 2007 (WA)<br>Petroleum (Submerged Lands) (Environment) Regulations 2012 (WA) |
| South Australia | Petroleum (Submerged Lands) Act 1982 (SA) | Petroleum (Submerged Lands) Regulations 2005 (SA) |
| Tasmania | Petroleum (Submerged Lands) Act 1982 (Tas) | Petroleum (Submerged Lands) Regulations 2008 (Tas)<br>Petroleum (Submerged Lands) (Management of Environment) Regulations 2008 (Tas) |
| Victoria | Offshore Petroleum and Greenhouse Gas Storage Act 2010 (Vic) | Offshore Petroleum and Greenhouse Gas Storage Regulations 2011 (Vic) |
| New South Wales | Petroleum (Offshore) Act 1982 (NSW) | Petroleum (Offshore) Regulations 2006 (NSW)<br>Pipelines Regulation 2005 (NSW) |

*Source:* Hunter and Chandler (2013)

As outlined in Figure 3.5, there are two possible environmental approval processes that a project may have to undergo – OPGGSA approval and Environmental Protection and Biodiversity Conservation Act 1999 (Cth) (EPBCA) approval. All projects in Commonwealth Waters are required to undergo environmental approval under the OPGGSA, with this approval conferred by NOPSEMA. Only projects that trigger a referral under the 'Matters of National Environmental Significance' (MNES) as defined under Chapter 2, Part 3, Division 1 of the EPBCA will be referred for assessment under the EPBCA. As of February 2014, the Commonwealth has delegated the assessment of EPBCA referrals to NOPSEMA.

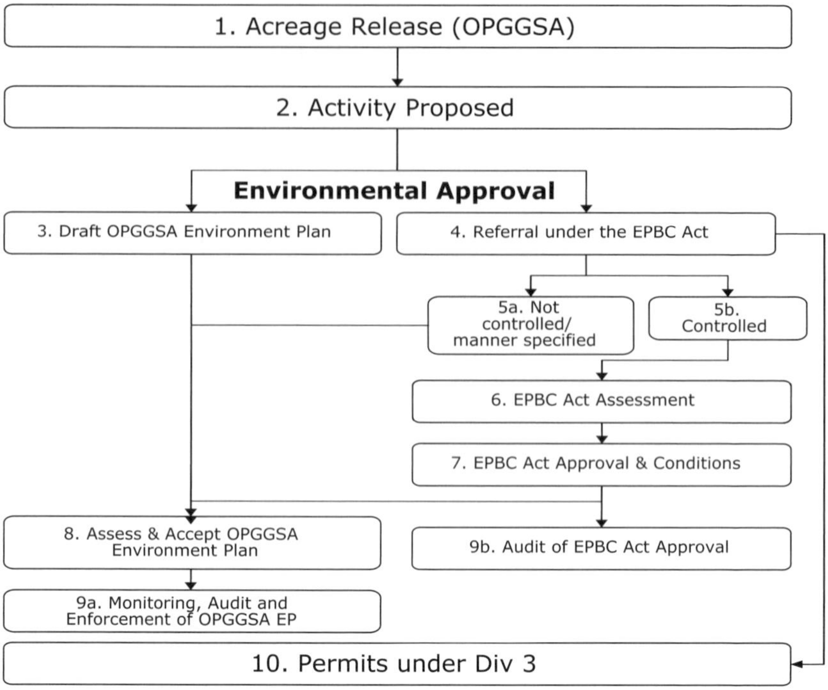


*Source:* NOPSEMA

*Figure 3.5 Legislation regulating the environmental impacts of offshore petroleum activities in Coastal Waters*

*Environment approvals by NOPSEMA under the OPGGSA* The object of the Environment Regulations, as outlined in r 3, is to ensure that petroleum operations are carried out in a manner consistent with the principles of ecologically sustainable development and in accordance with an environment plan (EP) that has appropriate environmental performance objectives, standards and measurement criteria from determining whether the objectives and standards have been met.[142] Thus the Environment Regulations require an operator to have an accepted EP in place for a petroleum activity. The Environment Regulations, which regulate the EP, are based on four principles:

142 NOPSEMA, 'Environment Plan Core Concepts', Guidance Note N4700-GN1074, 2012, 17 January 2013 at www.nopsema.gov.au/assets/document/N-04700-GN1074-Environment-Plan-Content-Requirements-Guidance-Note.pdf.

1. objective-based, setting out general requirements, where the operator is then responsible for determining their approach to meet these objectives;
2. risk-based, where the central basis of the EP is the assessment and management of the environmental impacts and risks of the petroleum activity to ALARP;
3. performance-based, where appropriate environmental performance objectives and environmental performance standards are in place that have measurement criteria to determine if they are being met; and
4. systems-based, where a structured risk management process is applied to developing the EP which shows the continual and iterative steps that are followed to assess and manage environmental impacts and risks effectively.[143]

The required content of an EP is detailed within the Environment Regulations and must include all the elements specified in the Environment Regulations which require an operator to have an accepted EP in place in order for a petroleum activity to occur. The operator is therefore required to submit a draft management plan to NOPSEMA for the activities that are going to be undertaken. NOPSEMA then assesses the plan, using an objective-based system utilizing the concepts of ALARP. NOPSEMA will then either accept the EP, or will send the plan back to the project proponent for redrafting.

Given that the EP process is a systems-based approach, the EP submitted by the operator is required to set out the following elements appropriately:

- describe the activity, environment, regulatory and other requirements, and acceptable levels for the impacts and risks;
- detail environmental impacts and risks for the activity;
- evaluate and demonstrate that impacts and risks are of an acceptable level;
- evaluate and demonstrate that the impacts and risks are reduced to ALARP; and
- define environmental performance objectives, environmental performance standards and measurement criteria that capture performance and monitoring requirements.[144]

---

[143] Ibid.
[144] Ibid.

The role of NOPSEMA in regulating the environment under the OPGGSA is twofold. NOPSEMA's first role is that of an assessor, where NOPSEMA either accepts or refuses to accept the operator's EP in the planning phase. NOPSEMA accepts an EP if there are reasonable grounds for believing that the EP meets the criteria outlined in rr 10 and 11 of the Environment Regulations. NOPSEMA's second role is that of monitoring. Once the EP is accepted by NOPSEMA, petroleum activities can commence, and will be monitored by NOPSEMA to ensure the operator's continued compliance with the EP during the operational phase.

*Environment approvals by NOPSEMA under the EPBCA* An additional form of regulatory approval that may be required for a petroleum activity is approval under the EPBCA, where an *action* will require additional approval from the Environment Minister's delegate (NOPSEMA) if the action has, will have or is likely to have a significant impact on a matter of national environmental significance (MNES).

Under s 523 of the EPBCA, an *action* is defined broadly to include a project, a development, an undertaking, an activity or a series of activities, or an alteration of any of these things.[145] A significant impact is defined as:

> an impact which is important, notable, or of consequence, having regard to its context or intensity. Whether or not an action is likely to have significant impact depends on the sensitivity, value and quality of the environment which is impacted, and on the intensity, duration, magnitude and geographic extent of the impacts.[146]

The MNES include:

- listed threatened species and ecological communities;
- migratory species protected under international agreements;
- Ramsar wetlands[147] of international importance;
- the Commonwealth Marine Environment;
- world heritage properties;
- the Great Barrier Reef Marine Park; and
- nuclear actions.

---

145 EPBCA, s 523.

146 Department of Environment, 'Glossary' (2013), available 15 August 2014 at www.environment.gov.au/epbc/about/glossary.html#significant.

147 These are wetlands designated under Art 2 of the United Nations Convention of Wetlands of International Importance Especially as Waterfowl Habitat, concluded at Ramsar, Iran in 1971.

To determine whether an action requires approval under the EPBCA, the project is referred to NOPSEMA if the project includes one of the MNES. The project can be referred by either the project proponent themselves, or a third party such as a minister, or a government agency. Generally, the project proponent will refer the project, to determine whether the referral requires assessment under the EPBCA.

The process for approval is outlined in Figure 3.5. An application for referral is assessed to determine whether it requires assessment by the minister (or delegate) as a 'controlled action', on the grounds of posing a significant risk to an MNES. If it does not pose a significant risk (an uncontrolled action), the project will not be referred for assessment. Where an action is deemed to be a controlled action, it is referred to the minister for assessment. The action is then assessed under the EPBCA, and will either be approved or not. If approved, the action will also be assessed under the normal EP assessment and approval process by NOPSEMA under the OPGGSA.

## 5. CONCLUSION

Conventional petroleum resources in Australia are largely confined to the offshore area of Australia, save for large gas deposits in the Cooper-Eromanga Basin in South Australia, which has been producing gas for consumption in the densely populated eastern Australian States for over 40 years. Given the concentration of conventional petroleum resources in the offshore zone within the Australian jurisdiction, this chapter has focused on the regulation of offshore petroleum.

Australia's offshore petroleum resources are regulated in three distinct zones: State Waters, where the waters lie wholly within the jurisdiction of the state/NT; Coastal Waters, an area from the baseline seaward to 3 nautical miles conferred under the OCS in 1979; and Commonwealth Waters, which is the remaining offshore area seaward of 3 nautical miles. With the exception of WA, there are no activities in State Waters, and only Victoria and WA have any petroleum activities of importance in Coastal Waters. As such, this chapter has focused on the grant of access to petroleum resources and the regulation of petroleum activities in Commonwealth Waters. Grant of access to petroleum resources and the regulation of petroleum activities in Coastal Waters mirrors that of the Commonwealth, but by state regulatory agencies.

A restructured regulatory framework for the regulation of offshore petroleum activities in Commonwealth Waters commenced in January

2012. It created a new titles administrator, NOPTA, who awards exploration permits under licensing rounds and administers the various access and operational titles required for petroleum exploration and production. The restructured framework also enlarged the powers of the existing safety authority NOPSA to include environment and well integrity. The new operational regulator, NOPSEMA, utilizes an objective-based framework to regulate all aspects of petroleum operations: safety (platform and personal), well integrity and environment. The framework is premised upon a permissioning regime, where the appropriate (safety, well integrity or environment) plan is submitted to NOPSEMA, who either approve or reject the plan. In order for a plan to be approved, the operator is required to demonstrate in the appropriate plan that they have identified the risks associated with the activity, addressed the risks, and reduced the risk to ALARP. Alongside this approval role, NOPSEMA also maintains a role in monitoring and enforcing the approved plan, to ensure that risks remain ALARP, thereby reducing the risk of personal, facility or environmental harm.

## BIBLIOGRAPHY

APPEA, 'Submission No 16 to Productivity Commission', *Review of the Regulatory Burden on the Upstream Petroleum (Oil and Gas) Sector*, 19 September 2008, available 12 June 2014 at www.pc.gov.au/__data/assets/pdf_file/0019/83422/sub016.pdf.

Australasian Parliamentary Council Committee, 'National Uniform Legislation: Acts of Jurisdictions Implementing Uniform Legislation [as at June 2012]' (2012), available 25 June 2013 at www.pcc.gov.au/uniform/National%20Uniform%20Legislation%20table.pdf.

Australian Academy of Technological Sciences and Engineering, *Technology in Australia, 1788–1988* (Australian Academy of Technological Sciences and Engineering, Melbourne, 1988).

Australian Maritime Safety Authority, 'Major Oil Spills in Australia' (2009), available 12 January 2014 at www.amsa.gov.au/Marine_Environment_Protection/Major_Oil_Spills_in_Australia/.

Bills, K and Agostini, D, *Offshore Petroleum Safety Regulation: Marine Issues* (Australian Government, Canberra, 2009).

Bills, K and Agostini, D, *Offshore Petroleum Safety Investigation: Varanus Island Incident Investigation* (Government of Western Australia, Perth, 2009).

Borthwick, D, 'Report of the Montara Commission of Inquiry' (2010), Department of Resources, Energy and Tourism, available 12 January 2014 at www.ret.gov.au/Department/Documents/MIR/Montara-Report.pdf.

Bradbrook, A, 'The Relevance of the Cuius est Solum Doctrine to the Surface Landowner's Claims to Natural Resources Located Above and Beneath his Land' (1988) 11 *Adelaide Law Review* 462.

Brazil, P, *Offshore Constitutional Settlement 1980: A Case Study in Federalism* (Centre for International and Public Law, Faculty of Law, Australian National University, Canberra, 2001).

Crommelin, M, 'The Legal Character of Petroleum Production Licences in Australia' in T Daintith, *The Legal Character of Petroleum Licences: A Comparative Study* (University of Dundee, Dundee, 1981).

Daintith, T, 'A Critical Evaluation of the Petroleum (Submerged Lands) Act as a Regulatory Regime' (2000) *AMPLA Yearbook 2000* 91.

Daintith, T, *Discretion in the Administration of Offshore Oil and Gas* (AMPLA, Melbourne, 2005).

Daintith, Terence, 'Research Report: Discretion and Prescription in the Administration of Petroleum Policy' (2005), available 6 January 2015 at www.esrc.ac.uk/my-esrc/grants/RES-000-22-0329/outputs/Download/8c52e4da-4130-46b5-a25b-3e009 01126ee.

Dam, Kenneth, *Oil Resources: Who Gets What How?* (University of Chicago Press, Chicago, 1976).

Department of Environment, 'Glossary' (2013), available 15 August 2014 at www.environment.gov.au/epbc/about/glossary.html#significant.

Department of Industry, *Bid Assessment Criteria* (2014).

Department of Industry, 'Cash Bid Exploration Permit Guidelines' (2014), available 12 October 2014 at www.petroleum-acreage.gov.au/2014/exploring-investing-and-bidding.

Department of Industry, Tourism and Resources, *Cash Bidding System: A Guide in Relation to the Australian Government's Petroleum (Submerged Lands) Act 1967* (RET, Canberra, 2004).

Department of Resources, Energy and Tourism, 'Competitive Cash Bidding' (2013), available 14 December 2014 at www.industry.gov.au/resource/Documents/upstream-petroleum/Cash-Bidding-Fact-Sheet.pdf.

Department of Resources, Energy and Tourism, *Commonwealth Government Response to the Productivity Commission Review* (DRET, Canberra, 2011).

Department of Resources, Energy and Tourism, 'Good Oilfield Practice: A Guideline in relation to the Offshore Petroleum and Greenhouse Gas Storage Act 2006' (2013), available 14 April 2014 at www.nopta.gov.au/_documents/guidelines/Good-Oilfield-Practice-Guideline.pdf.

Department of Resources, Energy and Tourism, 'Offshore Acreage Release 2008: An Overview for Applicants' (2008), available 12 March 2014 at www.ret.gov.au/resources/Documents/Acreage%20Releases/2008/site/page2.htm.

Fraser Institute, *Global Petroleum Survey, 2010* (Fraser Institute, Vancouver, 2010).

Geoscience Australia, 'Australia's Maritime Jurisdiction Extended' (2008), available 12 December 2013 at www.ga.gov.au/ausgeonews/ausgeonews200806/inbrief.jsp.

Hunt, Constance, *The Offshore Petroleum Regimes of Canada and Australia* (Calgary Institute of Resources Law, Calgary, 1989).

Hunt, Michael, *Minerals and Petroleum Law* (Lexis Nexis, Melbourne, 1996).

Hunter, Tina and Chandler, John, *Petroleum Law in Australia* (Lexis Nexis, Sydney, 2013).

Maritz, Athol, 'Work Program Bidding in Australia's Upstream Oil and Gas Industry, 1985–1999' (2003), ABARE Report 03.14, prepared for the Department of Industry, Tourism and Resources, available 12 January 2014 at www.abare.gov.au/publications_html/energy/energy_03/er03_work_program.pdf.

NOPSEMA, 'Environment Plan Core Concepts, Guidance Note N4700-GN1074' (2012), available 17 January 2013 at www.nopsema.gov.au/assets/document/N-04700-GN1074-Environment-Plan-Content-Requirements-Guidance-Note.pdf.

NOPSEMA, 'Management of Offshore Well Operations: Interim Guideline' (2011), available 14 April 2014 at www.nopsema.gov.au/assets/Guidelines/N-04600-GL 0936-Management-of-Offshore-Well-Operations-Plans-Interim-Guideline.pdf.

NOPSEMA, 'Safety Case' (2013), available 12 April 2014 at www.nopsema.gov.au/assets/Uploads/document/QandA%20-%20Safety%20Case%20in%20Context.pdf.

Productivity Commission, *Review of Regulatory Burden on the Upstream Petroleum (Oil and Gas) Sector* (Productivity Commission, Melbourne, 2009).

Rodgers, Trudi and Webster, Stewart, 'Resource Rent Mechanisms in Australian Primary Industries: Some Observations and Issues' (2007), paper presented at the 51st Annual Conference of the Australian Agricultural and Resource Economics Society, 13–16 February 2007, Queenstown, New Zealand.

White, M, *Australia's Offshore Laws*, (Federation Press, Sydney, 2009).

# 4. Licensing the exploration for and production of petroleum on the UK continental shelf

**Greg Gordon and John Paterson**

## 1. INTRODUCTION

Production of hydrocarbons began on the UK continental shelf (UKCS) in 1975 and since then some 41 billion barrels of oil equivalent have been produced. While it is widely accepted that the majority of recoverable oil and gas originally *in situ* has been extracted, even conservative estimates suggest that 12 billion barrels of oil equivalent remain and some put the figure higher. Whatever the total remaining to be produced, however, there can be no denying that the rate of production is now well past its peak. The peak year for oil production was 1999 with around 130 million tonnes extracted, and for gas 2000 when 108 billion cubic metres were produced. Production is now substantially lower, at around 40 million tonnes of oil and 37 billion cubic metres of gas per annum.[1]

Oil and gas reserves in the North Sea area of the UKCS are therefore becoming progressively depleted to the extent that it is now described as a mature basin. This refers to the probability that all the major discoveries have now been made, the infrastructure is ageing and the viability of smaller, more marginal discoveries may often depend on the ability to tie them back to that infrastructure before it is decommissioned. Meanwhile, the area of the UKCS west of Shetland, which was for many years regarded as a promising but technologically challenging frontier region, is becoming better established and some of the most exciting projects are under way there. The extent to which the remaining reserves are located

[1] Production data is available from DECC, 'Oil and Gas: Field Data' (2014), available 7 December 2014 at www.gov.uk/oil-and-gas-uk-field-data.

and exploited will, however, depend upon the attractiveness of the province to ongoing investment.[2]

This chapter will consider the way in which the licensing regime that allows exploration for and production of oil and gas offshore has developed and currently operates. After dealing with the fundamental issue of ownership and control of hydrocarbons on the UKCS, the chapter will go on to look at the key features of the UK's licensing regime. It will then consider the different forms of licence before looking in greater detail at the key clauses of the production licence and the variant forms of that licence that have evolved over the years to meet specific challenges. The chapter then considers the financial dimension of the UK's licensing regime, addressing in brief the fiscal regime that complements it. The treatment of offshore installations under the licence is then examined before discussion of the UK's approach to decommissioning. The considerable challenge of third party access to infrastructure is then considered, followed by health, safety and environmental regulation on the UKCS, which is currently subject to ongoing reform in the aftermath of events in the Gulf of Mexico in 2010. The conclusion to the chapter addresses the current debate about the reforms to the licensing regime that have been identified in a recent government-commissioned review as necessary to ensure that there is maximum economic recovery of the remaining hydrocarbons.

## 2. OWNERSHIP AND CONTROL

The question of who owns hydrocarbons *in situ* has been contentious historically. The property law regimes in place within each of the constituent legal systems in the UK stipulate that ownership of minerals ordinarily follows ownership of the overlying land. Early efforts to develop hydrocarbons onshore were obstructed by landowners claiming mineral rights or seeking compensation for them. As a consequence, the initial licensing regime simply ignored the issue of ownership and gave the state the right to prevent hydrocarbon operations without a licence – albeit that this did not in turn lead to any significant developments being taken forward.[3] The position was fundamentally altered with the passage

---

[2] For a discussion, see A Kemp, 'North Sea Oil and Gas', in *Scotland's Future: The Economics of Constitutional Change* (A Goudie, ed, Dundee University Press, Dundee, 2013), p. 243.

[3] For a fuller account of the historical development of this area, see T Daintith, G Willoughby and A Hill (eds), *UK Oil and Gas Law* (3rd ed, 2003,

of the Petroleum (Production) Act 1934, section 1(1) of which stated that 'property in petroleum existing in its natural condition in strata in Great Britain is hereby vested in [the Crown]'. The Act contains no provisions for compensation on the basis that the Crown's view was that this was simply a statement of the pre-existing legal fact. Perhaps against expectation – and perhaps because few sizeable conventional onshore finds were made – this position has been accepted without demur. The 1934 Act remains the source for the Crown's ownership of hydrocarbons *in situ* onshore.

As regards to offshore, however, the Crown does not claim ownership over oil and gas deposits located within the continental shelf, mirroring the position under international law whereby the coastal state enjoys 'sovereign rights' in relation to the continental shelf in contrast to sovereignty in relation to the territorial sea.[4] Thus, when the Continental Shelf Act 1964 extended the onshore licensing regime under the 1934 Act to the continental shelf, reference to section 1 was deliberately omitted.[5] In its place, section 1(1) of the 1964 Act provides that 'any rights exercisable by the United Kingdom outside territorial waters with respect to the sea-bed and subsoil and their natural resources' are vested in the Crown. It can be argued, of course, that from the perspective of international law, at least, insofar as these sovereign rights can be defended by the coastal state against other parties, it does not much matter precisely what they are. However, such an approach is somewhat reductive. The legal nature of the right may bear, for instance, upon the kind of security that can be granted in order to fund exploration and production projects and upon the remedies open to the state in the event that a person attempts to extract petroleum otherwise than in accordance with a licence. Differing views have been expressed on the legal nature of the state's right (and therefore, the right taken by the state's licensees). Daintith contends that, although not a full right of property, the right is essentially proprietary in nature;[6] others have argued that petroleum resources *in situ* are ownerless and that original ownership is acquired at

---

looseleaf, Sweet and Maxwell, London, updated to 2013) ('Daintith, Willoughby and Hill'), Chapter 1.

[4] Convention on the Continental Shelf 1958, Art 2(1). See now United Nations Convention on the Law of the Sea (UNCLOS) 1982 Art 77(1).

[5] The other principal sections of the 1934 Act were expressly referred to: see the Continental Shelf Act 1964 s1(3).

[6] Daintith, Willoughby and Hill, paras 1-345–6.

the point of extraction by any party authorized by the coastal state to extract them.[7]

## 3. KEY FEATURES OF THE UK'S OFFSHORE PETROLEUM LICENSING REGIME

The UK experimented with a state oil company (BNOC) briefly during the 1970s, but otherwise has not directly participated in the industry, preferring to license private parties to explore for and produce hydrocarbons in return for fees, royalties and tax on profits. The Petroleum Act 1998, a consolidating statute, continues the essential features of the licensing regime first set out in the 1934 Act, most strikingly in the extent to which it is discretionary in nature. Section 3 of the 1998 Act empowers the Secretary of State to grant licences to such persons as he or she thinks fit, on such terms and conditions as he or she thinks fit, for such financial consideration as he or she, with the consent of the Treasury, may determine. The apparent scope of the Minister's discretion has, however, been significantly limited both by EU law and by the government's willingness to respond to the industry's desire for greater transparency. Thus, the EU Hydrocarbons Licensing Directive provides that the Secretary of State may only utilize specific objective and non-discriminatory criteria in deciding between competing licence applications (such as their technical and financial capability, and the manner in which they propose to work) and has removed previous government efforts to promote UK commercial interests in the award of licences.[8] In

---

7 See the discussion in G Gordon, 'Petroleum Licensing', in G Gordon, J Paterson and E Usenmez, *Oil and Gas Law, Current Practice and Emerging Trends* (2nd ed, Dundee University Press, Dundee, 2011) ('Gordon, Paterson and Usenmez'), paras 4.8–4.9.

8 Other EU laws also have the effect of curtailing the Minister's discretion: see e.g. Council Directive 92/43/EEC of 21 May 1992 on the Conservation of Natural Habitats and of Wild Fauna and Flora ('the Habitats Directive'); Parliament and Council Directive 2011/92/EU of 13 December 2011 on the Assessment of the Effects of certain Public and Private Projects on the Environment (Codification), as amended by Parliament and Council Directive 2014/52/EU of 16 April 2014 amending Directive 2011/92/EU on the Assessment of the Effects of certain Public and Private Projects on the Environment ('the Environmental Impact Assessment Directive'); Parliament and Council Directive 2001/42/EC of 27 June 2001 on the Assessment of the Effects of certain Plans and Programmes on the Environment ('the Strategic Environmental Assessment Directive').

the same way, a marks scheme has been introduced by means of ministerial guidance to indicate clearly to applicants which activities will attract the highest scores in the decision-making process on the award of licences.[9]

As to the precise legal nature of the petroleum licence, it may be said to be partly regulatory and partly contractual. The form is contractual, insofar as both parties sign it and consideration passes, but the effect is regulatory insofar as it imposes obligations substantially only on the licensee and confers various powers on the relevant government Minister. The conditions which the Minister may impose on licensees are generally to be found in Model Clauses which section 4(1)(e) of the 1998 Act provides shall be prescribed by regulation and shall be incorporated into all petroleum licences. As a result of the contractual form of the licence, the set of Model Clauses in force at the time it is granted will govern the relationship between the Crown and the licensee and it will be impossible for the former to change the terms of the licence without the agreement of the licensee or recourse to retroactive primary legislation.[10]

## 4. FORMS OF LICENCE

The UK employs two principal forms of licence: the exploration licence and the production licence. The production licence in turn now has a number of variants: the traditional licence, the promote licence, the 6-year frontier licence and the 9-year frontier licence.

### Exploration Licence

The exploration licence is a non-exclusive licence with a term of three years, which in essence grants the licensee the right to prospect. The licensee, together with other exploration licence holders, may conduct exploration activities anywhere within the territorial sea or UKCS, subject to certain conditions. The licensee may not enter any area that is

---

[9] Higher marks are awarded e.g. for firm commitments to drill a well than for contingent offers. Where there is competition between a number of applicants for any given block, in the interests of transparency, the winning mark is published at the time of awarding the licence. See e.g. relative to the 28th offshore round, DECC, 'Winning Marks: By Block', available 7 December 2014 at www.gov.uk/government/uploads/system/uploads/attachment_data/file/370434/28R_MarksByBlock.pdf.

[10] For the current model clauses, see The Petroleum Licensing (Production) (Seaward Areas) Regulations 2008 (SI 2008/225).

already subject to a production licence without the consent of the production licensee. Seismic and other exploration activities may be carried out, but no drilling is permitted beyond 350 metres beneath the seabed.[11] The exploration licence holder may not carry out operations so as to interfere unjustifiably with navigation or fishing or with the conservation of the living resources of the sea. He or she must also notify the Ministry of Defence ahead of any proposed seismic activity, keep samples and records and appoint a fisheries liaison officer.[12]

## Production Licence

The production licence allows exploratory drilling and development work within, as well as production from, a defined area (or block) and is generally awarded by the Secretary of State in rounds. In recent years these rounds have occurred at intervals of around a year to eighteen months, with specific areas, divided into blocks, being offered to the industry.[13] The government thereby retains some control over the pace of development of its hydrocarbon resources. Whereas early licensing rounds were focused on specific areas, more recently the government has been willing to open all unlicensed areas of the UKCS for bids (unless there are environmental concerns as revealed by Strategic Environmental Assessment)[14] – a function of the maturity of the province discussed in the introduction to this chapter. In light of the Hydrocarbons Licensing Directive and the marks scheme, where there is competition for the same acreage the applicant offering to carry out the most sophisticated seismic survey and especially making firm commitments to drill exploration wells will generally be awarded the licence.

The production licence confers an exclusive right over the licensed area in relation to all phases, from exploration through development to production. It is, however, divided into three terms (the initial (exploration) term, the second (development) term and the production period) and can be revoked if the licensee fails, without good reason, to fulfil its

---

[11] The deposits sought lie at depths significantly greater than that: thus, despite what the terminology might suggest, a production licence will be required in order to drill an exploration well.

[12] For the Model Clauses for such licences, see The Petroleum Licensing (Exploration and Production) (Seaward and Landward Areas) Regulations 2004 (SI 2004/352) Schedule 1.

[13] At the time of writing there have been twenty-eight such rounds.

[14] See G. Gordon, 'Petroleum Licensing', in Gordon, Paterson and Usenmez, para 4.27.

obligations during any such term. Insofar as the licensee must surrender acreage at the end of the first and second terms, so that it is left only with such as it requires for production by the third period, there is both a clear incentive to explore efficiently and an early opportunity for the state to recycle acreage in the next round.

## 5. KEY CLAUSES IN THE PRODUCTION LICENCE

Among the most important Model Clauses in the production licence are the work programme approval and the acreage surrender obligation. The Minister must approve work (that is, exploration) programmes, meaning that he or she is able to control the pace of development and has oversight of exploration activity and the sums of money being spent. With the acreage surrender obligation, the Minister seeks to disincentivize any preference on the part of the industry to hold blocks for future exploration and possible development while resources are deployed in other jurisdictions. Whatever the internal commercial pressures on individual companies or joint ventures given commitments globally, the state is only concerned that its natural resources are efficiently explored and produced.

Generally speaking, the Minister will be keen on development sooner rather than later in order that taxable profits may be generated. Rent is charged on acreage, and although this remains (despite recent increases) relatively insignificant compared to income from a producing field, it does constitute another disincentive against sitting on a block without working efficiently. Royalty was payable in relation to fields which received development consent on or before 31 March 1982, but this has now been abolished.[15]

All licences are subject to the obligation that gas may not be flared without the consent of the Minister, but additional special conditions may be imposed where the licensed block is close to the shore, or is otherwise seen as environmentally sensitive.[16]

The consent of the Minister is required before a licensee sells or otherwise disposes of the ownership of the licence or of the beneficial interest in the hydrocarbons won and saved.[17] This form of control is

---

[15] For a discussion, see Emre Usenmez, 'The UKCS Fiscal Regime', in Gordon, Paterson and Usenmez, pp. 137–60.

[16] The Petroleum Licensing (Production) (Seaward Areas) Regulations 2008, Model Clause 23.

[17] Ibid, Model Clause 36.

standard in licences as the state has a clear interest in ensuring that the party holding the licence is one that has financial and technical resources to carry out the work it has approved.

The licensee must unitize the field if so instructed by the Minister.[18] Where the parties fail to agree, the Minister can impose unitization on them. There have apparently been no examples of this power being used, indicating that parties are keen to avoid the uncertainty that ministerial involvement would bring, even if one or other of them may be seen to be driving a hard bargain.

Finally among the licensee's key obligations is that of furnishing such information to the Minister as he or she may require.[19] This provides the state with ready access to considerable information about the geology of the UKCS and the nature and amount of production from it. This means the Minister has a wider and privileged view of the prospectivity of the province than any individual company.

## 6. VARIANT FORMS OF THE PRODUCTION LICENCE

The great majority of Model Clauses are common to all variants of the production licence. All, therefore, contain obligations, for example, to measure petroleum, retain records, keep accounts, to avoid harmful methods of work and to obtain the consent of the Minister before drilling wells. At first sight, the differences between the variant licences appear small (the duration of the initial term, the point at which the work obligation has to be carried out, the level of area rental) but are significant in practice.

---

[18] Ibid, Model Clause 23. This arises when two adjacent licensed areas are found to lie above the same field. Due to the migratory character of oil and gas, in such situations there is – in the absence of regulation preventing it – a risk that each licence group might seek to produce as much oil as possible, in order to maximize its own production from the shared reservoir. This situation – known as competitive drilling – can damage the production characteristics of the field, leading to reduced recovery rates in the long term. This is contrary to the state's interest, which lies in the maximum recovery of petroleum. Thus the Minister may order that the joint field be developed as a single unit (hence, 'unitization'). This requires both sets of licensees to work together in a unitization and unit operating agreement.

[19] Ibid, Model Clause 30.

The traditional production licence currently[20] has an initial term of 4 years (at the end of which the work programme must be completed and at least one-half of the initial licensed area must be surrendered), a second term of 4 years (at the end of which all parts of the licence, apart from those required for production, must be relinquished) and a production period of 18 years. Area rental for such a licence starts at the relatively low figure of £150 per km$^2$ during the initial term, rising to £7,500 per km$^2$ by the time the production period has been reached.

The promote licence was introduced after consultation with the industry in 2003 in order to make it easier for small companies, perhaps with considerable geo-technical ability but limited financial resources, to obtain licences. While there is a risk in allowing such companies to obtain production licences insofar as they may simply lack the wherewithal to meet their obligations, it is also the case that their knowledge may be crucial in helping the UK meet its objective of maximum economic recovery of hydrocarbons. The promote licence seeks to strike a balance between the risk and the reward. It abates the area rental fee by 90 per cent for the first two years, thus making it easy for small companies to enter the province, but sub-divides the initial term and its associated work obligation into Parts I and II, with a cut-off after Part I if certain obligations are not met, thus removing the risk that the state is left to tidy things up if a small company fails at a more advanced stage. Part I lasts two years during which time the work obligation will normally be to carry out desk-based work on existing data with the promote licensee thereby aiming to attract a co-venturer with the technical and financial capability required to carry out the more onerous and expensive physical exploration work in Part II.

The frontier licence was introduced in 2004 and restructured in 2008. It was initially designed, again in consultation with the industry, primarily to facilitate the exploration and development of the area to the west of Shetland, which has historically been less attractive to the industry due to its deeper water, difficult geology and lack of infrastructure. Whereas the promote licence is aimed at small companies, the frontier licence is very much intended for major companies who have the financial and technical resources to meet the challenges of this difficult environment. The rationale is that such companies will be encouraged to take on these challenges insofar as they are allowed to obtain the exclusive right to a very large area, to explore it for a longer period and to pay a lower rental

---

20 Note, however, that as earlier sets of Model Clauses prescribed different terms, many licences having a different term structure continue to exist.

than would be the case with a traditional licence. In return, the licensee must relinquish the great majority (seven-eighths) of the area for recycling into subsequent licensing rounds at the end of the initial period. There are now two variants of the frontier licence: one has a 6-year initial term followed by a 6-year second term and an 18-year production period, while the other, which is available only for the Atlantic margin area lying to the west of the Outer Hebrides,[21] has a 9-year initial term but is otherwise identical to the 6-year variant.

The promote and frontier variants of the production licence are generally regarded as success stories insofar as they have indeed respectively encouraged new entrants to the North Sea and major company activity to the west of Shetland. They may thus be seen to indicate that the Department of Energy & Climate Change (DECC) and its predecessors have adopted an approach to the regulation of the offshore industry that is both suitably proactive and suitably responsive. Whether this assessment is justified given the criticisms of DECC contained in the Wood Review is another matter. For all that the department may have done a good job in relation to the licence variants, there would appear to be questions about whether its broader role has indeed been conducive to the achievement of the stated objective of maximizing economic recovery of hydrocarbons.[22] This point will be returned to below when the Wood Review is considered in greater detail.

## 7. FINANCIAL PROVISIONS

As discussed above, the means by which the Minister chooses between competing bids for the same acreage are essentially the financial and technical capacity of the bidders and especially the work programme each is offering to carry out. As such, auctions based on cash premium bidding are not a feature of the UK licensing system. The state's take from offshore hydrocarbon operations is accordingly made up of a combination of licensing fees, area rental fees and various forms of

---

[21] The licence is available only for blocks lying beyond the 7°W line in Strategic Environmental Assessment area 7. See DECC, 'Types of Licence on Offer for Each of the Respective Strategic Environmental Assessment (SEA) Areas', available 7 December 2014 at www.gov.uk/government/uploads/system/uploads/attachment_data/file/273694/28R_licence_types_SEA.pdf.

[22] For a discussion, see Gordon, 'Petroleum Licensing', above, paras 4.54–4.69.

taxation.[23] Licensing fees are a relatively insignificant part of the state's take, while area rental fees are somewhat more substantial, though still small when compared to tax on the profits of production. Given fluctuations in the price of oil and the impact on production of unscheduled shutdowns, the revenues received by the state in this way are highly volatile.[24] In the fiscal year 2011–12, for example, the UK's total revenue from oil and gas production was £11.3 billion, while in 2012–13 it was £6.5 billion.[25]

There are currently three main components to the UK's offshore tax regime: Corporation Tax (CT), Supplementary Charge to Corporation Tax (SC) and (for fields developed prior to 1993) Petroleum Revenue Tax. The rate for CT is currently 30 per cent with capital allowance being made for major investment costs on a 100 per cent first year basis. SC was initially introduced as a windfall tax to capture excess profit at a time of high oil prices and levied at a rate of 10 per cent. This was subsequently increased to 20 per cent as prices continued to rise. When the rate was increased once again to 32 per cent in 2011, however, few even outside the industry saw it as anything other than a cash grab by a government struggling with the ongoing effects of the financial crisis. A number of allowances deductible against SC are offered to encourage a variety of especially economically marginal and/or technically challenging developments.[26] CT and SC are both subject to a ring-fence, meaning that losses incurred outside upstream oil and gas operations may not be set off against profits from upstream UKCS production. Significantly, however, the ring-fence is not (for these taxes) applied at a field-by-field level, and so losses from one field may be offset against profits from another. Decommissioning expenses are recoverable in part against

---

23 Royalty payments were initially a feature of the UK's regime, but have now been abolished. It is important to note, however, that the state does obtain significant indirect gains from the oil and gas industry in the form of employment, tax on employees' income, tax on the profits of UK-based service sector companies, and so on.

24 The maturity of the UKCS and the consequent ageing of the infrastructure means that the impact of unexpected shutdowns may increase in future. This problem is all the more significant when one considers that the interconnectedness of the infrastructure means that the failure of a particularly important nodal point will not only result in one platform being shutdown, but perhaps a substantial part of the UKCS.

25 HMRC, 'Statistics of Government Revenues from UK Oil and Gas Production', December 2013, p. 6.

26 E.g. small fields, high pressure/high temperature fields, heavy oil fields, remote gas fields, large deep-water fields, high cost brownfield projects.

payments of CT and SC going back to 2002. Petroleum Revenue Tax (PRT) is chargeable only in relation to fields developed before March 1993, but was abolished for fields given development consent after that time. PRT is levied at 50 per cent and, insofar as it is conceptually a field tax, the ring-fence does operate on a field-by-field basis. There is a specific set of allowances for PRT and decommissioning costs are allowable against PRT without limit of time. Note also that PRT is deductible when calculating profits for the purposes of CT and SC.[27]

The basic objective of the UK tax system for offshore hydrocarbon operations is to strike a balance between maximizing the state's take while incentivizing ongoing investment in a maturing province with challenging frontier areas. To some extent it may be said that this objective is met, but there must be doubts about whether the system is transparent and simple enough to encourage especially smaller companies and also those unfamiliar with the UKCS to enter. This is because the means by which the government seeks to take account of specific challenges encountered by individual developments is by way of tailored allowances which are contained in general tax legislation. While this approach succeeds in meeting the needs of sophisticated and experienced investors, it produces a complex and convoluted appearance to potential new entrants who may not appreciate the fact that dedicated allowances may be available even if none of those currently on offer is relevant to their circumstances. The overall objective could perhaps be met by getting rid of the complex allowances and introducing instead a progressive tax system in which, for example, in the case of a marginal field, an Investment Return Allowance could be claimed against the Supplementary Charge.[28] The ongoing fiscal review announced at the 2014 Budget might be an opportunity for such a change to be effected.[29]

---

[27] For further details, see Usenmez, 'UKCS Fiscal Regime', above.

[28] The Scottish Government, 'UK Continental Shelf Tax Regime: Options for Reform' (2011), available 7 December 2014 at www.scotland.gov.uk/Resource/Doc/919/0121778.pdf, p. 5. Other possibilities were also outlined but Investment Return Allowance was the favoured option.

[29] See HM Treasury, 'Review of the Oil and Gas Fiscal Regime', available 7 December 2014 at www.gov.uk/government/consultations/review-of-the-oil-and-gas-fiscal-regime-a-call-for-evidence.

## 8. OFFSHORE INSTALLATIONS

In recent years, the means by which oil and gas are extracted from the UKCS has shifted increasingly away from the use of large new platforms, which characterized the early developments, and towards the use of floating production, storage and offtake vessels, semi-submersibles and sub-sea completions tied back to existing platforms. Whichever approach is employed, however, wellheads and associated equipment and pipelines will require to be installed on the seabed. As such, a range of permissions is required before any such installation may be carried out. The petroleum license prohibits the carrying out of 'relevant work', which includes 'any structures and any other works whatsoever which are intended by the Licensee to be permanent', without the consent of the Minister.[30] Such consent cannot be given unless the Minister has considered the environmental impact of the proposal.[31] Equally, the operator of a proposed production installation must provide design information to the Health and Safety Executive (HSE) prior to submitting an application to DECC for field development consent[32] and may not operate the installation before a safety case has been submitted to and accepted by the HSE.[33]

### Decommissioning of Offshore Installations

With the UKCS now a mature province, there is every prospect that the relatively slow pace of decommissioning so far experienced will pick up substantially in the coming years. The precise outcome in any given field will depend on a variety of factors including the age of the infrastructure and the extent to which remaining reserves justify the cost of new investment to extend its life to allow further recovery. This in turn will depend upon further factors such as the fiscal environment and the impact of the implementation of the Wood Review discussed further below.[34] Current estimates put the total cost of decommissioning the

---

[30] The Petroleum Licensing (Production) (Seaward Areas) Regulations 2008, Model Clause 17.

[31] The Offshore Petroleum Production and Pipe-Lines (Assessment of Environmental Effects) Regulations 1999.

[32] HSE, *Offshore Installations (Safety Case) Regulations 2005*, HSE's involvement in the design and construction process (including processing of design notifications), Offshore Information Sheet No 11/2007.

[33] Offshore Installations (Safety Case) Regulations 2005, Reg. 7(1)(c).

[34] See especially the conclusion to this chapter below.

UKCS at some £35 billion, although it is not unusual now to hear those in the industry warning of a substantially higher figure. This presents a potential headache for the Treasury insofar as decommissioning costs are allowable against profits for tax purposes, but equally offers a significant opportunity for the contracting community (whose profits will of course be subject to CT).

In terms of the legal obligations facing operators in relation to decommissioning, these stem from the obligations the UK is subject to under international law as a consequence of its being a contracting party to, in particular, UNCLOS 1982,[35] the London Convention 1972[36] and the regional instrument which guides international cooperation on the protection of the marine environment of the North-East Atlantic, the OSPAR Convention 1992.[37] The net effect of their provisions is that there is a general presumption that offshore installations should be entirely removed when they are no longer required. Derogations from this position, meaning that an installation could be left wholly or partly in place, are available only in a limited range of circumstances. There is a strong presumption against installations being dumped offshore, with a preference instead for their reuse, recycling or final disposal on land.

The UK's implementation of these obligations is difficult to discern in either primary or secondary legislation.[38] Such references as there are to decommissioning are focused in particular on who will be responsible rather than on the means by which it will be carried out or the end result that must be achieved. This level of detail is to be found instead in guidance produced by DECC.[39]

---

[35] UNCLOS Art 60(3) as applied to the continental shelf by Art 80.

[36] London Convention on the Prevention of Marine Pollution by Dumping of Wastes and Other Matter of 1972. The 1996 Protocol to the Convention is of particular significance as it provides for a stricter regime than that stipulated in the Convention in its original form.

[37] Convention for the Protection of the Marine Environment of the North-East Atlantic of 1992. The Convention provides for a commission which may take decisions which are binding on all parties to the Convention. The provisions of OSPAR Decision 98/3 are of particular importance in the decommissioning context.

[38] The power in the hands of the Minister under s39 of the Petroleum Act 1998 to make regulations in relation to decommissioning has not so far been used.

[39] DECC, Guidance Notes, *Decommissioning of Offshore Oil and Gas Installations and Pipelines under the Petroleum Act 1998*, version 6, March 2011.

The key statutory provision is section 29(1) of the Petroleum Act 1998, as amended by the Energy Act 2008, which provides that the Secretary of State may serve a written notice requiring the submission of a programme detailing the measures proposed to be taken in relation to the abandonment of an offshore installation or submarine pipeline. The Act also provides that the abandonment programme must contain a cost estimate, specify the times at or within which the proposed measures are to be taken or indicate how those times are to be determined and, if it proposes that an installation or pipeline is to be left wholly or partly in place, include the details of any continuing monitoring and maintenance. The Minister may reject the programme or approve it, with or without conditions. Each of the parties submitting the programme is jointly and severally liable for its execution. In the event of non-compliance, the Minister can serve a default notice, failure to comply with which may give rise to criminal penalties. In such circumstances the Minister may arrange any necessary remedial action necessary, and recover the costs from the defaulting party.[40]

The precise way in which the UK goes about ensuring the development of an appropriate abandonment plan has evolved over the years, not least as a result of the problems surrounding the decommissioning of the Brent Spar oil storage buoy in the mid-1990s. Whereas the UK was content that its approach, involving detailed consideration of a variety of options by the operator and the regulator in order to determine the best practicable environmental option, was wholly consistent with international law, Greenpeace took the view that the process was flawed and that the proposed dumping of the buoy in deep water would set a dangerous precedent. The international furore that ensued persuaded Shell to abandon the original plan and, much to the annoyance of the UK government, to set up in essence a parallel policy process seeking stakeholder input with a view to determining a socially acceptable as well as a technically feasible solution. The net effect of this event is that while the same broad approach is adopted, DECC guidance now requires that where it is proposed that a derogation will be sought from the general duty to remove an installation in its entirety the operator will involve stakeholders from an early stage in the process.[41]

Given the costs involved in decommissioning, as well as the technical challenges and the potential for public concern, it is not surprising to find

---

40 Petroleum Act 1998, s37.

41 For a discussion, see John Paterson, 'Decommissioning of Offshore Oil and Gas Installations', in Gordon, Paterson and Usenmez, pp. 285–329.

that the UK government is particularly keen to ensure that it may turn to a wide range of parties to take responsibility for the process in relation to any individual platform or other piece of infrastructure. The notice requiring the preparation of an abandonment plan may, therefore, be served not only on the operator of the installation, but also on licensees, participants in a joint operating agreement, persons who own an interest in the installation and associated companies of any of the primary duty-holders.[42] The notice may also be served prospectively on parties who intend to take up any of these roles, in recognition of the fact that a field may not perform as expected and require decommissioning at a much earlier stage than had been anticipated, but also raising an issue of relevance in the context of a transfer of assets.[43] Where a party disposes of its interest in a platform, the Minister may withdraw a section 29 notice,[44] but it is important to note that such a party may be effectively called back at any time in the future to fulfil decommissioning obligations[45] – an indication of the government's desire to ensure that it has maximum protection in the event that an incoming party proves unable to meet the costs of abandonment or to lack the technical capacity. The power to reimpose a section 29 notice exists notwithstanding the fact that full use is made by the Minister of the power to require appropriate financial security from an incoming party in the event of an asset transfer.[46] It may be wondered quite what advantage is enjoyed by a party which has had their section 29 notice withdrawn. There is certainly none at the level of statute, but in guidance the Minister has indicated that he would utilize the power in section 34 only as a last resort.[47] Parties divesting themselves of assets and seeking withdrawal of section 29 notices are nevertheless aware that the risk of call-back exists and accordingly are as keen as the government to require incoming parties to provide security for future decommissioning liabilities. As understandable as this approach undoubtedly is, it has served to complicate and in some cases to kill asset transfers.

Asset transfer has also been complicated by the fact that although, as mentioned previously, relatively generous allowances are available for

---

42 Petroleum Act 1998, s30.

43 Ibid, s30, as amended by the Energy Act 2008, s72(4).

44 Ibid, s31(5).

45 Ibid, s34.

46 Ibid, s38(4).

47 DECC, Guidance Notes, *Decommissioning of Offshore Oil and Gas Installations and Pipelines under the Petroleum Act 1998*, version 6, March 2001, para. 3.11.

decommissioning costs in the fiscal regime, the industry has been concerned that future governments might make unfavourable changes. As a consequence, decommissioning security agreements have been drawn up as if the allowances available at the time of signature will not be available at the point of decommissioning, thereby increasing the cost of such security. Lobbying by the industry and recognition finally by the Treasury that this problem was holding up desirable asset transfers between, for example, exiting majors and incoming late-stage specialists, led to the unprecedented offer by the state in the 2012 Budget to guarantee the value of future decommissioning allowances. This is achieved by the Chancellor of the Exchequer entering into legally binding deeds ('Decommissioning Relief Deeds'), the first of which were signed in October 2013. So long as no change is made to the existing tax regime, there will be no need to claim on the deed. Should there be a detrimental change, however, then the company can use the deed to claim a 'difference payment' from the government. This in essence constitutes a circumscribed form of stabilization arrangement.[48]

## 9. THIRD PARTY ACCESS TO INFRASTRUCTURE

A crucial challenge confronting the mature North Sea area of the UKCS in particular is that of ensuring that third parties seeking to tie small marginal developments back to existing infrastructure are able to do so in a timely and cost-effective manner. It has been recognized from a very early stage of the development of the UK upstream industry that such access to infrastructure might not be readily forthcoming. Thus, since 1975, legislation has been in place that allows a thwarted third party to approach the Minister who has the powers to impose a determination, setting out the terms and conditions on which access may be taken. It was always intended, however, that this would be a last-resort solution, with the expectation that access would be negotiated on commercial terms between the parties. The system has not been regarded as successful and has been the subject of frequent amendment as deficiencies with the legislation have come to light.[49] There have also been problems at the

---

48 See HMRC, 'Decommissioning: Increasing Tax Certainty for Oil and Gas Investment in the UKCS' (2013), available 7 December 2014 at http://webarchive.nationalarchives.gov.uk/20130403010244/http://www.hmrc.gov.uk/budget2013/tiin-1118.pdf.

49 See e.g. the discussion on the technical deficiencies that formerly attended the legal definition of 'infrastructure' and 'pipelines' in Uisdean Vass, 'Access to

level of commercial negotiations, with companies seeking access (often relatively small and with relatively modest developments) manifesting a reluctance to approach the Minister on the basis that this is not the 'done thing'. Even were an access-seeker to achieve a successful outcome following a referral to the Minister, they may fear that their willingness to involve the government would lead to subtle reprisals from others in the industry. There may also be a concern on the part of such small players that the Minister may impose a deal on them that they would not ultimately be able to afford.

Efforts have been made to solve these problems, but it remains the case that this is an area of concern on the UKCS. DECC has issued guidance notes detailing how its statutory power would be used as a means of reducing the uncertainty felt by the industry.[50] This Guidance is good in parts: it summarizes relevant legislation and explains the *procedure* DECC would use in making a determination; but it does not set out the *norms* that DECC would use. It certainly states which factors DECC would have regard to in making its decision, but provides little information about how DECC would weigh those factors. The guidance has accordingly had little effect in encouraging third parties to approach the Minister in the event of a dispute on access. The industry itself has also been active in trying to address the problem of third party access. Oil and Gas UK and its predecessor, the United Kingdom Offshore Operators Association have published and updated an Infrastructure Code of Practice (ICOP) since the mid-1990s.

The most recent ICOP was published in 2012 and provides that all members of Oil and Gas UK are expected to abide by the Commercial Code of Practice, a brief (one-page) document which is composed of ten principles, drawn at a high level of abstraction, of good negotiation practice.[51] The ICOP also requires infrastructure owners to provide relevant technical and commercial information in a timely manner. In this last regard, access-seekers can now very easily obtain information on a pipeline's spare capacity or ullage as well as its entry specifications (that

---

Infrastructure', in G Gordon and J Paterson, *Oil and Gas Law: Current Practice and Emerging Trends* (Dundee University Press, Dundee, 2007), para 5.7.

[50] DECC, 'Guidance on Disputes over Third Party Access to Upstream Oil and Gas Infrastructure', (2013), available 7 December 2014 at www.gov.uk/government/uploads/system/uploads/attachment_data/file/212482/Guidance_on_Disputes_over_Third_Party_Access_to_Upstream.pdf.

[51] See Oil and Gas UK, 'Infrastructure Code of Practice' (2012), available 7 December 2014 at www.oilandgasuk.co.uk/cmsfiles/modules/publications/pdfs/ICoP%20revised%202013.pdf.

is, the quality of hydrocarbons admissible). In addition to this high-level information, the ICOP also requires that bona fide enquirers have ready access to more detailed technical information. The ICOP has also led to the key features of successful commercial deals being published so that there is now some clarity as to the commercial norms applying in this otherwise opaque realm.

Given the reluctance of third parties to make use of statutory powers in the event of an obstructive owner, the ICOP also provides for the so-called Automatic Referral Notice (ARN) procedure. In essence this provides that if the parties have not concluded a deal within six months, then the Minister will automatically become involved, thus removing the stigma that appeared to be deterring referrals previously. As simple as this solution appears, the ARN system has not been successful in practice. It has proved difficult for third parties to determine when they are six months away from concluding a deal, at which point the ARN should be issued by the access-seeker. This uncertainty has provided fearful access-seekers with an excuse for foot-dragging. There is also the problem that there is no effective sanction in the event that the third party fails to issue an ARN. In many respects, it is easy to conclude that the problem with this approach is that the ARN is anything but automatic, but rather leaves a key decision in the hands of a party who remains worried about industry attitudes to involving the Minister and about the terms the latter may ultimately impose. It is perhaps not surprising to find, therefore, that ten years on from the introduction of ARN, only two referrals to the Minister have been proceeded with, neither of which resulted in a formal determination being issued. Despite the apparent ingenuity and good intentions of the ARN, the procedure cannot be said to have been any more successful than DECC's efforts in relation to guidance.

The intractability of the access to infrastructure problem was the focus of yet further legislative efforts in the context of the Energy Act 2011, which introduced a new provision designed to address the problem of the nervous access-seeker. Section 83 of the 2011 Act provides that the Minister may now issue a determination on his or her own initiative – in other words, even if the matter has been referred by a party seeking access. While this does constitute a shift in emphasis, the Minister must still wait until the parties have had a reasonable time to conduct a commercial negotiation. Furthermore, DECC appears to be at pains to reassure the industry that it will not make use of the section 38 power except in the last resort. This reticence, surprising in light of the continuing problems with access to infrastructure, may appear more understandable when one considers both the complexity of the task facing DECC in the event that it elected to use the power and the relatively limited resources it has at its disposal.

It is significant that among the key observations in the Wood Review[52] is the fact that the current practice of designing new infrastructure only for specific projects has led to inefficiencies. This is seen as a result both of self-interest on the part of infrastructure owners and of weak governance by DECC. The Review consequently concludes that there is a need for 'a stronger Regulator, able to become pro-actively involved and minimise disruption and delays'.[53] Quite how this will be achieved in practice is not clear at the time of writing, but it is significant that the government has accepted the conclusions of the Wood Review and is committed to implementing the proposed reforms.[54]

## 10. HEALTH, SAFETY AND THE ENVIRONMENT OFFSHORE

The regulation of health and safety at the outset of operations offshore was a matter covered briefly only in the petroleum licence. The licensee was required to comply with any written instructions in this regard issued by the Minister. The instructions that were forthcoming required licensees to follow the Institute of Petroleum's Code of Practice, a document that was drafted very much as guidance but which now had regulatory effect thrust upon it. Insofar as the Minister was only able to sanction licensees for breach of this Code (notwithstanding the many other parties involved in the industry) and then only by way of revocation of the licence, the whole approach to health and safety offshore appeared incoherent and incomplete at best. These problems were soon revealed when in late 1965 the Sea Gem jack-up drilling rig collapsed with the

---

[52] This was initiated by an invitation from the Minister in June 2013 to review the regulation and stewardship of the UKCS from the perspective of maximizing economic recovery of offshore oil and gas. The final version of the review was published in February 2014. See Sir Ian Wood, 'UKCS Maximising Economic Recovery: Final Report' (2014), available 7 December 2014 at www.woodreview.co.uk/documents/UKCS%20Maximising%20Recovery%20Review%20FINAL%2072pp%20locked.pdf.

[53] Ibid.

[54] DECC, 'Government Response to Sir Ian Wood's UKCS Maximising Economic Recovery Review' (2014), available 7 December 2014 at www.gov.uk/government/publications/government-response-to-sir-ian-woods-review-of-the-uk-continental-shelf-ukcs.

loss of 13 lives. The inquiry into the disaster recommended the introduction of a code of statutory authority with credible sanctions. This recommendation was implemented by the Mineral Workings (Offshore Installations) Act 1971, which in turn provided the framework for the establishment of a set of detailed prescriptive regulations, 11 in all by the time it was complete in 1980.[55]

This 'command and control' phase of regulation, in which the state took responsibility for the achievement of offshore health and safety, continued until the Piper Alpha disaster in 1988 in which 167 men lost their lives. The Report of the Inquiry chaired by Lord Cullen that followed the disaster was highly critical of the prescriptive regulatory approach. It recommended its replacement by a new goal-setting approach in which the regulator lays out general objectives but the means of achieving them is a matter for the individual operator at the level of each installation. In greater detail this requires the operator to identify the hazards facing the installation, to quantify the risks and then to indicate the ways in which these risks will be reduced to a level that is as low as is reasonably practicable. The key document in this approach, beyond the initial goal-setting regulations,[56] is the Safety Case, a document in which the operator essentially makes the case to the regulator that the design and operation of installation is safe in the terms outlined above.[57] A breach of any of these regulations, and indeed of the procedures that the operator has set out in the safety case, may result in the prosecution of any responsible party, whether the operator or another.

Given the shift in the degree of responsibility from the regulator to the operator in the post-Piper Alpha era, it would be easy to characterize the goal-setting, safety case approach as essentially self-regulation. This, however, would be inaccurate. The safety case must be submitted to the HSE, in the case of a production facility, at least six months before operations begin,[58] and operations must not commence until the regulator has accepted the safety case.[59] Once accepted, the operator must ensure that all procedures and arrangements described therein which might

---

[55] For a discussion of these issues and for a full listing of the instruments in question, see John Paterson, 'Health and Safety at Work Offshore', in Gordon, Paterson and Usenmez, para 8.15.

[56] For a discussion of these issues and for a full listing and short commentary on each of these Regulations, see ibid, paras 8.38 to 8.43.

[57] The Offshore Installations (Safety Case) Regulations 2005, (SI 2005/3117).

[58] Ibid, Reg. 7(1)(b).

[59] Ibid, Reg. 7(1)(c).

affect health or safety are followed.[60] Thereafter, the safety case must be thoroughly reviewed by the duty holder every five years, or sooner if required by the HSE.[61] Insofar as the safety case must be understood to be a 'living' rather than a static document, it must also be revised by the duty holder when it is 'appropriate' to do so,[62] for instance, if changes have been made 'that merit reappraisal of the risk control arrangements, whether or not they require the adjustment of measures to be taken'.[63] Other regulations, too, require acceptance by the HSE and/or external verification, for example, the obligation for independent examination of the well design concept.[64]

The UK's approach to health and safety offshore appeared well-established until the Deepwater Horizon disaster in the Gulf of Mexico in April 2010 prompted a review of existing arrangements not only by the relevant authorities in the UK, but also by the European Commission. This led in due course to the passage of the Offshore Safety Directive,[65] which Member States must transpose into national law by July 2015.[66] Transitional provisions apply which mean that in relation to installations and well operations at a planning stage, Member States do not have to apply the Directive until July 2016, and in relation to installations which already exist, Member States do not have to apply the Directive until July 2018. While the Directive is in no small measure influenced by the UK's approach, it will not be without its implications for the content of the safety case and indeed for the identity of the regulator. As regards the content of the safety case, this document will now have to consider environmental risks as well as those affecting health and safety. As regards the identity of the regulator, the complication here is that the Directive requires there to be no conflict of interest between the body

---

60 Ibid, Reg. 16.

61 Ibid, Reg. 13.

62 Ibid, Reg. 14.

63 HSE, 'Material Change to a Safety Case: Offshore Information Sheet No. 3/2011' (2011), available 7 December 2014 at www.hse.gov.uk/offshore/infosheets/is3-2011.htm.

64 Offshore Installations and Wells (Design and Construction, etc.) Regulations 1996, Reg. 18.

65 Directive 2013/30/EU of the European Parliament and of the Council of 12 June 2013 on Safety of Offshore Oil and Gas Operations and Amending Directive 2004/35/EC [2013] OJ L 178/66.

66 For a discussion, see G Gordon, 'Offshore Safety: The European Commission's Legislative Initiatives', in *European Energy Law Report* X (M Roggenkamp and H Bjornebye, eds, Intersentia, Cambridge, 2014) pp. 139–72.

responsible for economic development of offshore hydrocarbons (licensing) and that responsible for health, safety and environmental regulation. While the problem of a potential conflict between licensing and health and safety in the UK was solved after Piper Alpha by shifting responsibility for the latter away from the Department of Trade and Industry (DTI) (DECC's predecessor as the licensing authority) and to the HSE, responsibility for much environmental regulation remained with the DTI. In the aftermath of Deepwater Horizon, which was of course a significant environmental disaster as well as one that resulted in injury and death and which revealed a deeply conflicted regulator in the US, the European Commission is keen to avoid any such problem with the EU. At the time of writing, the manner in which this problem will be resolved in the UK is not yet clear.

For the time being therefore, responsibility for a significant body of rules, some contained in the licence, but the overwhelming majority of which are found in delegated legislation, governing various aspects of environmental protection remains in large part with DECC. These rules can be divided into various groups. First of all, at the point at which the decision is made as to whether a particular area of the UKCS can be opened for licensing, a Strategic Environmental Assessment must be carried out so that DECC, which is the responsible authority, is aware of the environmental consequences.[67] In addition, Environmental Impact Assessments will normally be required before consent will be given for a variety of hydrocarbon operations including drilling.[68] Exploration activities are also regulated insofar as there are restrictions on the licensee's right to conduct seismic surveying as the acoustic disturbance involved in that process may adversely affect marine wildlife.[69] During operations, the licensee is obliged to avoid harmful methods of working. The operator of an offshore installation is also obliged to produce an Oil Pollution Emergency Plan (OPEP), which must be submitted to DECC

---

[67] The Environmental Assessment of Plans and Programmes Regulations 2004 (SI 2004/1633). For details, see DECC, 'Offshore Energy Strategic Environmental Assessment (SEA): An Overview of the SEA Process', available 7 December 2014 at www.gov.uk/offshore-energy-strategic-environmental-assessment-sea-an-overview-of-the-sea-process.

[68] The Offshore Petroleum Production and Pipelines (Assessment of Environmental Effects) Regulations 1999 (as amended) (SI 1999/360). For details, see DECC, 'Oil and Gas: Offshore Environmental Legislation', available 7 December 2014 at www.gov.uk/oil-and-gas-offshore-environmental-legislation.

[69] See, inter alia, The Offshore Petroleum Activities (Conservation of Habitats) Regulations 2001 (SI 2011/1754) (as amended).

for approval and must be renewed every five years or where there is a material change of circumstances. This plan must set out the measures that would be implemented in the event of an oil spill,[70] both in relation to tracking oil pollution and containing and dealing with it.[71] The operator is also subject to regulations controlling the volume of petroleum and chemicals that may be lawfully discharged from offshore installations.[72] Further, the Secretary of State also has a range of powers where a major pollution event has occurred or appears likely to occur. These range from the ability to issue directions up to quite draconian powers where the Minister considers that issuing directions is going to be or has proven to be inadequate. These latter powers include taking charge of the emergency response operation, and taking whatever action the Minister sees fit in relation to an offshore installation or its contents up to and including the destruction or sinking of the installation.

## 11. CONCLUSION

The foregoing chapter has offered a relatively brief overview of the UK's approach to the licensing of offshore hydrocarbon operations and discussed some of the key features, not least those which have particular significance at the moment as a result of recent events, notably the treatment of health, safety and the environment. In the months ahead, however, it is likely that relatively significant changes will occur. These may affect the licence itself – including existing licences – and will certainly affect the identity of the responsible authority. These changes arise as a result of the Wood Review, which has been referred to at several points above. In essence, this review stems from a request from DECC in mid-2013 to Sir Ian Wood, a respected figure in the offshore

---

[70] For a discussion of the civil liability issues arising from a spill of oil from an installation on the UKCS, see G Gordon, 'Oil, Water and Regulation Don't Mix: Environmental Liability for Offshore Operations in the UK', Parts 1 and 2: (2013) 25 ELM 3–11 and (2013) 25 ELM 121–32. For a comparative discussion of the position within the EEA, see BIO by Deloitte (2014) *Civil Liability, Financial Security and Compensation Claims for Offshore Oil and Gas Activities in the European Economic Area, Final Report prepared for European Commission – DG Energy*, available 7 December 2014 at www.ec.europa.eu/energy/oil/offshore/doc/201408_offshore_oil_and_gas_activities.pdf.

[71] The Merchant Shipping (Oil Pollution Preparedness, Response and Co-operation Convention) Regulations 1998.

[72] For example, the Offshore Petroleum Activities (Oil Pollution Prevention and Control) Regulations 2005 (SI 2005/2055) (as amended).

industry on the UKCS, to consider what would be required in order to ensure the achievement of the state's objective of maximizing economic recovery in what must be the late phase of the North Sea (even as the west of Shetland continues to present certain features of a frontier province). A full examination of the Review is beyond the scope of this chapter, but it is sufficient to note in concluding that Sir Ian is by no means convinced of the continued relevance of DECC as the key regulator in relation to the petroleum licence. While DECC would presumably continue to play a role in relation to policy under his proposals, the key regulatory role would pass to a new independent regulator at arm's length from the department. This regulator would possess the powers necessary to ensure that those licensed to operate on the UKCS (and thus entrusted with the stewardship of the state's hydrocarbon resources) are always doing so in such a way as would maximize the economic recovery of the remaining hydrocarbons *in situ*. The language of the Review at times suggests a facilitative role for the new regulator, and at others indicates that the regulator would have stronger powers of intervention and enforcement. Whereas all indications in the immediate aftermath of the Review's publication in February 2014 were that the industry was fully behind Sir Ian's recommendations (which were perhaps not unsurprisingly warmly welcomed by the government), the passage of months since then has revealed a certain trepidation that a newly empowered independent regulator would change the manner in which business has traditionally been done on the UKCS in ways which might not be to the liking of all the commercial actors. At the time of writing, there is still no clarity on the shape and form of the new regulatory authority, and only relatively vague indications as to the legislative approach to the powers. A period of intense discussion is under way involving key stakeholders as the government seeks to ensure that this key opportunity to shape the future direction of offshore developments in the long-term interests of the state is not wasted and the industry seeks to ensure that the government understands where the balance lies between protecting state interests and creating disincentives to the investments that will be required if maximum economic recovery is to be achieved. Fifty years after the passage of the Continental Shelf Act in 1964 one might have been forgiven for suspecting that the UK would by now have a very stable, and perhaps even slightly boring, licensing regime. Nothing could be further from the truth.

# 5. Regulation of the Norwegian upstream petroleum sector[1]

**Ernst Nordtveit**

## 1. HISTORY AND OVERVIEW OF NORWEGIAN PETROLEUM ACTIVITIES

Norway became a petroleum producing country in the early 1970s. Up to the early 1960s it was not expected that petroleum could be found in Norwegian land or sea areas. The discovery of natural gas in Groningen in the Netherlands in 1959, however, led to an interest from some oil companies to explore for petroleum further north in the North Sea.

In October 1962 the Norwegian authorities, to their surprise, received an application from the American oil company Phillips Petroleum for an exploration license in the Norwegian part of the North Sea. The Norwegian authorities regarded this as an application for an exclusive license. This was unacceptable to the Norwegian authorities, but the interest from the oil companies made them think that it might be timely to establish a legal regime and a policy for handling the interest in petroleum activity. A small group of people in the Ministry of Foreign Affairs led by Jens Evensen took the initiative to establish a legal framework for possible future activity. In May 1963, the Norwegian government proclaimed sovereignty for Norway over the Norwegian continental shelf, based on the Continental Shelf Treaty from 1958.[2] An Act on submerged areas was also passed by the Norwegian parliament,

[1] Most of the literature on Norwegian petroleum law is in Norwegian. In this chapter, I have only made references to literature or material that is available in English.

[2] Convention on the Continental Shelf 1958. Done at Geneva on 29 April 1958. Entered into force on 10 June 1964. United Nations, *Treaty Series*, vol. 499, p. 311

Stortinget.[3] The main provisions in the act were that any natural resources on and under the ground of the subsea areas belongs to the Norwegian State and that exploration and production will be subject to a license from the King.[4] The same year some companies were granted licenses to carry out preparatory exploration by seismic surveys, but not drilling.

The process of clarifying the borders of the Norwegian continental shelf took place over the next 45 years. The borders in the North Sea were settled by agreements with Denmark and with the United Kingdom in March 1965 in accordance with the median line principle.[5] These agreements were of great importance to Norway as several of the major Norwegian oil and gas fields are located near the borders of the English or Danish continental shelves, or even crossing the border towards England.[6]

An agreement on the border towards the Icelandic continental shelf, in the area between the island of Jan Mayen and Iceland, was reached in October 1981. According to this agreement the two countries will cooperate on petroleum activity in a defined area between Iceland and Jan Mayen, covering 45 470 km², of which 32 750 km² is on the Norwegian side of the border and 12 720 km² on the Icelandic side. Each country has the right to participate with a 25 per cent share of the activity that takes part on the other country's side within the defined area.

---

3 Act of 21 June 1963 No. 12 relating to scientific research and exploration for and exploitation of subsea natural resources other than petroleum resources (1963 Act). English version available 14 December 2011 at www.npd.no/en/Regulations/Acts/Scientific-research-act/. This act prevailed until the first petroleum act in 1985 became the formal basis for regulation of Norwegian petroleum activity, when it was changed to only regulate other natural resources than petroleum. The 1985 Act was repealed by the Act of 29 November 1996 No. 72 relating to petroleum activities (Petroleum Act), see English version available 26 June 2012 at www.npd.no/en/Regulations/Acts/Petroleum-activities-act/.

4 Norway is a monarchy, but the king has had no real political power since 1905. The decisions are madc by the Government, led by the prime minister, but formally the decisions are made in a meeting with the King presiding and in the form of a royal declaration. Any reference to the King later on will mean the Government.

5 See United Nations Convention on the Law of the Sea (UNCLOS), Article 83.

6 The Ekofisk field is located near the border towards the Danish continental shelf and the Norwegian Statfjord field is a part of a structure going into the UK sector, where it is called Murchison.

Norway presented its claim for delimitation of the Norwegian continental shelf towards the Arctic Ocean area in 2006. An agreement was made with the Commission on the Limits of the Continental Shelf (CLCS) in April 2009. Norway is so far the only country to have had its claims for continental shelf in the Arctic accepted. By the agreement, an area of 235 000 $km^2$ was added to the Norwegian continental shelf.

In April 2010 an agreement was finally reached on the maritime border with Russia in the Barents Sea, after nearly 40 years of negotiations. The agreement was a compromise between the Norwegian claim based on the median principle and the Russian claim based on the 'sector-principle'.

The borders of the Norwegian continental shelf are consequently defined towards all neighbour countries and towards the deep sea areas along the Norwegian continental shelf. The Norwegian continental shelf is about four times the size of the Norwegian mainland, and covers about one third of the total European continental shelf.[7]

After the borders in the North Sea were defined, the first Norwegian licensing round was announced in April 1965. Based on the applications, production licenses for 78 blocks[8] in the North Sea were awarded to oil companies or groups of companies. The first major discovery of petroleum was made by Phillips Petroleum on the Ekofisk field in December 1969. This was one of the largest offshore petroleum deposits ever found, containing oil and natural gas.

Production from Ekofisk commenced in 1971, and in the following years a number of major discoveries were made, like the natural gas field Frigg in 1971 and Statfjord in 1974, which is one of the largest offshore oil fields in the world. Later the natural gas field Troll, containing 1300 billion cubic metres of natural gas,[9] and oil fields like Gullfaks and Oseberg and several smaller fields added considerably to Norwegian petroleum production.

---

[7] In addition to the continental shelf adjacent to the Norwegian mainland, Norway also has jurisdiction over the Bouvet Island in the southern Atlantic, and its continental shelf. Norway also has claims in Antarctica (Queen Maud Land) and its continental shelf. This area is regulated by the Antarctic treaty from 1959.

[8] The continental shelf is divided into 'blocks' of 15 latitude minutes and 20 longitude minutes and numbered. A production license might cover one or several blocks or parts of blocks, Petroleum Act, s 3-3(1).

[9] From the start Troll was considered a natural gas field only, but due to technical innovation it became possible to extract the oil from the field, also turning Troll into an important oil field.

Until the end of the 1970s, Norwegian petroleum activity was limited to the areas south of the 62nd parallel, but from around 1980 the activity was gradually moved north with the granting of licenses in the Norwegian Sea and later the southern part of the Barents Sea. Discoveries have been made in the Norwegian Sea in fields such as Heidrun and Ormen Lange, and in the Barents Sea with the field Snøhvit (Snow white), which is a natural gas field, and lately oil fields like the Goliat field and the Skrugard field. The gas from Snøhvit is transported by pipeline to an LNG plant at Melkøya from where it is transported by ship to the customers.[10]

The petroleum industry has become the largest economic activity in Norway, with around 42 per cent of the expected petroleum to be found on the Norwegian continental shelf being produced so far. Currently the Norwegian petroleum production represents 23 per cent of the country's GDP. Petroleum exports count for 52 per cent of Norwegian exports. The total value created from petroleum activity until today is more than 12 000 billion kroner in today's value, while 3000 billion kroner have been invested in petroleum activity. Around 43 000 people are working in the petroleum industry or in a directly connected activity. Indirectly it is estimated that petroleum activity creates 250 000 jobs in Norway.[11]

The Norwegian production of petroleum has varied over the years and changed from mainly oil production in the first years to a larger proportion of natural gas. The oil production peaked in the years from 1996–2002 with about 180 million $Sm^3$ of oil equivalents annually and is now down to around 100 million $Sm^3$.[12] Around 100 billion cubic metres of natural gas[13] and 1.7 million tons of NGL[14] and condensate, all at a value of around 200 billion NOK, is exported from the Norwegian continental shelf to the continent and the UK every year, supplying 18 per cent of the European consumption of natural gas. The Norwegian gas

---

[10] Detailed and updated information on Norwegian petroleum activity can be found on the Fact Sheet that is published every year by the Ministry, see Facts 2014 Norwegian Petroleum Sector, 22 August 2014, available 22 August 2014 at www.regjeringen.no/en/find-document/dep/OED/Reports/2014/Facts-2014–All-you-need-to-know-about-Norwegian-petroleum-activities/id757846/?regj_oss =20. The NPD has established a new service called 'Factpages' and 'Factmaps', available 12 January 2015 factpages.npd.no/factpages/Default.aspx. An application 'Oil facts' for iPhone and androids can also be downloaded for free.

[11] Facts 2014. The Norwegian Petroleum Sector, Ministry of Petroleum and Energy and Norwegian Petroleum Directorate, (Oslo 2014), page 13.

[12] Ibid, 13.

[13] 107 million cubic metres in 2013.

[14] Short for Natural Gas Liquefied (propane and butane).

exports accounted for approximately 24 per cent of EU gas consumption in 2013, with almost all Norwegian exports going to countries in the EU. An extensive infrastructure for processing and transportation of natural gas from the Norwegian continental shelf to the markets in the UK and on the European continent has been developed, with a network of nearly 8000 km of submarine pipelines and two large processing plants in Norway (Kollsnes and Kårstø). The main importers of Norwegian gas are Germany, France, UK and Belgium.

Petroleum activity generates great revenues for the state. From around 2000 a new system for managing the state revenue from petroleum activity was introduced. All revenue from petroleum activity is channelled into a special fund called the Government Pension Fund Global, and is invested in stocks, bonds and property abroad.[15] Only up to 4 per cent of the capital can be spent annually to cover state expenses. This limitation is set in order to avoid inflation arising from overspending of petroleum revenue.

Norway is regarded as a country that has succeeded in using the petroleum resources as a tool in order to develop the whole society, and has developed a good system for managing the resources as well as the revenue from the activity.[16] The Norwegian petroleum policy and system for governance of the petroleum sector has been developed over time and has sometimes caused conflict and strong political debate in Norwegian society. Today there seems to be a general consensus on the main elements of the system for governance of petroleum activity in Norway.[17]

---

[15] Information on the value of the fund, investment strategy, etc. can be found on the website available 15 February 2010 at www.nbim.no/en/. In January 2014 the value was more than 6500 billion NOK. The value is constantly updated.

[16] Tina Hunter, *Legal Regulatory Framework for the Sustainable Extraction of Australian Offshore Petroleum Resources* (University of Bergen, Bergen 2010), 30–33.

[17] An updated presentation of the Norwegian petroleum policy can be found in the White Paper: Meld. St. 28 (2010–2011) (Report to the Storting) 'An Industry for the Future: Norway's Petroleum Activities' (hereafter White Paper 2011). The English version is available (24 June 2011) at www.regjeringen.no/nb/dep/oed/kampanjer/petroleumsmeldingen.html?id=636866.

## 2. LEGAL FRAMEWORK FOR OFFSHORE PETROLEUM ACTIVITIES IN NORWAY

Norway has developed extensive legal regulation and administration of its petroleum activity and related areas. The governance of petroleum activity in Norway also has to be adapted to the legal frameworks established in international law and European law that are binding on Norway.

All Norwegian petroleum deposits are located on the continental shelf outside the Norwegian territorial waters. The Norwegian jurisdiction over these areas has to be exercised within the limits of the international law of the sea, especially UNCLOS Articles 77–81, and international treaties on protection of the environment like the OSPAR convention.

Norway is not a member of the European Union, but the Agreement on the European Economic Area (EEA) which was entered into in 1992 in reality integrates the three EFTA countries Iceland, Norway and Lichtenstein into the European internal market, except for agriculture and fisheries. The EEA agreement is an international agreement which obliges Norway and the other EFTA partners to implement legal acts determined by the European Union in its internal law, based on specific procedures, in order to create equal market conditions for all partners. EU legal acts that are considered to be relevant to the EEA shall be implemented in the EEA agreement through specific procedures.

The EEA agreement contains some general principles that have a strong impact on petroleum activity. Article 4 of the EEA agreement bans any discrimination based on nationality between the partners, and the agreement also establishes the free movement of capital, goods and labour. This has an impact on licensing as well as the regulation of procurement for petroleum activity.

The EU has also adopted several legal acts which directly regulate or are of great relevance to petroleum activity. Council Directive 93/38/EEC of 14 June 1993 coordinating the procurement procedures of entities operating in the water, energy, transport and telecommunications sectors also regulates procurement procedures in the petroleum industry.[18]

Directive 94/22/EC of the European Parliament and of the Council of 30 May 1994 on the conditions for granting and using authorizations for the prospection, exploration and production of hydrocarbons (Licensing directive) regulates the procedures and criteria for granting licenses for

[18] The directive is implemented in Norwegian law by regulation of 7 April 2006 no. 403.

petroleum activity. Directive 2009/73/EC of the European Parliament and of the Council of 13 July 2009 concerning common rules for the internal market in natural gas regulates the transport and marketing of natural gas, in the upstream market as well as in the downstream market.

Directive 2013/30/EU on the safety of offshore oil and gas operations has not yet been integrated in the EEA agreement, and Norway claims that this directive is not a part of the EEA agreement.

In the Norwegian constitution the legislative power is vested in Stortinget, while the executive power is vested in the King. The central public administration is organized in ministries, headed by a minister who is a member of the government. The Ministry for Petroleum and Energy (hereafter the Ministry) is responsible for governing the petroleum sector, within the acts, budget decisions and other decisions made by Stortinget. Norway has a parliamentary system that means that the government needs the support of Stortinget. A minister who does not have the confidence of Stortinget would have to resign. Stortinget also has the right to instruct the government or minister on how to act in major cases that formally would be under the government's or minister's power to decide.

Petroleum activity is regulated by several acts (enacted by Stortinget) and a large body of detailed regulations.[19] The Petroleum Act regulates the right to the natural resources, licenses for exploration and production of petroleum, resource management, utilization of installations, plans for field development, transport of petroleum and termination of the activity, disposal of installations after termination of the activity, organization of the state participation in the activity, liability for pollution damage and compensation to fishermen for occupation of earlier fishing grounds for petroleum activity, and so on.

In accordance with Norwegian tradition, the Petroleum Act regulates the main questions and leaves technical details to more specific regulations. The power to make such regulations is delegated to the government, the Ministry or administrative bodies under the Ministry like the Norwegian Petroleum Directorate (NPD) or the Petroleum Safety Agency (PSA).

A great number of regulations have been adopted for petroleum activity with a basis in the Petroleum Act or other relevant acts. Some regulate very concrete questions of technical or practical character, others

---

[19] An English version of the most central acts and regulations can be found on the website of the Norwegian Petroleum Directorate, available 7 July 2008 at www.npd.no/en/Regulations/.

have a more general scope, but are still more detailed than the Act. Regulation of 18 May 2001 no. 516 on delegation of authority to the Ministry of Oil and Energy, regulating management of the State's Direct Financial Interest (SDFI) in petroleum activity, and Regulation of 18 May 2001 no. 749, regulating resource management in petroleum activity, are important general regulations.

Petroleum activity is also governed by more individual, but to a great extent standardized, conditions for the individual licenses for exploration, field development, transport or termination of activity.

Other Norwegian law than the Petroleum Act, including customary law, will apply to activity on the continental shelf, unless provision is made in another act, is decided by the King or follows from international law or an agreement with another state. This does not, however, apply to removable installations that are not permanently placed on the Norwegian continental shelf, unless it is decided in an act, or by the King in accordance with an act, that Norwegian law will apply.[20]

In addition to the Petroleum Act, the most important acts for petroleum activity are the Pollution Control Act,[21] regulating ordinary pollution as well as from offshore activity, and the Working Environment Act,[22] regulating working conditions and safety measures for personnel. The Marine Resources Act[23] will also apply to petroleum activity. The Act on Biodiversity[24] does not apply fully outside the territorial waters but the general aim of sustainable development, the aim for the management of biological species, and the general principles of management as evidence-based management, the precautionary principle, ecosystem-based management, polluter pays principle and use of best possible technology, will also apply to the continental shelf.

The executive branch of the government is responsible for the management of petroleum as a state asset. Within the framework of the Constitution, legislation and decisions by Stortinget, the authorities have

---

[20] Petroleum Act, section 1-5.

[21] Act 13 March 1981 no. 6, relating to protection against pollution and relating to waste, see section 4.

[22] Act 17 June 2005 no. 62, see section 1-3.

[23] Act from 6 June 2008 no. 37 on management of wild living marine resources. Also the Petroleum Act regulates the obligations for the petroleum authorities to take consideration for fisheries and marine resources into account in decisions allowing petroleum related activities, but there are some discussions to what extent the fisheries authorities on the basis of the Marine Resources Act could interfere also in petroleum activities.

[24] Act 19 June 2009 no. 100, which regulates management of diversity in biology, landscape, geology and ecological processes.

the obligation to manage the petroleum as a state asset and maximize the value creation of the petroleum in order to increase the tax revenue for Norwegian society.[25] Some decisions like the granting of licenses are made by the King in Council, but the management is to a large extent delegated to the Ministry of Energy and Petroleum.[26]

Two separate administrative bodies have been established to control and regulate petroleum activity, the Norwegian Petroleum Directorate (NPD)[27] and the Petroleum Safety Authority (PSA),[28] both located in Stavanger.

The NPD is a governmental specialist directorate and administrative body under the Ministry. Its main objective is to gather and analyze data from the continental shelf and contribute to prudent resource management based on safety, emergency preparedness and safeguarding of the external environment. The NPD secures the realization of the resource potential by long-term resource management and, in cooperation with other authorities, ensures comprehensive follow-up of petroleum activities. The NPD is an adviser to the Ministry, but has also been delegated the authority to make regulations and individual decisions in several areas. The NPD is also responsible for conducting metering audits and collecting fees from the petroleum industry.

The responsibility for regulating and supervizing the safety and emergency preparedness and for the working environment in petroleum activity has been given to the Ministry of Labour. The PSA is responsible for the technical and operational safety, emergency preparedness and working environment of petroleum activity. The activity is divided into six areas: drilling and well technology, process integrity, structural integrity, logistics and emergency preparedness, occupational health and safety, and HSE management. An important task is safeguarding the legal quality of regulations, orders and other products where the formal basis must be clear and unambiguous. The PSA is more of an independent authority with responsibility for safety and less of a political body than the NPD, but the PSA reports to the Ministry of Labour.

The Ministry of Climate and Environment has the responsibility for managing environmental protection and the external environment. The Norwegian Environment Agency is, for example, also responsible for enforcing the Pollution Control Act in relation to petroleum activity.

---

25 The Constitution 1814, section 19 and the Petroleum Act, section 1-2.
26 See www.regjeringen.no/en/dep/oed.html?id=750.
27 See www.npd.no/en/.
28 See www.ptil.no/?lang=en_US.

Taxation and fees from petroleum activity are the responsibility of the Ministry of Finance, and the Petroleum Tax Office is subordinated to the Ministry of Finance.

## 3. PETROLEUM LAW: HISTORICAL DEVELOPMENT, REFORM AND CURRENT LAWS

### Development of Legislation

The Norwegian petroleum regime must be understood within the background of the political situation in Norway and the economic and industrial development in the first decades of the Norwegian petroleum era. Norway was, prior to the discovery of petroleum, a modern industrialized country, which to a large extent was based on natural resources like forests, minerals, fish and hydro-energy. Norway also had a large merchant fleet and a well-developed ship-building industry. The introduction of petroleum activity in the early 1970s coincided with a downturn in the shipping industry as a result of the oil embargo in 1974, the general international economic downturn from around 1977 and competition from emerging ship-building countries in Asia. At the same time, the Arab oil embargo in 1974 and the rapid increase in oil prices due to OPEC's strong position throughout the 1970s and the first part of the 1980s created a strong bargaining position for Norway as a promising new petroleum province.

Norway was an established democracy with a developed legal system and a well-developed welfare system with free education, free health care, and social services. Since the early 1900s Norway had developed a system of a mixed economy and a tradition for strong state involvement in management of natural resources. The petroleum resources were seen as an opportunity for the industrial, social and economic development of Norwegian society, and it was important to secure the benefits for Norwegian society.[29]

---

[29] The principle that the petroleum resources should be used to develop a better quality society for the benefit of all regions and all people was formulated early and is now expressed in the Petroleum Act, section 1-2, 2:

> Resource management of petroleum resources shall be carried out in a long-term perspective for the benefit of the Norwegian society as a whole. In this regard the resource management shall provide revenues to the country and shall contribute to ensuring welfare, employment and an improved environment, as well as to the strengthening of Norwegian trade and industry

Norway also had long experience in managing natural resources and in using foreign capital and technical competence to develop natural resources from the mineral industry dating back to the 1500–1600s and from the development of hydro-electric energy from the late 1800s. Petroleum activity represented different challenges than earlier activities and new models for the regulation and governance of petroleum activity had to be developed. An important basis for the Norwegian license system was a report on possible models for governing the petroleum resources based on a survey of models used in different countries, made by Jens Evensen.[30] His advice to the Norwegian authorities was to use a licensing system in combination with joint ventures or joint operating agreements. This system was adopted and is still the main system for management of the petroleum resources in Norway, but the regulations and administrative practice have been altered substantially throughout the Norwegian petroleum era.

Norwegian petroleum law has been developed gradually as the government gained experience and with the development of petroleum activity. The basis for development of Norwegian petroleum law was the 1963 Act, until the more detailed regulation in the first petroleum act in 1985. The 1963 Act delegated the authority to give more detailed rules to the King, and regulation was developed by revised regulations several times before the 1985 Act.

Based on the 1963 Act, petroleum activity was regulated by the government in the form of regulations contained in 'royal declarations'.[31] A special committee of experts was established in 1972 with the task of developing the Norwegian petroleum legislation. The committee presented a draft for a general petroleum act in 1979.[32] After a long process this led to the first petroleum act in 1985.

In 1995 the Ministry prepared a proposal for a new act. The reasons for the new act were the development on the continental shelf of more

---

and industrial development, and at the same time take due regard to regional and local policy considerations and other activities.

[30] Evensen, Jens Ingebret: Oversikt over oljepolitiske spørsmål bl. a. på bakgrunn av utenlandsk oljelovgivning og utenlandsk konsesjonspolitikk. Innstillinger og betenkninger nr. 21, Oslo 1971 (Survey of oil-political questions on the basis of foreign oil-legislation and foreign licensing policy).

[31] Royal declaration from 9 April 1965 which was repealed by royal declaration from 8 December 1972.

[32] NOU 1979: 43 Petroleumslov med forskrifter (Petroleum Act with regulations).

mature oil provinces and the need for regulation of cessation of petroleum activity. New technical developments making it possible to run parts of the petroleum activity from land and also the adaptation to EEA regulation were probably reasons as well. The new act is mainly based on the 1985 Act, but is in many respects more flexible and leaves more competence to the Ministry than the former act.

## License Policy and License Conditions

The Norwegian petroleum regime as demonstrated above consists of a framework of acts and regulations. To understand the Norwegian concession and licensing system it is equally important to look at the regulation developed in individual concession conditions, the use of license groups and joint venture agreements and the 'governance from within' through state participation in the activity. The principles developed for granting licenses have also been an important tool for governance of the activity.

Petroleum activity on the Norwegian continental shelf can only be carried out on the basis of a concession or license from the state.[33] This also gives the opportunity to grant licenses in certain conditions. Based on the tradition from the concessions for hydro-electricity, petroleum activity was from the start in reality to a great extent regulated by individual conditions for the licenses.[34] These conditions were mainly standardized, but could be altered from concession round to concession round and thus made it possible to develop regulation based on experience and the development of the industry, without changing the law. There has also been a clear development in the conditions from licensing round to licensing round, but the basic structure has been relatively stable over the years.

Licenses were granted in licensing rounds where a restricted number of new fields were made available for applications. The areas available for application were selected by the authorities, often based on suggestions by the companies on the basis of information obtained by exploration. The authorities were interested in information about the geology in

---

33 In Norway the concept of concession (*konsesjon*) has traditionally been used for public permits to exploit natural resources, also in other areas than petroleum. In line with international terminology in the petroleum industry I will use the word 'license' in this chapter. As far as I can see there is no difference in reality between the two concepts.

34 The Petroleum Act, section 3-1 now also states that licenses can be granted on certain conditions, but this is not necessary under general Norwegian administrative law.

different areas of the continental shelf and followed a systematic strategy in order to investigate different parts of the continental shelf.

Norway early developed a policy where licenses were given to a group of companies composed by the state. Even if the companies applied individually for a license in different fields in which they were interested, the authorities would grant each license to a group of several companies.[35] It would then be a condition for the license for each company that they accepted to share the license with the other companies in the group and that they entered into a joint venture agreement with them, for the purpose of cooperating in the petroleum activity based on the license. After the fully state-owned company Statoil was established in 1972, Statoil would get a share in all licenses. Statoil was not only established with the purpose to generate economic revenue for the state, but was also an important part of the state governance of the petroleum industry 'from within'. Statoil had a veto right in the license groups in cases of great national importance and also played an important role in giving Norwegian industry an opportunity to compete for contracts for goods and services for the petroleum industry.

Until the EEA agreement Norwegian companies could be given priority in the licensing process. In addition to Statoil, it had long been Norwegian policy to bring in two other Norwegian companies. Hydro Petroleum (Hydro) was a part of the major industrial company Norsk Hydro ANS, in which the Norwegian State had been a majority shareholder since the 1960s, and Saga Petroleum AS was an independent privately owned Norwegian company. These two companies were granted shares in many licenses, but they did not have any special position in the joint ventures. This structure lasted for almost 30 years, but when Saga ran into economic problems in 1999 due to low oil prices and some strategically unfortunate transactions, it was taken over by Hydro in cooperation with Statoil. Saga was merged with Hydro, while Statoil took over 25 per cent of Saga's shares in Norwegian petroleum licenses and its role as operator on two fields (Snorre and Tordis). This process seems to have been mainly run by the companies without government intervention. Statoil was partly privatized in 2001, and in 2006 Statoil merged with Hydro Petroleum.

---

[35] See section 4 on the criteria for selecting the members of the group.

## State Participation and Governance 'from Inside'

The most important license condition has been that the companies sharing a license entered into a joint venture agreement between themselves in accordance with a draft attached to the license, formulated by the state. In the earliest licensing rounds these agreements were called 'state participation agreements', emphasizing that the main purpose was to regulate the relation between the state and the private partners in licenses where the state took part through Statoil. The state participation agreements had provisions that the other partners in the license should carry Statoil's exploration costs and about the transfer of technology and knowledge.

The agreement would formally be an agreement between the partners in the license and the state would not sign the agreement. The agreement regulated the commercial and practical aspects of cooperation on general terms, but it also contained some regulations of great importance for the public governance of the activity and the joint venture agreements were considered part of the concession system for the petroleum industry. Only the upstream activity was regulated by the agreement, and the crude oil produced was sold individually by each partner. The agreements had provisions on the joint sale of natural gas and the gas was until 2000 sold jointly in field contracts; see below on the development of the sale of natural gas.

The agreements were developed from concession round to concession round and gradually became more standardized. From the 10th and the 11th round in 1987 the agreement was called a 'cooperation agreement' or 'joint venture agreement', which reflects their more neutral character.[36] One important change was that the earlier carried interest arrangements, under which Statoil's exploration costs carried by the other joint venturers were totally excluded. The royalty provisions were also abolished from the 11th round.

The system with different joint venture agreements in each licensing round created a situation with many different contracts in the different joint ventures. The development in different areas and especially the

---

36 See A. Frihagen, 'The Joint Operating Agreement and Licence Terms in the Norwegian 11th Round 1987' in 11. Runde avtalene, Foredrag og arbeidsnotater til Det niende og tiende bergenske oljerettssymposium redigert av A. Frihagen. Offentligrettslige skrifter 1988: 1, Bergen 1988 (The 11th round agreements, lectures and working notes for the ninth and tenth Bergen Petroleum law symposium, edited by A. Frihagen, Papers in Public Law 1988: 1, Bergen 1988).

changes that were necessary due to changes in the state participation and organization of the sale of natural gas, made it necessary to make several changes in the contracts. In order to establish an updated and more unified regulation the state and the license holders in 2006 agreed on a new standard joint venture agreement that would repeal all former agreements.[37] For new production licenses new agreements are being developed; see section 4 below.

## Procurement and Industrial Development

An important part of the Norwegian petroleum policy from the start was to use petroleum activity as a basis for industrial development. This had also been done in relation to the hydro-electric industry in Norway with great success during the early 20th century.

From the start of the petroleum activity it was a requirement that Norwegian goods and labour should be used whenever they were competitive in price and quality. The Petroleum resolution from 1972, Section 53, stated that Norwegian goods and services should be used whenever they were competitive in price and quality. A special office ('the goods and service office') was established in the Ministry to ensure that this provision was followed. Also Statoil had the task of making sure Norwegian industry was given the possibility to participate in competition for contracts with the petroleum industry, using its influence as an operator of some fields and as a member of the steering committees in the joint ventures.

Another tool in developing Norwegian industry was making it a part of the criteria for granting licenses for petroleum activity that the applicants had contributed to the development of Norwegian industry. If the applicant could demonstrate investments or support for Norwegian industrial development, this would be taken into consideration in deciding which companies should be granted petroleum licenses. It turned out that this incentive led to some not very rational investments, and Norway decided to abolish this system around 1990.

Norway has developed an industry for the production of goods and services for the petroleum industry that also exports to other countries. The export industry for petroleum related products is now actually the second largest export industry in Norway after the export of petroleum itself.

---

[37] See 'Enclosure to Production Licenses 1972–2006', 27 February 2007 available at www.regjeringen.no/en/dep/oed/documents-and-publications/Laws-and-rules-2/Rules/2007/framework.html?id=455506.

The EEA agreement led to a shift in Norwegian petroleum legislation and practice. To some extent this was also in line with a shift in Norwegian policy that had already started prior to the EEA agreement. The EEA agreement meant that Norway could no longer treat companies from other EEA countries differently than Norwegian companies, due to Article 4 of the EEA agreement and the Licensing directive. Licenses now had to be granted on the basis of objective criteria; see under section 4 below for the current regulation and practice. Also the obligation to use Norwegian goods and services had to be abandoned and special procedures in order to make sure such discrimination does not take place were introduced.

The most important change in Norwegian policy as a result of EU/EEA regulation was the change in the system for the sale of natural gas. After trying different models, in 1986 Norway established a model where all natural gas from Norwegian fields was sold in field contracts. All partners in the joint ventures were obliged to deliver the gas jointly under the joint field contract. Statoil and the other Norwegian companies negotiated the contract on behalf of the license groups. The purpose was to sell gas from the most profitable fields from a national economic point of view first, and to construct the infrastructure as cost effectively as possible. It seems that all companies agreed to this system, as they at the time were more interested in the oil, while the Norwegian authorities also wanted to create a benefit from the sale of natural gas.

In 2001 the EU Commission started formal proceedings against approximately 30 Norwegian gas producers, claiming that the organization of the sale of natural gas was violating EU competition law. Norway abolished the system of the joint sale of natural gas from 1 January 2002. The joint venture agreements were changed so that the license partners on the Norwegian continental shelf were given the right to sell their own gas. The EU Commission cancelled the proceedings in 2002.

Norway negotiated an agreement with the owners of the natural gas pipelines where the pipelines were merged into one joint venture, and a fully state-owned company Gassco AS was established with the purpose of managing the pipeline network and making decisions on access to the network.[38]

---

[38] See for more detail on the development of the system for transportation and sale of natural gas, E. Nordtveit, 'Between Market and Public Interest: Organization and Management of the Norwegian System for Sale and Transportation of Natural Gas', in K. H. Søvig and S. Schütz, eds., *Undring og*

The development of the Norwegian petroleum regulation and licensing system can be said to have developed from a system with strong government interference and priority for Norwegian industry towards a more market-based system, open to international competition. In the next sections I will analyze the current regulations on some central issues.

## 4. GRANT OF LICENSES FOR PETROLEUM ACTIVITIES

### Introduction

The Norwegian license system has from 1972 been based on a system with licenses for petroleum activities in combination with joint venture or joint operating agreements with state participation. Still there have been substantial changes and development in the licensing process and in the criteria for granting licenses. An important driving factor for the changes has been the EEA agreement and the development in EU law in the energy sector and in national and international environmental law.

### Opening of Areas for Petroleum Activity

Licenses for petroleum activities can only be granted in areas that are open to petroleum activity. The procedure and the criteria for the decision to open a new area for petroleum activity are determined in the Petroleum Act, section 3-1. Before opening a new area for petroleum activity an evaluation of the different interests in the area shall take place. The procedure is similar to that of an environmental impact assessment (EIA), but not only the effects on the environment, but also the commercial, social and economic consequences of petroleum activity in the area shall be evaluated, especially the impact on fisheries. Local authorities and organizations with interests in the matter will be asked their opinion and a public announcement of which areas that might be opened and the nature and extent of the intended operations, shall be issued.

In order to establish a holistic, ecosystem-based and sustainable management of the sea areas under Norwegian jurisdiction, special cross-sectorial management plans have been decided for the Norwegian

---

*erkjennelse: Festskrift til Jan Fridthjof Bernt 70 år* (Fagbokforlaget, Bergen 2013) p. 469–82.

Sea, the Barents Sea and the North Sea and Skagerak.[39] These plans establish the basis for management of all activity from the base lines to the outer limit of the Exclusive Economic Zone.

In the White Paper for the North Sea and Skagerak the following is said about the role of the management plans:

> The management plans clarify the overall framework and encourage closer coordination and clear priorities for management of Norway's sea areas. They increase predictability and facilitate coexistence between industries that are based on the use of these sea areas and their natural resources. The management plans are also intended to be instrumental in ensuring that business interests, local, regional and central authorities, environmental organisations and other interest groups all have a common understanding of the goals for the management of the area in question. The Government will continue and further develop the system of management plans, and make it more effective.

The management plans will be the basis for public decisions on the use of the sea areas, also on the opening of new areas to petroleum activity, granting of petroleum licenses, and so on. One aim is to establish frameworks for petroleum activity and measures for better co-existence of fisheries and petroleum activity.

The North Sea with the exception of Skagerak and some areas close to the coast, the Norwegian Sea and large parts of the southern Barents Sea are open to petroleum activities. The question of opening for petroleum activity in the Jan Mayen area is under consideration. The question of opening the 'Lofoten and Vesterålen' areas which are the breeding place for the Norwegian-Russian cod stock, one of the largest remaining fish stocks in the world, has created strong political tension in Norway and has been a central issue in the last two general elections. The same is the case for the northern part of the Norwegian shelf in the northern Barents Sea.[40]

---

[39] The White Paper on the management plan for the North Sea and Skagerak (Meld. St. 37 (2012–2013)), English version, available 26 April 2013 at www.regjeringen.no/en/dokumenter/meld.-st.-37-2012-2013/id724746/.

[40] The management of the unopened areas is discussed in the White Paper 'An Industry for the Future: Norway's Petroleum Activities', Chapter 6, see footnote 17 above.

### Exploration License

In areas open to petroleum activity the Ministry may grant licenses to explore for petroleum within limited areas of the seabed or its subsoil. The exploration license gives a non-exclusive right to exploration in the areas covered by the license. This means that several companies can have exploration licenses in the same area. The license does not give any preferential right for a production license in the area at a later stage. The exploration license does not give the right to exploration in areas covered by a production license unless otherwise decided by the Ministry; see Petroleum Act, section 2-2.

### Production License

The central license for petroleum activity is the production license which gives an exclusive right to exploration, exploration drilling and production of petroleum deposits in the area covered by the license; see Petroleum Act, section 3-3(3).[41] The holder of a production license will become the owner of produced petroleum from the license area. The question of which company will be selected and on which conditions the license will be given is therefore of high importance as the license system is a central element in the governance of petroleum activity. Above I have described the development of the Norwegian licensing system. Here I will present the current system as it has developed.

A production license is granted by the King in Council (Government).[42] Production licenses are 'as a rule' granted in 'license rounds' where specific areas are announced as open for applications; see the Petroleum Act, section 3-5(1). In some cases licenses can be granted without announcement (Petroleum Act, section 3-5(4)). The announcement shall take place in *The Norwegian Gazette* (*Norsk Lysingsblad*) and the *Official Journal of the European Communities*, and stipulate a time limit for the filing of applications of not less than 90 days. The Norwegian policy for the last few years has been to announce a licensing round every second year. Before a licensing round is announced all companies who already have a license on the Norwegian shelf and all prequalified companies[43] are invited to nominate areas which they would

---

[41] About 700 licenses have been granted on the Norwegian continental shelf so far.

[42] Petroleum Act, section 3-1(1).

[43] See about prequalification below.

be interested in. Based on this nomination and other considerations the authorities will decide which areas shall be announced.[44]

Licenses in the traditional licensing rounds are announced in new areas, where licenses have not been granted earlier. This is important in order to investigate the geology in new areas. As the Norwegian shelf becomes more 'mature' in the sense that some areas have been investigated earlier, it has also been of interest for the authorities as well as some companies to search for smaller deposits in areas that have been explored earlier. In order to achieve this, an arrangement with award of licenses in predefined areas (TFO) was introduced in 2003 parallel to the ordinary licensing rounds for new areas. The TFO scheme gives companies the possibility to apply for licenses in mature areas of the continental shelf with a well-known geology.[45] In these areas the possibility of discoveries are high, but often the discoveries are small and dependent on the use of existing infrastructure. The TFO scheme has proved to be important and has given a basis for new discoveries in the last few years.[46] The government now announce one TFO every year.

Selection of the companies that shall be given the right to explore for and produce petroleum is of crucial importance for the development of the petroleum industry and the management of the petroleum resources. The ability of the companies to find petroleum and to produce it in an effective and safe manner will decide whether the activity will be a failure or a success.

Norway has always granted production licenses on the basis of a broad discretionary evaluation of the applicant's qualifications in order to select companies that are able to combine effective exploration and production of petroleum with safety and environmentally sound practice. The challenges in the Nordic areas due to climate and other factors and the risk of environmental damage in vulnerable areas, have made it necessary to choose companies with strong technical capability, financial resources and a proven record for responsible conduct. Norway has never considered auctioning of licenses in any form as a possible option.

---

[44] The process leading up to the 23rd licensing round are taking place as this is written.

[45] See the announcement for TFO 2014, available 3 April 2014 at www.npd.no/en/Topics/Production-licences/Theme-articles/Licensing-rounds/APA-2014/APA-2014 – announcement-/. Here the predefined area was extended with six blocks in the Norwegian Sea and three blocks in the Barents Sea.

[46] Most of the discoveries under this scheme are small, but the John Sverdrup field in the North Sea that was discovered in 2011 was the largest offshore field discovered that year.

As I have pointed out above, until 1992 Norway used licensing to develop national oil companies and a national industry for the fabrication of installations and for services to the petroleum industry. This is no longer possible. The Licensing directive, Article 5(a) and (b) demands that the decision shall be made on the basis of 'the technical and financial capability of the entities' and 'the way in which they propose to prospect, to explore and/or to bring into production the geographical area in question'. The Petroleum Act, section 3-5(3) uses the wording 'factual and objective criteria'. Nationality will not be used as a criterion for granting licenses.

The invitation to apply for a license contains a detailed list of which factors will be emphasized in the decision. The applicant's competence in all relevant areas, that means technical competence as well as competence in research and development, health and safety, environment and the ability to explore and produce from the field cost effectively, will be relevant. Especially important is that the operator of the field has strong technical capacity. Sufficient financial capacity is a requirement and also a geological understanding of the area, earlier experience of the Norwegian shelf, experiences with the applicant, and so on. The provision in the Licensing directive that financial capability shall be taken into consideration does not necessarily mean that the company with the strongest financial capability will be chosen. If there is another company with sufficient financial capability and a stronger record in other relevant areas, this might be decisive.

It is possible for interested companies to be evaluated for prequalification, in order to demonstrate that they fill the criteria for obtaining a license before they spend resources on an application process. This is of course not a guarantee that the company will be granted a license.

An important part of the Norwegian licensing policy has been that licenses are granted to a group of companies. It was of great importance for Norway during the negotiations on the content of the Licensing directive that the option to use government constructed license groups could be maintained; see the Licensing directive, Article 5(d).

The selection of participants in each license as well as the operator is crucial, and strong emphasis is placed on finding the best qualified companies for the task and putting together a group of companies with diversified competence and interests. The companies in each license have an obligation to participate actively in the management of the joint venture, and it is important that they have the capacity to make an independent and qualified evaluation of the questions before the management committee.[47] The Norwegian authorities see it as important to avoid passive participants, and therefore are inclined to avoid applications from groups composed by the applicants as they often tend to be dominated by the operator. It is also an aspect in selecting partners that some of them will be interested in keeping the costs down. The selection of partners must also be seen in relation to the fact that the voting rules are designed so that small partners might play an important role as well.

The use of government-composed license groups including the appointment of the operator is a key factor in what is regarded as the successful management of petroleum activity in Norway. It makes a basis for good decisions in the partner groups and also provides the authorities with information and different opinions when they are evaluating proposals for field development plans, transport solutions, and so on.

It is a license condition that the companies sharing a license will enter into a joint venture agreement and an accounting agreement between themselves. The license will also set the time limit for the initial period for the license, usually 10 years, Petroleum Act, section 3-9(1). A licensee that has fulfilled his work obligation pursuant to section 3-8 and other individual conditions for the license, can apply for prolongation of the license for a part of the area as stipulated in the license, section 3-9(2) and (3). The prolongation period is set individually. As a general rule it is 30 years, but might be extended up to 50 years and in some cases even longer.

## 5. ASSESSMENT OF PETROLEUM FIELD DEVELOPMENT (FOR EXAMPLE, FIELD DEVELOPMENT PLANS)

If a licensee makes a discovery there is a special procedure in the joint venture agreement in order to declare the field commercial. All partners in the joint venture must accept the commerciality declaration expressly

[47] See the Norwegian standard joint venture agreement (Agreement concerning petroleum activities) (available at freepdfs.net/k-verk-vedlegg-1-2-engpdf-regjeringen/9f1ca30600b230dca926510100e4cbe8/) article 1.3: 'The management committee is the supreme body of the joint venture. Each Party shall contribute to the management and control of the joint venture activities.' This is obviously a clause that is more aimed at safeguarding the public interest than the interests of the other partners.

and in writing on the basis of a proposal for field development from the operator. This is not decided by a vote in the management committee.

Development of the field and the start of production cannot take place before the authorities have approved a plan for development and operation of the petroleum deposit (PDO), Petroleum Act, section 4-2. A separate approval is also needed in order to install and operate facilities (PIO) under the PDO. The authorities in February 2010 issued guidelines for the preparation of PDO and PIO. The production license gives the license group the right to produce petroleum and become the owner of the produced petroleum, but a PDO and PIO that fulfils the demands for good resource management, and protection of health, safety and environment is needed.[48]

The plan shall contain 'an account of economic aspects, resource aspects, technical, safety related, commercial and environmental aspects, as well as information as to how a facility may be decommissioned and disposed of when the petroleum activities have ceased'.

The Ministry may in situations when 'particular reasons so warrant, require the licensee to produce a detailed account of the impact on the environment, possible risks of pollution and the impact on other affected activities, in respect of a larger defined area'.

The plan shall also comprise information on facilities for transportation or utilization required by the Petroleum Act, section 4-3.

## 6. GOVERNMENT INVOLVEMENT IN THE EXTRACTION OF PETROLEUM

Even if exploration for and production of petroleum resources is left to national and foreign oil companies through a system of licenses, the state has a strong interest in how petroleum activity is organized and conducted. The state is interested in the highest possible tax revenue from the petroleum activity as well as in its impact on the general economic, regional and social development of Norway. The Norwegian State has used different instruments in order to obtain the best possible total management of the Norwegian petroleum resources as a whole and as high a market value of Norwegian petroleum as possible. It has also been

---

[48] See the detailed 'Guidelines for Plan for Development and Operation of a Petroleum Deposit (PDO) and Plan for Installation and Operation of Facilities for Transport and Utilization of Petroleum (PIO)', 4 February 2010, issued by NPD and published on and available 4 February 2010 at www.npd.no/Global/Engelsk/5-Rules-and-regulations/Guidelines/PDO-PIO-guidelines_2010.pdf.

an important goal that petroleum activity should be conducted with a strong consideration for other industries, especially fisheries, and for the environment.

It was considered at an early stage that the state also should participate directly in petroleum activity, not only from an economic and commercial point of view, but also in order to obtain information and to govern the activity 'from within'. The state has combined the role as regulator with the position as owner in managing the resources and regulating the activity.

Statoil was established as a fully state-owned company in 1972 and is seen as an important tool for safeguarding the state's interests in petroleum activity. The company developed a strong economy and played an important role in the development of the Norwegian supply industry to petroleum activity. In the early 1980s the company's strong economic position and political role raised concern among some political parties that the company might become too dominant in Norwegian society, and would be difficult to keep under political control.

The new government that came into power in 1981 wanted to reduce the economic and political power of Statoil, and a committee was appointed to develop a proposal which was presented in 1983.[49] A compromise was reached between all major political parties represented in the Storting in 1984. The state's participation interest in the licenses was divided into one part belonging to Statoil and one part organized in the State's Direct Financial Interest (SDFI). The revenue from the SDFI goes directly into the state treasury. Statoil's veto-rights in the license groups were abolished and transferred to the state in order to reduce Statoil's dominant position in the license groups. The division of the income did, however, not apply to the Statfjord field, which would be the most important source of income for Statoil in the years to come.[50]

The political compromise in this case became a basic element in the future organization of the state's involvement in the petroleum industry and it also established an understanding in the Norwegian political environment for the need for broad political consensus as a basis for the long-term governance of petroleum activity.

Statoil continued to manage the SDFI. The voting rights could, however, only be exercised on the basis of special authorization from the shareholder meeting of Statoil, which was the Minister of Petroleum and

---

[49] NOU 1983: 16 Organiseringen av statens deltagelse i Petroleumsvirksomheten (Organization of the state's participation in the petroleum activity).

[50] The SDFI is now regulated in the Petroleum Act, Chapter 11.

Energy. Statoil also had, and still has, the task of selling the state's share of the petroleum from the different license groups. This arrangement lasted until the part privatization of Statoil in 2001 made it necessary to reorganize the management of the SDFI.[51]

As a result of the part privatization of Statoil, a new fully state-owned shareholder company Petoro AS was established, with the purpose of managing the state's interests in the petroleum fields. The company is regulated by the Limited Companies Act, but some special regulation of its position is determined in the Petroleum Act, Chapter 11. The state is liable for any obligations incurred by the company and the company cannot go bankrupt, Petroleum Act, section 11-3. Petoro acts as participant in the licenses, but the revenue from the sale of petroleum goes directly into the State Treasury. The Storting provides the budget to cover Petoro's costs of running the company and for investments. Petoro must be considered as a special form of company developed in order to carry out the task of managing the SDFI.

In line with overall development in the general political doctrine, there might be a development towards less public interference in pure commercial decisions than in the earlier stages of petroleum activity, especially in decisions on investment, in cases where the companies are against this. The state generally leaves the commercial decisions to the companies. Also Petoro AS is supposed to take commercial decisions independently and to act like a petroleum company. Still the authorities have played an active role in influencing the activity and the company's conduct in many respects through administrative decisions aimed at safeguarding good resource management. The public involvement in the reorganization of the sale and distribution of natural gas as a result of the conflict with the EU Commission, as presented above, is one example of this.

The Petroleum Act gives the Ministry authority to make decisions on joint use of infrastructure, coordinated field development or unitization of two or more fields, in order to lower the costs. The state can also instruct a license holder to take special measures in order to increase the depletion rate. How far this authority will be used is a political question at any time.

---

[51] The proposal to partly privatize Statoil was presented to the Storting in St.prp. nr. 36 (2000–2001).

## 7. ENVIRONMENTAL REGULATION FRAMEWORK

Norway has always aimed at a high level of environmental standards in its petroleum activity. Extensive legislation and management systems have been developed to protect the external environment in all phases of petroleum activity, from the question of opening an area to petroleum activity, to approval of plans for development and operation and through to cessation of petroleum activity. The introduction of management plans for the ocean areas have established the basis for integrating petroleum activity into a more holistic or ecosystem-based decision system for the utilization of the offshore areas. Even if these plans are not legally binding they will govern the decisions made by public authorities under the Petroleum Act and other acts regulating the utilization of the sea areas. The Norwegian Biodiversity Act[52] does not apply outside the territorial sea, except for the general principles, like the precautionary principle and the principle of use of best possible technology.

The licensees have an unlimited liability for pollution damage from the activity, irrespective of fault, Petroleum Act, section 7-1. The liability will not only cover ordinary economic loss, but also consequential damage and the cost of restoration of damaged areas. Under Norwegian general law pure economic loss is also covered, with some limitation.

## 8. SAFETY FRAMEWORK

Petroleum activity is governed by a strict safety regime. The Petroleum Act, section 9-1 states that the petroleum activities 'shall be conducted in such manner as to enable a high level of safety to be maintained and further developed in accordance with the technological development'.

The licensee and other participants are also obliged at all times to 'maintain efficient emergency preparedness with a view to dealing with accidents and emergencies which may lead to loss of lives or personal injuries, pollution or major damage to property'; see section 9-2.

The Norwegian safety regime has moved away from the traditional 'command and control' paradigm towards a functional system based on good governance and best practices and with the involvement of industry, government and trade unions representing the people working in the industry. The readjustment towards goal-oriented and performance-based regulation, have also influenced how the regulations are formulated. The

---

[52] Act 19 June 2009 no. 100.

functional responsibilities for the safe conduct of activities have been transferred from regulators to companies and the establishment of cooperative approaches by different stakeholders. Legal, industrial and professional standards have replaced the detailed legal texts and the main responsibility rests with the companies.

The aim is to involve the companies and their employees in developing a safe practice based on the best possible knowledge and technology at all times, not just to comply with detailed and written legislation that easily gets outdated. Also this places incentives for innovations in technology and organizational solutions on the stakeholders themselves. The burden of proof for compliance with the requirement of the legal standard in the Petroleum Act, section 9-1 lies on the licensee.

The unlimited liability also creates strong incentives for the licensee to maintain a high standard of safety for environment and personnel.

Norway objects to the Offshore Safety Directive (OSD), which the European Union has issued as a response to the Macondo accident.[53] Norway considers the OSD as a step backwards that would reduce the safety level on the Norwegian continental shelf. Norway claims that the OSD is not a part of the EEA agreement and has decided not to implement the directive in Norwegian law.

[53] Directive 2013/30/EU of the European Parliament and of the Council of 12 June 2013 on safety of offshore oil and gas operations and amending Directive 2004/35/EC. See also G. Gordon, 'On offshore safety: The European Commission's legislative initiatives', in M. Roggenkamp and H. Bjørnebye (eds.), European Energy & Law report, Volume X, (2014) pages 139–169.

# 6. The offshore petroleum licensing regime in the United States

**Owen L. Anderson and Christopher Kulander***

## 1. INTRODUCTION

The United States government claims the exclusive right to control exploration and exploitation of oil and gas resources to most of the Outer Continental Shelf (OCS)[1] and other submerged lands bordering the United States seaward for 200 nautical miles (or more where continental shelf extends further seaward), subject to state title claims to certain territorial waters.[2] The federal OCS includes all submerged lands that lie seaward of state-owned submerged lands ceded to the states by the Submerged Lands Act of 1953[3] and that are subject to the jurisdiction and control of the United States. Due to a longstanding leasing moratorium, the only areas of active federal offshore leasing are the Gulf of Mexico off the coasts of Texas, Louisiana, Mississippi, and Alabama and limited areas offshore of Alaska.

* This chapter is an adaptation of a portion of Chapter 7 prepared by the same authors for the oil and gas law casebook: John S. Lowe, Owen L. Anderson, Ernest E. Smith, David E. Pierce, and Christopher S. Kulander, *Cases and Materials on Oil and Gas Law* (6th ed. West, St. Paul, Minnesota, 2012) [hereinafter Lowe, *et al*].

[1] The 'continental shelf' may be defined as a gently sloping extension of the continental landmass that is submerged beneath shallow coastal waters. Legally, the 'outer continental shelf' is that portion of the continental shelf that lies seaward of state territorial waters and which generally extends to the continental slope, a steeper slope where the continental land mass gives way to the abyssal plain of the ocean.

[2] Although the United States has yet to ratify the United States Convention of the Law of the Sea (UNCLOS), it does claim an 'exclusive economic zone' or 'EEZ' and continental shelf, consistent with UNCLOS.

[3] 43 U.S.C. §§ 1301–15.

Under the Submerged Lands Act, coastal states hold title to the inlets and offshore lands within three nautical miles of their respective shorelines.[4] However, the States of Florida and Texas hold title to offshore submerged lands off their respective Gulf coasts within three marine leagues of their shorelines, approximately ten standard miles out to the landward OCS boundary.[5]

The first documented offshore wells were drilled from piers off the coast of southern California in the 1890s. The first freestanding offshore well was drilled in 14 feet of water off the coast of Louisiana in 1938. Modern offshore wells in the Gulf of Mexico have achieved production through almost 10,000 feet of water.[6]

Modern 3D seismic technology has lowered the risk of drilling dry holes in high-cost deep-water areas. The Outer Continental Shelf Deep Water Royalty Relief Act[7] improved the economics of drilling in deep water by enacting provisions for suspending the royalty ordinarily owed the federal government for certain volumes of production from wells drilled in deep water with the aim of ultimately increasing production or encouraging development on certain producing or nonproducing leases, respectively.

The Outer Continental Shelf Lands Act of 1953 (OCSLA), as amended, governs the leasing of federal offshore lands.[8] From 1982 to 2010, the offshore leasing program was administered by the Mineral Management Service (MMS). After the Deepwater Horizon disaster, the MMS was replaced by the Bureau of Ocean Energy Management, Regulation and Enforcement (BOEMRE). On October 1, 2011, BOEMRE was, in turn, replaced by the Bureau of Ocean Energy

---

[4] 43 U.S.C. § 1312.

[5] Under the Act, coastal states and territories were given the right to prove greater historical boundaries. Only Texas, Puerto Rico, and Florida have succeeded. For background, see *United States v. Louisiana, et al.*, 364 U.S. 502 (1960). A marine league is equal to three geographical or nautical miles. A geographical or nautical mile is about 6,075 feet, whereas a statute mile is 5,280 feet.

[6] Xiong Tong, 'Shell Produces Oil from World's Deepest Subsea Well in Gulf of Mexico,' *EnglishNews.CN*, November 18, 2011.

[7] Pub. L. 104–58, Title III, §§ 301 *et seq.*, 109 Stat. 563–6 (November 28, 1995).

[8] 43 U.S.C. §§ 1331–56. A host of other federal laws also apply to certain offshore oil and gas operations. These include the Jones Act, 46 U.S.C. § 688; the Longshore and Harbor Workers' Compensation Act, 33 U.S.C. §§ 901–50; the Coastal Zone Management Act, 16 U.S.C. §§ 1451–64; and the Rivers and Harbors Act, 33 U.S.C. § 403.

Management (BOEM), the Bureau of Safety and Environmental Enforcement (BSEE), and the Office of Natural Resources Revenue (ONRR) as part of a major reorganization stemming from the Deepwater Horizon disaster. These three new organizations each took control over one or more important functions of MMS/BOEMRE in response to criticism that MMS/BOEMRE faced conflicts of interest across the spectrum of their duties. BOEM is responsible for managing offshore leasing, resource evaluation, the review and administration of oil and gas exploration and development plans, renewable energy development, National Environmental Policy Act (NEPA) analysis, and environmental studies. BSEE is responsible for safety and environmental oversight of offshore oil and gas operations, including permitting and inspections of offshore oil and gas operations. ONRR collects revenues from leasing and production of federal offshore lands.

A major focus of the OCSLA was to establish a leasing and development program for offshore lands. In 1978, amendments were adopted to address environmental concerns and to provide for state involvement in the formulation of leasing plans and in the conduct of offshore operations that affect coastal zones.[9] Under the 1978 amendments to OCSLA, the Secretary of the Interior may suspend lease operations and cancel leases when the potential harm outweighs benefits and where operators are not in compliance with environmental requirements.[10]

## 2. DEVELOPMENT OF OFFSHORE FIELDS

Under the amended OCSLA, there are four steps in the development of an offshore oil well on federal offshore lands: (1) development of a five-year leasing plan by the federal government; (2) the solicitation of bids for discrete parcels of acreage and issuance of offshore leases to the highest bidder; (3) exploration operations; and (4) drilling, development, and production operations.

---

[9] In general, offshore activities must be consistent with an affected state's coastal zone management plan as promulgated pursuant to the Coastal Zone Management Act of 1972, as amended. 16 U.S.C. §§ 1451–64.

[10] 43 U.S.C. § 1334(a)(1).

## Five-Year Leasing Plan

Under OCSLA, the Department of the Interior (DOI) must first prepare a five-year leasing plan for offshore leases within a region.[11] This plan, which is a schedule of the proposed lease sales, must indicate the size, timing, and location of the leasing activity that the Secretary of the Interior determines will best meet the nation's energy needs for the following five years.[12] In making this determination, the Secretary must consider the economic, social, and environmental impact of the plan, including an equitable sharing of developmental benefits and environmental risks among the various regions.[13] The Secretary must also solicit comments from all interested federal agencies and the governors of states that might be affected by the proposed operations – typically those states landward of the OCS acreage being contemplated for leasing.[14] The Secretary must submit the proposed final program and any comments to the President and the Congress[15] and must give final approval to the plan. This process can take as long as two years.

Eight-seven percent of federal offshore acreage is closed to oil and natural gas exploration and development. Currently, federal leases for oil and gas development have been issued only for certain Alaskan waters and off the coast of Texas, Louisiana, Mississippi, Alabama, and the western portion of Florida despite strong evidence that commercial oil and gas deposits may be found off both the Atlantic and Pacific coasts and off the western coast of the Florida peninsula. Prior to the Deepwater Horizon disaster, leasing of areas off the Atlantic coast of the United States that had been off limits for exploration were being considered, but these plans were quashed in the aftermath of the accident.

## Bid Solicitation and Lease Issuance

After the final five-year leasing plan has been approved, the next step in the offshore development process is the solicitation of sealed bids and the issuance of offshore leases. Under the OCSLA, the Secretary of the Interior must set the timing of lease sales and the location of leased tracts

---

[11] Id. § 1344(a).

[12] Id.

[13] Id.

[14] Id. § 1344(c)(1). Under the Coastal Zone Management Act of 1972, any federal offshore leasing activities 'directly affecting the coastal zone' must be consistent with each coastal state's approved coastal zone management plan.

[15] Id. § 1344(d)(2).

to strike a proper balance between environmental protection and the potential for the discovery of oil and gas.[16] Since the 1990 amendments to the Coastal Zone Management Act (CZMA), lease sales are subject to CZMA section 307(c)(1)'s 'consistency provisions' for federal agency activities.[17]

Table 6.1 enumerates the steps involved in issuing an offshore lease and the anticipated time schedule for each step.

*Table 6.1 Steps for leasing on the Outer Continental Shelf (OCS) of the United States*

| Steps | Lower 48 | Alaska |
|---|---|---|
| Identify areas of hydrocarbon potential, Call for information and nomination, and Publish notice of intent to prepare an EIS Month 1 | Month 1 | |
| Area identification Month 4 | Month 4 | |
| Draft EIS | Month 12 | Month 15 |
| Public comment period | Month 14 | Month 17 |
| Final EIS | Month 18 | Month 21 |
| Proposed notice of sale | Month 19 | Month 22 |
| Governors' comments | Month 21 | Month 24 |
| Final notice of sale | Month 22 | Month 25 |
| Sale | Month 23 | Month 26 |
| Bid review | Month 24 | Month 27 |
| Lease issued | Month 25 | Month 28 |

*Note:* EIS – Environmental Impact Statement.

Following the call for information and nomination, BOEM identifies areas of potential hydrocarbon resources.[18] The oil and gas industry is invited to comment on these areas and to nominate areas for leasing. The governors of affected states may also make comments regarding environmental concerns[19] and consult with the Secretary to determine whether

16 Id. § 1344(a)(3).
17 16 U.S.C. § 1456(c)(1).
18 30 C.F.R. § 256.23.
19 Id. § 256.25.

leases in close proximity to the coast may contain oil or gas reservoirs underlying state lands.[20]

Next, a draft environmental impact statement (EIS) is prepared unless a multi-sale EIS has previously been issued for an area with recurring sales. The EIS identifies the planning area, possible environmental impacts, potential coastal zone management conflicts, and alternatives to the proposed leasing program. The release of the draft EIS is followed by a public comment period. Comments on the proposal and on the draft EIS are evaluated in the final EIS. In the case of a subsequent lease sale in an area with recurring sales, BOEM does an environmental assessment (EA) for each such leasing episode to determine whether a supplemental EIS needs to be prepared or whether BOEM may make a Finding of No New Significant Impact (FONNSI). An EA is typically less rigorous than an EIS.

After the final EIS is released, a proposed notice of sale is published in the Federal Register. This announcement contains the proposed terms and conditions of sale, the bidding protocol, the blocks of offshore acreage to be leased, special lease stipulations, and any required mitigation measures. The notice of sale is also sent to the governors of affected states, who are invited to comment upon the size, timing, and location of the proposed lease sale. The Secretary must respond in writing to their comments. In the final notice of sale, which is an invitation to submit bids, the Secretary may include any comments that are deemed 'valid.' The final notice is published in the Federal Register at least 30 days before the actual date of bidding and sale.

Leasing of federally controlled offshore land is by competitive sealed bid.[21] After the bids have been gathered, the winning bid for each block of leased acreage is publically announced live over the internet and radio. The BOEM, with approval of the Secretary, may use alternative bidding systems to increase income to the government or to give smaller oil and

---

[20] Id. § 256.25(b). Since coastal states hold title to offshore lands within three miles (Texas and Florida (on the Gulf coast) hold title to three marine leagues) of their respective shorelines, a three mile 'buffer zone' of federal offshore land lies seaward of each state's offshore limits. This area is commonly referred to as the '8(g) Zone', because Section 8(g) of OCSLA (43 U.S.C. § 1337(g)) governs federal leasing of submerged lands within three nautical miles of a coastal state's seaward boundary. In other words, there are two separate zones of federal offshore lands: (1) the 8(g) Zone, the innermost three-mile strip of federal offshore land lying immediately seaward of state-owned offshore lands; and (2) all federal offshore land lying seaward of the 8(g) Zone. *Texas v. Secretary of Interior*, 580 F.Supp. 1197, 1198 (E.D.Tex.1984).

[21] 43 U.S.C. § 1337(a)(1).

gas companies a better opportunity to acquire offshore leases.[22] To achieve these somewhat inconsistent goals, the Secretary is authorized to use any of the following bidding and royalty systems:[23] a cash bonus with a royalty of 12.5 percent or more; a variable royalty bid with either a fixed work commitment or a fixed cash bonus; a cash bonus bid or a work commitment bid and a diminishing sliding scale royalty beginning at 12.5 percent; a cash bonus with a fixed net profits share of not less than 30 percent; a net profits share bid with a fixed cash bonus; a cash bonus bid with a royalty of at least 12.5 percent and a fixed net profits share of at least 39 percent; a work commitment bid with a fixed cash bonus and a fixed royalty; or any other bidding system with variables, terms, and conditions that the Secretary determines to be useful to accomplish the purposes and policies of OCSLA. OCSLA provides generally for a royalty rate provision of not less than 12.5 percent of the amount or value of the production saved, removed, or sold.[24] Some more recently issued offshore leases have a royalty rate as high as 18.75 percent. Nearly all offshore sales have been conducted using the variable cash bonus and fixed-rate royalty system.

The Secretary may accept or reject any bid within 90 days of the sale.[25] Tied highest bids are rejected, unless the tied bidders file an agreement with the BOEM to accept the lease jointly within 15 days of being notified of the tie.[26] Before any bids are accepted, the Attorney General and the Federal Trade Commission review the highest bid for each lease to determine whether any bids violate antitrust laws and to assure that fair market value has been realized.[27] Before a lease is issued, the successful bidder must post a bond with the BOEM of at least US $50,000; the BOEM may later require additional bonds in the event that offshore operations are eventually commenced.[28]

Because the costs and risks of offshore oil and gas development are high, several operators may submit joint bids to acquire joint interests in

---

[22] 30 C.F.R. § 256.32(e).

[23] 43 U.S.C. § 1337(a)(1)(A)–(H).

[24] 43 U.S.C. § 1337(a)(1).

[25] 30 C.F.R. § 256.47(e)(2). See also *Kerr-McGee Corp. v. Morton*, 527 F.2d 838 (D.C.Cir.1975) (holding that the Secretary may reject the highest bid if the Secretary believes that the bid is too low).

[26] 30 C.F.R. § 256.47(c).

[27] 43 U.S.C. § 1337(c).

[28] Id. § 256.58(a).

offshore leases.[29] Joint bidders enter into agreements that name the parties, identify targeted lease blocks, address allocation of interests in the lease, require disclosure of certain information, provide for confidentiality of that information, set procedures for determining a bid, provide for termination of the agreement regarding certain parcels for which there is no mutual interest, provide for execution of a joint operating agreement,[30] and provide that the agreement shall not be construed as a partnership or joint venture.

Although no case law addresses the antitrust aspects of joint bidding for oil and gas leases, parties are advised not to submit a solo bid that is lower than the highest suggested bid proposed by a fellow member of the joint bidding group.[31] Pursuant to federal regulation, the Department of the Interior periodically publishes a list of parties who are prohibited from entering into joint bidding arrangements with other listed parties.[32] This list consists of companies whose daily production exceeded 1.6 million barrels of oil or natural gas or liquefied petroleum equivalent in the preceding six months.[33]

Offshore tracts are awarded to the 'highest responsible qualified bidder'.[34] To be deemed 'responsible', a bidder must be financially capable of fulfilling the lease conditions and have fully complied with the conditions of any prior leases.[35] To be deemed 'qualified,' a bidder must meet strict conditions, including the requirement to be a United States citizen, an alien lawfully admitted for permanent residence, or a private,

---

[29] See generally, A. Hoppe, 'Offshore Agreements,' Paper 3, 18 Annual Oil, Gas & Min.L.Inst., (The University of Texas School of Law 1992) and Terry E. Hogwood, 'Bidding Agreements: Tailoring the Right One for Your Company,' Monograph Series No. 3, 1986 ABA Sec. on Nat. Res., Energy, and Envt'l L.

[30] Joint operating agreements covering offshore wells are lengthy and detailed documents, but they contain many of the same provisions commonly found in onshore agreements. One major difference, however, is that offshore agreements commonly contain lengthy and onerous nonconsent penalties for a party's refusal to participate in the drilling of an offshore well. Due to the high cost of offshore operations, offshore operating agreements often contain detailed provisions concerning the construction, ownership, and operation of platforms and other offshore facilities, including detailed provisions on preferential purchase rights in the event any party wishes to sell its interest.

[31] Terry E. Hogwood, 'Bidding Agreements: Tailoring the Right One for Your Company,' Monograph Series No. 3, 1986 ABA Sec. on Nat.Res., Energy, and Environmental L.

[32] See, e.g., 56 Fed.Reg. 55, 928.

[33] 30 C.F.R. § 256.41.

[34] 43 U.S.C. § 1337(a)(1).

[35] 43 U.S.C. § 1337(d).

public, or municipal corporation organized under United States law, or associations thereof.[36] All leases become effective on the first day of the month following the date of execution by the Secretary.

Generally, federal offshore leases, a form of concession, include the following terms and provisions: a leasehold consisting of a compact area not exceeding 5,760 acres;[37] a lease primary term of five years (up to ten years for certain deep water leases) and a habendum clause providing for a secondary term for so long as oil or gas is produced in paying quantities or approved drilling or well reworking operations are being conducted;[38] payment of the amount or value specified by the bidding system;[39] the right to explore, develop, and produce oil and gas conditioned upon due diligence requirements and upon approval of a plan of development;[40] suspension or cancellation of the lease;[41] annual rentals;[42] payment of royalties;[43] and offers of production to small or independent refiners under the Emergency Petroleum Allocation Act of 1973.[44] The lease is only four pages long, but it is subject to detailed regulations. A United States oil and gas lease/concession differs from concessions in nearly all other countries in that the term may perpetuate indefinitely by production 'in paying quantities', which means that the production revenues obtained by the lessee must exceed only the day-to-day costs of well operations; the well does not have to be profitable.

Historically, coastal states do not get a share of federal revenues from offshore leases. An exception is where federal leaseholds are located within the 'buffer' or 'Section 8(g) Zone', section 1337(g) of OCSLA. The buffer or Section 8(g) Zone is an interval of federal OCS marine acreage three nautical miles in width parallel to the shore and extending seaward from the state/federal boundary located three miles or three

---

[36] 30 C.F.R. § 256.35(b). As for other requirements, see International Investment Survey Act of 1976, 21 U.S.C. §§ 3101–8 (1988) as amended by the Trade and Tariff Act of 1984, Pub. L. No. 98–573, § 306(b)(1), 98 Stat. 2948, 3009 (1984); Pub. L. 101–533, § 6(A), November 7, 1990, 104 Stat. 2348.

[37] 43 U.S.C. § 1337(b)(1). Larger leaseholds are allowed if the leasing of a larger tract is necessary to secure a reasonable economic production unit. Id.

[38] Id. § 1337(b)(2). To encourage development in deep water or under adverse conditions, the lease may provide for a primary term of up to ten years.

[39] Id. § 1337(b)(3).

[40] Id. § 1337(b)(4).

[41] Id. § 1337(b)(5).

[42] Id. § 1337(b)(6).

[43] The royalty rate is commonly one-sixth; however, pursuant to alternate bidding systems, the rate could be higher, lower, or not reserved.

[44] Id. § 1337(b)(7).

marine leagues from the coast. If a federal leasehold lies wholly within the buffer, the federal government must remit to the state 27 percent of the revenues generated by the lease.[45] Where a federal lease lies only partially within the Section 8(g) Zone, the amount remitted to the coastal state must equal the percentage of the leasehold lying within the Section 8(g) Zone multiplied by 27 percent of the revenues generated by the lease.[46]

In 2006, largely in response to Hurricanes Rita and Katrina, Congress passed the Gulf of Mexico Energy Security Act (GOMESA),[47] which requires that 50 percent of 'qualified OCS revenues' be disbursed to (1) Gulf producing states (Alabama, Louisiana, Mississippi, and Texas) and their coastal political subdivisions, and (2) as financial assistance to states under Section 6 of the Land and Water Conservation Fund Act of 1965 (16 U.S.C. sections 4601–8). The revenues to which these requirements apply are limited to those derived from leases issued after the date of GOMESA's enactment in two relatively small areas made available for leasing under the Act for fiscal years (FYs) 2007 through 2016. For FY 2017 and thereafter, the disbursement requirements will apply to leases issued after the date of GOMESA's enactment in the entire federal area subject to leasing in the Gulf of Mexico except for areas under leasing moratorium and leases in the Section 8(g) Zone.[48] GOMESA also establishes areas within the Gulf of Mexico that are prohibited for oil and gas leasing, including areas off Florida.

---

[45] Id. § 1337(g)(2). In *Louisiana ex rel. Guste v. United States*, 832 F.2d 935 (5th Cir.La.1987), the Court of Appeals for the Fifth Circuit ruled that the State of Louisiana was not entitled to relief against drainage by federal wells or to unitization since the share the state received from 8(g) Zone leases was intended to settle such problems between sovereigns permanently (thus, this case could be viewed as upholding the 'rule of capture' on the federal OCS).

[46] 43 U.S.C. § 1337(g)(2).

[47] Pub.L.No. 109–432, 120 Stat.2922.

[48] The 75 percent of the 50 percent paid to Gulf producing states is allocated among those states under a formula that the Secretary must establish by rule, under which the allocated amounts are inversely proportional to the distances between the closest point on the state's coastline and the geographic centre of each leased tract (with no state receiving less than ten percent of the total) for FY 2007–2016. A similar formula is prescribed for FY 2017 and thereafter. The Secretary must pay 20 percent of the amount allocated to each Gulf producing state to the coastal political subdivisions, according to the complicated inverse distance allocation prescribed for the Coastal Impact Assistance Program enacted by section 384 of the Energy Policy Act of 2005, 43 U.S.C. § 1356a. In addition, the total amount of revenues to be disbursed may not exceed US $500 million for each of FY 2016–FY 2055.

## Exploration

After being granted a federal OCS lease but before exploration activities begin, an operator must file an exploration plan with the DOI.[49] This plan is subject to an environmental analysis that could lead to an EIS. The plan must be consistent with any affected state's approved coastal zone management plan and must set forth a schedule of anticipated exploration activity, a description of the equipment to be used, the general location of each well to be drilled, and other pertinent information.[50] The DOI must approve or reject the plan within 30 days.[51] Plans that are rejected may be revised. If revision is necessary, the Secretary may suspend the running of the lease term and any lease activity until the plan is approved.[52] The lease may be canceled if the plan cannot be altered to avoid harm to the environment,[53] or if the lessee commences exploration without an approved exploration plan or fails to comply with an approved plan.[54]

## Development and Production

Before beginning this phase of operations, the lessee must file a development and production plan with the DOI.[55] The plan must be consistent with any affected state's coastal zone management plan and must set forth all facilities and operations proposed by the lessee, including the location of the operations, all environmental and safety standards to be implemented, and other pertinent information.[56] Within ten days of receipt of this plan from the lessee, the Secretary must forward the plan to the governor of any affected state.[57] An environmental analysis must also be done and, if approval of the plan is a 'major federal action,' the Secretary must prepare an EIS.[58] The governor of an affected coastal state has 60 days from receipt of the plan to submit comments and recommendations that might influence the Secretary's

---

49 Id. § 1340(c)(1).
50 Id. § 1340(c)(3).
51 Id. § 1340(c)(1).
52 Id. § 1340(e)–(f).
53 Id. §§ 1340(c), 1334(c)–(d).
54 Id. § 1351(j).
55 Id. § 1351(a)(1).
56 Id. § 1351(a)(2).
57 Id. § 1351(a)(3).
58 Id. § 1351(f).

approval of the plan.[59] Affected states may veto any plan that is inconsistent with its state coastal zone management plan, but the Secretary of Commerce may override a state veto.[60] If a development and production plan is disapproved by the Secretary and cannot be altered to avoid harm to the environment, the lease may be cancelled.[61]

## 3. LEASE SUSPENSION AND CANCELLATION

### Suspension of Operations or Production

The BOEM Regional Supervisor may suspend production, including a requirement to produce, or any other operations on an offshore leasehold either on his own initiative or at the lessee's request for reasons that include the following: (1) to facilitate proper development; (2) to allow for construction or negotiation for use of transportation or production facilities; (3) to allow time to enter into gas or oil sales contracts; (4) to avoid operations that would result in premature abandonment of a producing well; (5) to allow a reasonable time to commence drilling; (6) to comply with the law or lease terms; (7) to prevent harm to the environment; (8) in the interest of national security; (9) to comply with NEPA; (10) to install equipment for safety or environmental reasons; (11) when lessees are subject to inordinate governmental delays in obtaining permits; or (12) to comply with judicial orders prohibiting production or other operations.[62]

A suspension may not exceed five years. If a lease is suspended, the term of the lease is increased for the period of the suspension, unless the suspension is the result of the lessee's gross negligence or willful violation of a permit, lease, or regulation.[63]

### Cancellation of Lease

If a lessee fails to comply with a law, regulation, or lease provision, the lease is subject to cancellation. If the owner of a *nonproducing* lease fails to comply with the law or with the lease terms, the Secretary of the

---

59 Id. § 1351(g).
60 Id. § 1351(d), 16 U.S.C. § 1456(c)(3)(B).
61 43 U.S.C. § 1351(h).
62 30 C.F.R. §§ 250.172–250.175.
63 30 C.F.R. §§ 250.170(a), 250.169.

Interior may cancel the lease, subject to judicial review.[64] If the owner of a *producing* lease fails to comply with the law or with the lease terms, only a federal District Court may cancel the lease.[65] Nevertheless, if the Secretary determines, after a hearing, that continued activity on the lease would probably harm the environment, that the threat of harm will not decrease, and that the advantages of cancellation outweigh the advantages of continuing the lease, the Secretary may cancel the lease whether or not the lease is productive.[66] The lease may also be cancelled if the lessee fails to submit an exploration plan or fails to comply with an approved plan. If an exploration plan or development and production plan is rejected and cannot be altered to avoid harm to the environment, the lease may be cancelled before exploration or development begins.

Any person adversely affected by a Secretarial decision to cancel a lease may appeal the decision to the Interior Board of Land Appeals (IBLA).[67] The DOI regulations further provide that the IBLA will consider the appeal under procedures applicable to onshore land appeals.[68]

## 4. ELIGIBILITY AND CONTINUING REQUIREMENTS FOR LEASING AND LEASE OWNERSHIP

The United States remains a popular target for foreign investment in general. For the period from 1996 to 2011, 2007 had the highest number of acquisitions by foreign individuals or entities acquiring an American firm.[69] The subsequent drop may be attributed partially to the more strict federal requirements covering foreign investment passed in 2007 as well as the economic downturn in the United States.[70] China, for example, accounted for 16 out of the 313 total mergers and acquisitions reviewed by the federal government between 2008–2010, behind such countries as

64 43 U.S.C. § 1334(c).

65 Id. § 1334(d).

66 Id. § 1334(a).

67 30 C.F.R. part 290, subpart A.

68 43 C.F.R. part 4, subpart E (30 C.F.R. 290.2).

69 James K. Jackson, Cong. Research Serv., RL33388, The Committee on Foreign Investment in the United States, 1, 16–17 (September 26, 2008).

70 Id., at 19.

the United Kingdom, Israel, France, and Canada.[71] Much of this investment is in offshore oil and gas assets. Such investments are subject to scrutiny and restrictions by the federal government.

## Citizenship and Corporate Control

Under the laws and regulations in the United States, various restrictions or controls exist on foreign nationals and foreign companies for holding a mineral right or to be a shareholder of a holder of a mineral right. Various federal laws touch upon foreign ownership of mineral rights. To own, hold, or control an interest in a federal lease, a person must meet citizenship[72] and other requirements.[73] Subject to certain exceptions, '[n]o person or entity shall take, hold, own or control more than 246,080 acres[74] of federal oil and gas leases in any one State [except Alaska] at any one time'.[75]

Mineral rights in oil and gas under public domain lands and lands returned to the public domain are subject to lease under the Mineral Leasing Act of 1920 (MLA),[76] and its amendments, and may be acquired and held by foreigners only through stock ownership, holding, or control, and only if the laws, customs, or regulations of their country do not deny similar or like privileges to citizens or corporations of the United States. Offshore federal mineral rights are also covered under this umbrella. Foreign corporations may not acquire or hold any direct or indirect interest in mineral prospecting permits or mineral rights, except that they may own or control stock in corporations holding such permit or mineral rights if the laws of their country do not deny similar or like privileges to citizens of the United States.[77]

---

[71] Id.

[72] A person qualifies if he is a citizen of the United States or an alien stockholder in a corporation that is organized under state or Federal law.

[73] 43 C.F.R. § 3102.5–1.

[74] Section 352 of the Energy Policy Act of 2005, Pub. L. No. 109–58, amended section 27(d)(1) of the MLA, 30 U.S.C. 184(d)(1), by excluding from the acreage limitation all acreage committed to a federally approved unit or communitization agreement or for which royalty was paid in the preceding calendar year.

[75] Id. § 3101.2–1(a). Alaska is divided into two leasing districts. A person or entity is limited to 300,000 acres of Federal leases in each district. Id. §§ 3102.5–1(b) and 3101.2–1(b).

[76] 30 U.S.C.A. §§ 181–96.

[77] 43 C.F.R. §§ 3472.1–2(d), 3502.13. Qualification Stipulations for alien ownership of mineral rights are described in 43 C.F.R. § 3502.10).

No joint venture is expressly required in the case where a foreigner or foreign company is a shareholder or owner of domestic mineral rights, but since foreign companies cannot directly own federally issued oil and gas leases or other mineral rights issued by the federal government, an American subsidiary is typically formed to own or control stock in corporations holding such permits, leases, or mineral rights. Foreign companies may own other non-federal minerals and leases in the United States, specifically those bought or issued from private owners or even states.

Generally, if the foreign corporation is based in a country hostile to the United States, ownership of mineral rights is disallowed or seriously curtailed.[78] Similarly, if ownership is sought by a foreign corporation from a country that, in turn, does not allow entities or individuals from the United States to own mineral rights in the foreign country, then ownership of American mineral rights is likewise curtailed. If any appreciable percentage of stock of a corporation is held by foreigners who are citizens of a country denying similar privileges to United States citizens, that corporation's application or bid for an oil and gas mineral right will be rejected, and that corporation's mineral right is subject to cancellation.[79]

## Survey Act of 1976

Ownership or effective control of ten percent or more of a corporation's voting securities activates the International Investment Survey Act of 1976 (Survey Act).[80] This information is collected by the Bureau of Economic Analysis (BEA).

---

78 The Trading with the Enemy Act of 1917 (12 U.S.C. § 95a *et seq.*), which contains the Foreign Assets Control Regulations and the Alien Property Custodian Regulations, curtails investment by and in countries deemed to be hostile to the United States (e.g., Cuba, Iran, and North Korea). An investment in mineral real property by foreign entities from these restricted countries is prohibited. The Office of Foreign Assets Control, an agency within the Department of the Treasury, administrates and enforces the regulations promulgated under this and similar acts.

79 43 C.F.R. § 3472.1-2(d).

80 See generally 21 U.S.C. §§ 3101–8 (1988), amended by the Trade and Tariff Act of 1984, Pub. L. No. 98–573, § 306(b)(1), 98 Stat. 2948, 3009 (1984); Pub.L. 101-533, § 6(A), Nov. 7, 1990, 104 Stat. 2348, overview of Survey Act accessed 12 January 2015 at www.bea.gov/papers.pdf/Mar04OverviewPaper3.pdf.

While the Survey Act does not impose restrictions on domestic investment by alien entities, it does generally require any alien investor (a) with more than 5 million dollars invested in domestic assets through direct investment, (b) that has more than 5 million dollars net income, or (c) owns more than 200 acres of land, to submit initial and subsequent periodic reports, as well as reports of certain business activities as they occur.[81]

More specifically, these reports require information about direct investment, financial and operating data of United States affiliates, and entity establishment and acquisition data. The reports also must contain the following disclosures: abbreviated income statements and balance sheets, employment and employee compensation, a separation of sales data into components (including bills of sale, sales of services, and sales to United States citizens or to foreign persons), structure and source of external financing, capital expenditures, changes in property, plant, and equipment, research and development expenditures, taxes paid, trade in goods, and numerous other data.[82]

In addition, every five years, the U.S. Department of Commerce issues a mandatory survey requesting information on the amount and type of foreign direct investment in American business enterprises. This survey is known as the Form BE-12 Benchmark Survey. All alien corporations, organizations, associations, branches, or ventures investing in United States real property must complete the report. The data obtained remains confidential and is used by the BEA for evaluating the effect of foreign investment in the American economy.

## Foreign Investment and National Security Act

Acquisition by an alien entity of control over strategically important domestic assets such as oil and gas may also trigger the Foreign Investment and National Security Act of 2007 (FINSA).[83] FINSA applies to foreign investors generally and thus is not limited to oil and gas activities.

The ‘Exon-Florio provision’, first implemented in 1988, gives the President the power to ‘suspend or prohibit any acquisition, merger, or takeover, of a person engaged in interstate commerce’ by a foreign entity,

---

[81] Id. (Note, however, that there are several exemptions mentioned in the statute that may exempt certain foreign entities from being required to file the reports).

[82] Id.

[83] 50 App. U.S.C.A. § 2170; 31 U.S.C. § 301; 5 U.S.C. 5315.

if the President determines that the acquisition will threaten the national security of the United States. FINSA codified the regulations and guidelines created by the Committee on Foreign Investment in the United States (CFIUS). CFIUS was created by Presidential order and is a federal inter-agency group of 12 members chaired by the Secretary of the Treasury. The law requires a mandatory investigation of any transaction involving an 'entity controlled or acting on behalf of a foreign government'.[84]

FINSA requires a foreign entity to address several concerns when seeking approval of an acquisition or merger. First and most importantly, the foreign entity must address the extent to which the transaction will affect the nation's 'critical infrastructure' in national security. Based on the extent to which the transaction will impact this critical infrastructure, government regulators may impose a 'mediation agreement' on the parties seeking CFIUS approval for the transaction.

Under FINSA, the CFIUS review process has been codified and made more rigorous. Although there is no requirement that a foreign entity seek CFIUS approval for a proposed transaction before it occurs, the federal government may suspend or unwind any transaction *after* the transaction closes if it determines that transaction represents a threat to the critical infrastructure described above. Before bringing in CFIUS, two threshold questions should be considered by an entity considering seeking CFIUS approval: (1) Whether the transaction gives an alien entity control over a U.S. firm? (2) Whether the acquisition implicates interests that could be characterized as important for national defence?

When considering the first question, the regulations promulgated under Exon-Florio define 'control' as the power to direct key matters affecting that firm, such as the power to sell off the corporation's assets, dissolve the firm, close its facilities, and terminate its contracts.

Since much of the review and investigatory criteria of CFIUS is confidential, intelligence regarding specific transactions remains sparse. Much of the available information has arisen from company and third-party sources after approval has been denied rather than from CFIUS itself, so it remains unclear exactly what particular factors may be considered detrimental to energy transactions. Prior actions by CFIUS do show, however, that mineral right purchases are considered to implicate interests that could be characterized as important for national defence and that oil and gas interests are considered strategic, along with the companies that own them. The only recent acquisitions of mineral rights

---

[84] 31 C.F.R. 800, *et al.*

by foreign entities that have given rise to CFIUS concerns have been proposed acquisitions of large corporations with extensive operations by foreign entities that are citizens of countries that represent a potential strategic challenge to the United States – namely China. In addition, CFIUS review generally is not required for asset acquisitions; therefore, the acquisition of a passive investment (for example, an overriding royalty interest) probably does not require CFIUS review.

Generally, in the case where an alien or alien company is the holder of a mineral right or a shareholder of a holder of a mineral right, the company is not required to supply the minerals produced by operations based on the mineral right to the domestic market.

## 5. ROYALTY VALUATION AND PAYMENT[85]

OCSLA gives the Secretary of the Interior general rulemaking authority to implement its provisions.[86] This authority takes on particular importance in the context of establishing the value of minerals produced for purposes of computing royalty.

Federal lease instruments are on standard forms that the Secretary revises from time to time. The lease terms include many statutorily required terms. This royalty rate is applied to the volume and value of the oil and gas the lessee produces to determine the amount of royalty owed:

$$\text{Volume} \times \text{Royalty Value} \times \text{Royalty Rate} = \text{Royalty Payable}$$

Lessees generally must pay royalties monthly.[87]

All federal leases contain two important general terms. One is that the lessee agrees to be subject to all reasonable rules and regulations in force when the lease is entered into or that thereafter may be promulgated. The second expressly reserves to the Secretary the authority to establish the reasonable value of production for royalty purposes.

The courts have uniformly upheld the Secretary's broad authority and discretion to establish the reasonable value of production for royalty purposes.[88] However, the Secretary's final determinations are subject to

---

[85] The casebook version of this segment was prepared by Peter J. Schaumberg, Beveridge & Diamond, P.C., Washington, D.C.

[86] 43 U.S.C. § 1334(a).

[87] 30 C.F.R. § 1210.53(a).

[88] See, e.g., *Independent Petroleum Association of America v. DeWitt*, 279 F.3d 1036 (D.C. Cir. 2002); *Marathon Oil Co. v. United States*, 604 F.Supp. 1375

judicial review under the Federal Administrative Procedure Act.[89] The Secretary's discretion therefore is not unfettered. The Secretary's authority has been delegated to the Director of the ONRR, one of the three successors to the MMS and BOEMRE mentioned above. Although most of the regulations were issued by MMS and the administrative and federal court decisions likewise involved that agency, for consistency, we refer to the revenue-collecting agency by its new name.

ONRR has issued regulations prescribing how lessees are to calculate the royalty value of production[90] of crude oil[91] and for federal gas.[92] The single most important threshold factor in applying the valuation rules is the relationship of parties to a sales transaction (particularly arm's length versus non-arm's length). Both the federal gas and federal oil royalty valuation rules contain definitions of 'arm's-length contract' and 'affiliate'.[93] An arm's-length contract is a contract between independent parties that are not affiliates and that possess opposing economic interests regarding the contract. An affiliate is a party controlled by or under common control with another party. Control is established in the case of ownership or common ownership of more than 50 percent of the voting securities or other forms of ownership. In the case of ownership or common ownership of 10–50 percent of the voting securities or other instruments of ownership, ONRR will consider a number of specified factors in determining whether control exists in a particular case. If the transaction is at arm's length, one set of valuation principles applies based on the lessee's 'gross proceeds'. If the transaction is not at arm's length, another set of valuation criteria applies using market-based indicia of value external to the specific transaction.

If lessees enhance the value of oil or gas production by transporting the production closer to market before selling it, the regulations allow the lessee to deduct a transportation allowance.[94] If the lessee incurs

---

(D.Alaska 1985), affirmed, 807 F.2d 759 (9th Cir.1986); *United States v. Ohio Oil Co.*, 163 F.2d 633 (10th Cir. 1947); *Shoshone Indian Tribe v. Hodel*, 903 F.2d 784 (10th Cir.1990); *Continental Oil Co. v. United States*, 184 F.2d 802 (9th Cir.1950); *California Co. v. Udall*, 296 F.2d 384 (D.C.Cir.1961).

89 5 U.S.C. §§ 701–6.

90 30 C.F.R. part 1206.

91 Id. §§ 1206.100–1206.120.

92 Id. §§ 1206.150–1206.160.

93 30 C.F.R. §§ 1206.101 and 1206.151, respectively.

94 Id. §§ 1206.156–1206.157.

expenses for 'processing gas' into more valuable constituent products, regulations allow the lessee to deduct a processing allowance.[95]

In natural gas and crude oil valuation under federal leases, there are fundamental valuation concepts and principles that have been the subject of considerable litigation and development over time. The following subsections highlight several of the most important principles and issues.

## Gross Proceeds

### Basic concept: Total consideration paid for production

Since the 1930s, and since offshore leasing began in 1954, the regulations for federal onshore and Indian leases have provided that under no circumstances may the value of production be less than the 'gross proceeds' accruing to the lessee for the disposition of the production.[96] The lessee's gross proceeds are the total consideration paid (in whatever form) for sale or disposition of the production.[97] Under the valuation regulations, with limited exceptions, if the lessee sells its production under an arm's-length contract, the royalty value is the gross proceeds accruing to the lessee under that contract.

Controversies arise as to whether various forms of payments between the lessee and its purchaser constitute part of gross proceeds. For example, gross proceeds include any reimbursements paid by purchasers to lessees for state severance taxes paid on the share of the production equal to the working interest, because the tax reimbursements are part of the consideration paid under the sales contract for all of the production.[98]

The consideration paid to the lessee when production is sold constitutes the lessee's gross proceeds even if the production is not sold in the local area. For example, in *Marathon Oil Co. v. United States*,[99] Marathon transported some liquefied natural gas from Alaska to where it was sold in Japan. The court held that royalty must be calculated based

---

[95] Id. §§ 1206.158–1206.159.

[96] Current codifications of the rule are found at 30 C.F.R. §§ 1206.152(h) and 1206.153(h) (federal unprocessed and processed gas); 1206.52(h) (Indian crude oil); and 1206.174(g) (Indian gas where an index-based method cannot be used). The gross proceeds rule applied uniformly to crude.

[97] See, e.g., 30 C.F.R. § 1206.151 definitions (federal gas); *Hoover & Bracken Energies, Inc. v. U.S. Dep't of the Interior*, 723 F.2d 1488 (10th Cir.1983).

[98] *Hoover & Bracken Energies, Inc. v. U.S. Dep't. of the Interior, supra*; *Enron Oil & Gas Co. v. Lujan*, 978 F.2d 212 (5th Cir. 1992).

[99] 604 F. Supp. 1375 (D. Alaska 1985), affirmed, 807 F.2d 759 (9th Cir. 1986).

upon the gross proceeds of sale less certain costs and not on the basis of the sales price that Marathon received for unliquefied gas sold under the provisions of a long-term sales contract in Alaska. When production is sold at a point remote from the producing lease, federal rules and case precedents have allowed the lessee to deduct the reasonable actual costs of transporting the production to the point away from the lease to arrive at the value of the production at or near the lease or unit.[100] The *Marathon* case reflects this principle. Value at the lease was determined by the landed sales price in Japan less the costs of liquefaction and tankering – that is, the transportation costs.

**What payments are payments for production?**

The way that sales transactions may be structured, particularly in light of regulatory requirements, has led to disputes between lessee-producers and the government about whether certain payments are payments for 'production' so as to be part of the lessee's gross proceeds. The most prominent disputes have arisen in the context of the former natural gas price regulations and the restructuring of the gas industry that occurred in roughly the same time frame as price deregulation.

Until the mid-1980s, most natural gas produced in the United States and shipped across state lines was sold under long-term contracts at or near Federal Energy Regulatory Commission (FERC) regulated ceiling prices. In most cases, the purchaser was a pipeline who then resold the gas to public utilities, industrial consumers, and others. Under these long-term regulated contracts, gas reserves were committed to the contract with the purchaser at fixed or increasing prices. The lessee-producer was not free to sell gas to someone else without a certificate of abandonment from FERC.

Most of these contracts contained 'take-or-pay' provisions, which obligated the purchaser to take a specified minimum volume of gas during an identified period or to pay for that quantity even if not taken in full. The purpose of a take-or-pay clause was to ensure a dependable revenue flow to the producer to cover operations, maintenance, and

---

[100] 30 C.F.R. §§ 1206.109–1206.112 (federal crude oil); 1206.156–1206.157 (federal gas); 1206.56–1206.57 (Indian crude oil); and 1206.177–1206.178 (Indian gas). Before the 1988 regulations, transportation costs were allowed under judicial and administrative cases. See, e.g., *United States v. General Petroleum Corp.*, 73 F.Supp. 225 (S.D.Cal.1946), affirmed; *Continental Oil Co. v. United States*, 184 F.2d 802 (9th Cir.1950); *Arco Oil and Gas Co.*, 109 IBLA 34 (1989); *Shell Oil Co.*, 52 IBLA 15 (1981); *Shell Oil Co.*, 70 I.D. 393, 396 (1963).

investment costs during periods of slack demand due to seasonal changes, and so on, because the producer's reserves were dedicated to that contract.

If a purchaser took less than the required minimum quantity during the period but paid for the minimum quantity under the take-or-pay clause, the payment for the difference between the minimum take volume and the lesser volume taken was called a 'take-or-pay' payment. The contract provision usually allowed the purchaser a period of time in which to recoup, or 'make up', a take-or-pay payment by later taking gas in excess of the required minimum quantity. Amounts previously paid as take-or-pay payments would be credited against the purchase price of the volume taken in excess of the contractual minimum take requirement ('make-up gas'). The purchaser typically would have to pay additional money if the market price had increased at the time of actual production. Under most contracts, if the purchaser never took make-up gas, the lessee kept the take-or-pay payment.[101]

The federal government assessed royalty on take-or-pay payments when received, which led to several lawsuits that resulted in conflicting district court rulings.[102] The losing parties' appeals were consolidated before the Fifth Circuit. In *Diamond Shamrock Exploration Co. v. Hodel*,[103] the government had argued that the take-or-pay obligation (and, therefore, a payment made under that obligation) was part of the consideration paid for all gas under the contract, not just make-up gas. In rejecting the government's argument, the Fifth Circuit held that 'royalties are not owed unless and until production, the severance of minerals from the formation, occurs'.[104] The 'production' to which the court referred was make-up gas. In other words, the court held that royalties were not due on a take-or-pay payment when received, but would become due later (as part of the lessee's gross proceeds) as and to the extent that the purchaser took make-up gas. To the extent that the purchaser did not take make-up gas, the lessee kept the take-or-pay payment royalty-free.

---

[101] Gas sales contracts currently in use are typically market price based contracts and no longer include take-or-pay clauses. However, the issue illustrates the potential complexity of the gross proceeds determination.

[102] *Diamond Shamrock Exploration Co. v. Hodel*, Civil No. 86–537 and consolidated cases (E.D. La. 1987) (upholding royalty assessment), and *Mesa Petroleum Co. v. U.S. Dep't. of the Interior*, 647 F.Supp. 1350 (W.D. La. 1986) (overturning royalty assessment).

[103] 853 F.2d 1159 (5th Cir. 1988).

[104] 853 F.2d at 1165.

The take-or-pay controversy was followed a few years later by the questions of whether and to what extent payments made to reform or terminate existing contracts are part of a lessee's gross proceeds. In the 1980s, due to both market and regulatory changes, the price of gas began to decline. Pipeline purchasers who could not resell the high-priced gas they were obligated to take under their long-term contracts began to violate the contract terms and sought to escape their price provisions. Because the pipelines could not resell the volumes their long-term contracts required them to take, they could not take those volumes from the producers and as a result also began to accrue large take-or-pay liabilities.

During the same period, FERC promulgated a series of regulatory changes with the objective of opening the industry to competition and changing the role of interstate pipelines from acting as purchasers to acting primarily as transporters.[105] The consequence of these developments was that producers and purchasers entered into agreements to revise or terminate their long-term higher-priced contracts. These agreements often were referred to as 'gas contract settlements' and generally included a lump-sum payment to the producer.

In *Independent Petroleum Association of America v. Babbitt*,[106] the D.C. Circuit Court held that a lump-sum payment made to compromise past accrued take-or-pay liabilities and to terminate a contractual relationship completely (often called a 'buyout') was not part of the consideration the lessee received for gas sold after the settlement to third parties at reduced prices and was not royalty-bearing. The court emphasized that the compromise payment for the take-or-pay liability and the buyout payment were 'non-recoupable' and therefore were not like make-up gas to which part of a take-or-pay payment is allocated.

The particular contract settlement at issue in *Babbitt* involved a settlement in which the producer and the purchaser terminated their relationship entirely. An entirely new and unrelated purchaser entered into a new contract with the producer.[107]

---

[105] See, respective FERC Order series numbered 380, 436, 500, 528, and 636.

[106] 92 F.3d 1248 (D.C. Cir. 1996).

[107] The *IPAA* decision was a 2-1 decision with an extensive dissent (by Judge Rogers). She found that it was reasonable for the Department to attribute the 'buyout' payment to gas that was subject to the contract between the lessee and the original purchaser and sold by the lessee to a different purchaser after the settlement but during the term of the original contract. She found that doing so

In contrast, *United States v. Century Offshore Management Corp.*,[108] involved a situation in which the purchaser made a lump-sum payment to replace the contract with a new contract with changed prices, and the same purchaser continued to take production after the settlement. In that case, the court held that the continued sale of production to the same purchaser at changed prices under the replacement contract provided a sufficient nexus between the payment and the production such that the lump-sum payment constituted part of the payment for production sold, even though the payment was not in a 'recoupable' form. The court distinguished the D.C. Circuit's decision in *Babbitt*, which involved a producer and purchaser that completely terminated their relationship.

**What is production subject to royalty?**

In *Diamond Shamrock*, *supra*, the federal Fifth Circuit Court of Appeals held that royalties are not owed unless and until production occurs. Following the explosion and sinking of the Deepwater Horizon rig in the Gulf of Mexico, some of the oil volumes discharged from the damaged Macondo well were recovered and sold by lease operator BP. The federal government issued an order to Anadarko Petroleum Corporation, 25 percent owner of the Macondo lease, requiring it to pay royalties on its share of the discharged and recovered oil volumes. Anadarko appealed the order to the IBLA, asserting that these volumes were not 'production' within the meaning of OCSLA; however, the IBLA held otherwise.[109]

**Duty to put production into marketable condition at no cost to the lessor**

The gross proceeds rule often operates in tandem with other requirements. For example, longstanding Interior Department rules for federal leases have required lessees to put production into marketable condition and to pay royalty on the value of the gas in marketable condition without deduction for the costs of treatment.[110]

Under the definition in the regulations, 'marketable condition' means production sufficiently free from impurities and otherwise in a condition that a purchaser will accept under a sales contract typical for the field or

---

was consistent with *Diamond Shamrock* because royalty was not assessed until after the gas to which the payment was attributed was produced.

[108] 111 F.3d 443 (6th Cir. 1997).

[109] *Anadarko Petroleum Corporation*, IBLA 2010-228 (January 20, 2012).

[110] Current rules are found at 30 C.F.R. §§ 1206.106 (federal crude oil); 1206.152(i) and 1206.153(i) (federal unprocessed and processed gas); 1206.52(i) (Indian crude oil); and 1206.174(h) (Indian gas).

area.[111] Thus, costs associated with removing water and base sediments from crude oil and with dehydrating, compressing, and desulphurization are not deductible when calculating royalty.

In *Amerada Hess Corp. v. Department of the Interior*,[112] the court upheld the Department's assessment of royalties on reimbursements for gas gathering costs under the gross proceeds rule and the marketable condition rule. The D.C. Circuit Court also considered the application of the marketable condition rule to natural gas produced from coal seams in *Amoco Production Co. v. Watson*.[113] Coal seam gas has higher levels of carbon dioxide ($CO_2$), an acid gas impurity. In *Watson*, the court held that the costs of removing $CO_2$ to the level of pipeline tolerances and the costs of transporting to the treatment plant the excess $CO_2$ were not deductible in calculating royalty value. The IBLA has held that a lessee may not avoid the requirements of the marketable condition rule by transferring the treatment or conditioning function to the purchaser in return for a lower price.[114]

Whether certain gas treatment costs are costs of placing production in marketable condition, which the lessee may not deduct in determining its royalty value, or whether they are allowable deductions for transportation or gas processing, is addressed by regulation.[115] In *Devon Energy Corp. v Kempthorne*,[116] the court upheld a DOI order providing that the marketable condition rule precluded Devon from deducting certain costs associated with compression and dehydration when calculating the 'gross proceeds' upon which royalties are owed on the ground that compression and dehydration were necessary to meet pipeline transmission tolerance requirements.

---

[111] See, 30 C.F.R. §§ 1206.101 (crude oil) and 1206.151 (gas). What condition is typical for the field or area may vary. See, *Branch Oil & Gas Co.*, 144 IBLA 304 (1998).

[112] 170 F.3d 1032 (10th Cir. 1999).

[113] 410 F.3d 722 (D.C. Cir. 2005), cert. denied in relevant part *sub nom., BP America Production Co. v. Burton*, 547 U.S. 1068 (2006).

[114] See, *Branch Oil & Gas Co.*, 144 IBLA 304 (1998); *Bailey D. Gothard*, 144 IBLA 17 (1998), affirmed; *Bailey D. Gothard v. United States*, No. CV 98–103 (D. Mont. 1999).

[115] See 30 C.F.R. §§ 1206.156–1206.157 (transportation); 30 C.F.R. §§ 1206.158–1206.159 (gas processing).

[116] 551 F.3d 1030 (D.C. Circuit 2008).

**Duty to market at no cost to the lessor**

Other regulations address whether lessees may deduct costs incurred to market their production (as opposed to costs incurred to put the production into marketable condition). Amendments to the ONRR gas valuation rules[117] provide that lessees must market production at no cost to the federal government. These 'duty to market' provisions were upheld in *Independent Petroleum Association of America v. DeWitt*.[118]

## Allowable Deductions in Determining Royalty Value (Allowances)

Under ONRR rules, two types of deductions are allowed when lessees determine royalty value: transportation allowances (applicable to oil and gas) and processing allowances (applicable only to 'wet' gas).

ONRR rules provide for the deduction of reasonable actual costs incurred to transport production to a point of sale that is not at or near the lease or unit where the oil or gas is produced.[119] The deduction recognizes that a lessee is not obligated to transport production to distant markets at its own expense. Thus the rule adjusts the value derived at a distant point of sale back to the lease.

In some circumstances, the distinction between transportation and 'gathering' may be disputed. As explained above, the ONRR does not permit any allowance for gathering because gathering is part of putting production into marketable condition. For federal royalty purposes, the regulations define 'gathering' as 'the movement of lease production to a central accumulation and/or treatment point on the lease, unit, or communitized area, or to a central accumulation or treatment point off the lease, unit, or communitized area as approved by BLM [Bureau of Land Management] or Bureau of Safety and Environmental Enforcement OCS operations personnel for onshore and OCS leases, respectively'.[120]

Under the cited provisions, if transportation arrangements are at arm's length, the lessee may deduct the costs paid to the transporter. If

---

[117] *See* 30 C.F.R. §§ 1206.152(i) and 1206.153(i) (1998–2001)) (federal unprocessed and processed gas) and 1206.106 (2001) (federal crude oil) (effective June 1, 2000).

[118] 279 F.3d 1036 (D.C. Cir. 2002).

[119] 30 C.F.R. §§ 1206.156–1206.157 (federal gas) and 1206.109–1206.111 (federal crude oil).

[120] 30 C.F.R. § 1206.151 (gas); see also the functionally identical definition for crude oil at 30 C.F.R. § 1206.101. Both the IBLA and federal courts have addressed the difference between gathering and transportation under ONRR rules. See *Amerada Hess Corp. v. Dep't. of the Interior*, 170 F.3d 1032 (10th Cir.1999); *Kerr-McGee Corp.*, 147 IBLA 277 (1999).

transportation arrangements are not at arm's length (e.g., the lessee also owns the pipeline or the pipeline is owned by an affiliated company), the lessee must calculate actual costs of constructing and operating the transportation system, and then allocate those costs to the production transported, according to requirements prescribed in the regulation.[121] The regulations[122] provide details on what cost elements may be included in the calculation of actual costs. Those include costs such as operations, labour, maintenance, and directly related overhead, and either depreciation or a return on undepreciated capital investment. In *Shell Offshore, Inc.*,[123] the lessee prevailed in an argument that the cost to enlarge an offshore platform to accommodate transportation-related facilities was allowable as a transportation cost because it was an integral part of a transportation system.

The other allowance mentioned above is a 'processing' allowance. A substantial portion of the natural gas produced is 'wet' gas – that is, gas that has heavier entrained liquid hydrocarbons such as ethane ($C_2H_6$), propane ($C_3H_8$), butane ($C_4H_{10}$), pentane ($C_5H_{12}$), and so on, as part of the gas stream flowing from the well. ('Dry' gas is pure methane ($CH_4$).) Wet gas usually has an energy content (measured in British thermal units (Btu)) that is too high for the gas to be useable without separating out the heavier hydrocarbons, which themselves are valuable products. Removal of the heavier hydrocarbons is called 'processing' or 'manufacturing' and is done at a gas processing plant, resulting in a product often called 'manufactured' gas. When gas is processed, the lessee must use the processed gas royalty valuation rules.[124]

Under ONRR rules,[125] the lessee may deduct the reasonable, actual costs of processing. The processing costs must be allocated among the gas plant products (i.e., the extracted liquid hydrocarbons). Unless the ONRR approves an extraordinary processing cost allowance,[126] no processing allowance may be taken against the value of the 'residue gas' (i.e., the dry methane remaining after the liquids are extracted). An

---

[121] In the natural gas context, the regulations provide for an exception to the requirement to calculate actual costs of constructing and operating a transportation system if the lessee has a tariff approved by the federal Energy Regulatory Commission. 30 C.F.R. § 1206.157(b)(5). (An analogous exception for crude oil was removed when the June 2000 rule was promulgated.)

[122] 30 C.F.R. §§ 1206.111 (crude oil) and 1206.157(b) (gas).

[123] 142 IBLA 71 (1997).

[124] 30 C.F.R. § 1206.153.

[125] 30 C.F.R. §§ 1206.158–1206.159.

[126] 30 C.F.R. § 1206.158(d)(2).

'extraordinary' processing allowance is permitted only when the lessee can show that the costs are 'extraordinary, unusual, or unconventional'.[127]

Under ONRR rules, if the processing arrangements are at arm's length, the lessee may deduct the amount it pays to have the gas processed.[128] If the processing arrangements are not at arm's length, (e.g., the lessee also owns the processing plant), the lessee must calculate the actual costs of construction and operation of the processing plant, and then allocate those costs to the production processed under requirements prescribed in the regulation.[129]

## Affiliated Transactions

While gross proceeds is a useful objective measure of the value of production when production is sold at arm's length, the premises underlying acceptance of the total consideration paid in the marketplace as the measure of value break down when the transacting parties are not dealing at arm's length. The valuation regulations effectively are bifurcated between arm's-length and non-arm's-length dispositions. In 30 C.F.R. sections 1206.152 and 1206.153 (federal unprocessed and processed gas), paragraph (b) of each section addresses arm's-length sales, while paragraph (c) addresses non-arm's-length dispositions. Under paragraph (b), the gross proceeds derived from an arm's-length sale generally are the value of the production, and not just the minimum value. For non-arm's-length dispositions, paragraph (c) establishes a hierarchical series of 'benchmarks' that are intended to look at more objective market indicia. Under those benchmarks, the non-arm's-length price may be the royalty value if it meets certain criteria of comparability with arm's-length prices. However, the 'benchmark' price must still be compared to the lessee's gross proceeds – the minimum value – under paragraph (h).

Until June 1, 2000, the federal crude oil royalty valuation rule contained analogous provisions.[130] In the rule that became effective on that date, however, the value of oil sold at arm's length is the lessee's gross proceeds. However, even under certain arm's-length conditions, the lessee may exercise a conditional option to use a spot market index-based value.[131]

---

127 Id. An unusual situation that led to the extraordinary processing cost allowance exception is addressed in *Exxon Corp.*, 118 IBLA 221 (1991).

128 30 C.F.R. § 1206.159.

129 Id. § 1206.159(b).

130 Former 30 C.F.R. § 206.102(b) and (c) (1999).

131 30 C.F.R. § 1206.102(a) and (d) (2001).

If crude oil production is not sold at arm's length before the oil is refined, the measure of value a lessee must use under the rule (as amended in 2004) depends on the area of the country in which the lease is located.[132] For example, for crude oil produced offshore from the Gulf of Mexico, the value is the average of the daily New York Mercantile Exchange (NYMEX) prices for the nearest month of delivery, plus the 'roll' (which takes rising or falling market trends into account), and further adjusted for location and quality.[133]

One of the principal disputes between the ONRR and industry that has arisen in the context of affiliate transactions is the proper measure of value when the lessee transfers to a wholly-owned affiliate (or to another affiliate that also is wholly owned by the same parent corporation), and the affiliate then resells at arm's length. The 2000 crude oil rule resolves this question for oil by using the affiliate's arm's-length resale proceeds to establish value,[134] but the issue remains for gas. The question first arose when lessees challenged ONRR orders to produce documents regarding the affiliate's arm's-length resale. In *Shell Oil Co. v. Babbitt*,[135] and *Santa Fe Energy Products Co. v. McCutcheon*,[136] the federal circuit courts upheld administrative decisions requiring lessees to produce documents regarding their affiliates' arm's-length resales.

The federal crude oil royalty valuation rule amendments promulgated in 2004 require that oil first sold to an affiliate and then resold at arm's length be valued based on the arm's-length resale price (unless the lessee opts to use a spot market index-based value).[137]

## Royalty in Kind

When the federal government leases oil and gas, the mineral leasing statutes provide for the payment of a royalty of a specified percentage of 'the amount or value' of production removed or sold from the lease.[138]

---

132 30 C.F.R. §§ 1206.102(a)(1) and (2) and 1206.103.
133 30 C.F.R. § 1206.103(c).
134 30 C.F.R. § 1206.102(a).
135 125 F.3d 172 (3d Cir. 1997).
136 90 F.3d 409 (10th Cir. 1996).
137 30 C.F.R. § 1206.102(a) (2007).
138 43 U.S.C. § 1337(a)(1) (OCSLA); 30 U.S.C. § 226(b) (Mineral Leasing Act).

The reference to 'amount' refers to the Secretary's authority to elect to take royalty in kind instead of in a cash payment.[139]

In a general sense, statutes provide for three principal types of disposition for royalty taken in kind. First, the Secretary may sell the product through a competitive sale process. Second, the product may be reserved for use by, or sold to, government agencies, including the Defense Department or the Department of Energy for storage in the Strategic Petroleum Reserve. Third, under circumstances specified in both statutes, crude oil may be made available at market price on a preference basis to small refiners who do not possess access to adequate supplies of crude oil.[140]

Congress added further provisions regarding royalty taken in kind from federal leases in section 342 of the Energy Policy Act of 2005.[141]

## Enforcement Statutes – FOGRMA and RSFA

In 1982, Congress enacted the Federal Oil and Gas Royalty Management Act of 1982 (FOGRMA).[142] FOGRMA does not address how much royalty a lessee is required to pay. Rather, Congress' principal purpose was to give the Secretary new and expanded authority to audit and enforce royalty payment obligations on federal and Indian oil and gas leases. Congress also imposed new fiscal accountability requirements on the Secretary regarding timely distribution of royalty revenues to designated recipients. Some of the major issues FOGRMA addressed include recordkeeping, civil and criminal penalties, and interest on late royalty payments. When Congress enacted the Federal Oil and Gas Royalty Simplification and Fairness Act (RSFA) in 1996, it amended certain of FOGRMA's provisions and added new provisions related to royalty management on federal oil and gas leases.[143] Some of RSFA's significant provisions address auditing, time limits on royalty collection and administrative adjudications, and interest on overpayments of royalty.

---

[139] 30 U.S.C. § 192 (Mineral Leasing Act) and 43 U.S.C. § 1353(a)(1) (OCSLA).

[140] The ONRR royalty-in-kind regulations at 30 CFR Part 1208 address only sales of royalty oil to small refiners.

[141] 42 U.S.C. § 15902.

[142] 30 U.S.C. §§ 1701 *et seq.*

[143] The RSFA amendments do not apply to Indian leases (RSFA section 9, 30 U.S.C. § 1701 note); FOGRMA's provisions as written and amended before August 1996 still apply to Indian leases.

**Administrative appeals and judicial review**

A lessee or a designee (a person designated by the lessee to make royalty or other payments on behalf of the lessee)[144] who receives an order from ONRR or a delegated state agency directing the payment of royalties, interest, or other monies, or directing the lessee or designee to provide information, may appeal that order administratively to the ONRR Director. The appeal procedures are set out in ONRR regulations.[145]

Any party to a case adversely affected by a decision of either the ONRR Director or a delegated state agency has the right to appeal that decision to the IBLA in the Office of Hearings and Appeals (OHA).[146] RSFA section 4 enacted a new FOGRMA section 115(h),[147] requiring the Department to issue a final decision in any royalty-related administrative proceeding within 33 months from when the administrative proceeding is commenced, unless the appellant agrees in writing to an extension. FOGRMA section 1724(h)(2) addresses the consequences if the Department fails to issue a final decision within the prescribed 33-month period.

Under the Administrative Procedure Act, judicial review of final royalty decisions is based on the administrative record, and generally is resolved through motions for summary judgment in the district courts. If an appellant paid the disputed royalties pending appeal, it would have the option of pursuing judicial review in a refund action in the U.S. Court of Federal Claims under the Tucker Act.[148]

## 6. EXPRESS 'IMPLIED' COVENANTS

The law of implied covenants developed to construe oil and gas leases covering privately owned oil and gas rights in the United States is applied to federal leases largely as a result of express regulations that are incorporated into the terms of a federal lease. These obligations include a duty by the lessee to protect a federal lease from drainage, to reasonably develop a federal lease, and to wisely manage and administer the lease, including a duty to pool or unitize a federal lease. In general, the basic obligation of a lessee is to act as a reasonable and prudent operator. For

---

144 30 C.F.R. §§ 1218.52 and 1290.102.
145 30 CFR part 1290, §§ 1290.100–1290.110.
146 Id. § 1290.108.
147 30 U.S.C. § 1724(h).
148 28 U.S.C. § 1491.

a case addressing a federal lessee's duty to protect against drainage, see *NGC Energy Co. v. Mono Power Co.*[149]

Lessees are required to protect the leasehold from drainage by drilling offset wells, or with permission of the Secretary, by paying compensatory royalty.[150] Consistent with the general implied covenant to protect against drainage, the federal government, as lessor, must ordinarily prove that the drilling of an offset well would be profitable to the lessee.[151] However, the burden shifts to the lessee to show that an offset well would not be profitable where the lessee is the draining party.[152]

Lessees are also obligated to develop a productive lease. After the federal government notifies the lessee in writing, the lessee must promptly drill and produce additional wells to assure that the lease is properly and timely developed if further development would be reasonably profitable.[153] In some cases, compensatory royalty may replace the need to drill new wells.

The lessee must also conduct all operations in compliance with applicable laws, regulations, lease terms, and other orders from the authorized officers. All operations must be conducted:

> in a manner which ensures the proper handling, measurement, disposition, and site security of leasehold production; which protects other natural resources and environmental quality; which protects life and property; and which results in maximum ultimate economic recovery of oil and gas with minimum waste and with minimum adverse effect on ultimate recovery of other mineral resources.[154]

## 7. EXPLORATORY UNITS, 'COMMUNITIZATION' AND UNITIZATION[155]

While federal leases do not contain pooling or unitization clauses, lessees may 'communitize' their leases with the consent of the Secretary of the

---

[149] 114 IBLA 141 (1990).

[150] 43 C.F.R. §§ 3100.2–2, 3162.2(a).

[151] *Nola Grace Ptasynski*, 63 IBLA 240 (1982).

[152] *Atlantic Richfield Co.*, 105 IBLA 218 (1988).

[153] 43 C.F.R. § 3162.2(c).

[154] Id. § 3162.1(a).

[155] For a thorough discussion of this complex topic, see Bruce M. Kramer and Patrick H. Martin, Chs 2 and 16, and Greg Nibert, Ch. 18, in Rocky Mtn. Min. L. Found., ed., Law of Federal Oil and Gas Leases (LexisNexis Matthew Bender, New Providence, NJ and San Francisco, CA, 2014).

Interior.[156] Communitization is the federal term for pooling. Without Secretarial consent, federal lands are not subject to state compulsory pooling or unitization laws.[157] Nevertheless, the DOI has long advocated unit operations as a conservation tool. The Secretary may require federal lessees to engage in unit operations when they are found to be in the public interest.[158] Accordingly, the Secretary will often consent to the pooling or unitization of federal acreage with private or state lands, and, on occasion, has used state compulsory pooling acts to force the pooling of federal acreage with other lands.[159] In the offshore context, such pooling might occur with state lands beneath territorial waters and, conceivably, state and private coastal lands.

Like pooling, communitization serves to perpetuate two or more leases, or parts thereof, by production from a single well. In general, however, only the communitized portions of federal leases will be held indefinitely by production from a unit well.[160]

Generally, unitization in the United States is done to facilitate secondary or enhanced recovery operations in a proven, developed field. While, subject to Secretarial approval, federal land may be unitized for enhanced recovery or for pressure maintenance in much the same manner as private lands, federal units may also be created to further exploration and primary development.[161] Indeed, a unique feature of federal unitization policy is the 'federal exploratory unit,' whereby federal lessees, with the

---

[156] 30 U.S.C. § 226(m). Lessees may also engage in unit operations on offshore and Indian lands with approval of the Secretary. 43 U.S.C. § 1334(a)(4) (offshore) and 25 U.S.C. § 396d (tribal lands under the Indian Mineral Leasing Act).

[157] *Kirkpatrick Oil & Gas Co. v. United States*, 675 F.2d 1122 (10th Cir.Okla.1982).

[158] 30 U.S.C.A. § 226(m). The offshore regulations specifically provide for both voluntary and compulsory unit operations. 30 C.F.R. §§ 250.190–250.194. In *Clark Oil Producing Co. v. Hodel*, 667 F.Supp. 281 (E.D.La.1987), the court found that the Secretary could compel unitization of offshore leases to protect correlative rights and to avoid the drilling of unnecessary wells.

[159] See, e.g., *Ohmart v. Dennis*, 196 N.W.2d 181 (Neb. 1972).

[160] 30 U.S.C.A. § 226(m) contains language similar to a 'Pugh clause' – a clause in a lease that provides for leased acreage not included in a spacing unit containing a producing well at the end of the primary term to be dropped from the lease. At the time of communitization, the non-unit portion of the lease will continue for the balance of its primary term or two years, whichever is longer, subject to continued payment of rentals. Id. During the balance of that period, the lessee could perpetuate the non-unit portion of the lease by obtaining production from that portion.

[161] Federal gas storage units may also be created. 43 C.F.R. § 3105.5.

consent of the Secretary, may combine multiple federal leases to conduct exploration and development drilling.

The objective of the federal exploratory unit is orderly exploration and development of the unit area, which is believed to be underlain by a common reservoir, without regard to the interior boundaries of the unit leases. Accordingly, the timely drilling of wells will hold all leases within the unit area for the duration of the unit. The lands are approved for inclusion in an exploratory unit on the basis of geological or geophysical inferences that the lands may overlie a common reservoir. Nevertheless, an exploratory unit may be created so as to encompass all formations within the unit area. However, all proposed exploratory wells must be drilled regardless of the results of earlier wells.[162]

Unwilling federal lessees may be forced to join.[163] A substantial majority of the lessees holding most of the leased acreage within the proposed unit area must approve of the basic unit plan of operations, however, before the Secretary will give approval. Once participants are included in a unit, they can also be forced to accept changes in the unit.[164] And the Secretary may expand and contract existing units as reasonably necessary to achieve the conservation objectives of the unit.

If oil or gas is not discovered in paying quantities within five years from the date of approval, the unit ordinarily terminates. If oil or gas is discovered in paying quantities, additional development drilling is usually required.

In an exploratory unit, the term 'paying quantities' means that the well will recover all costs of drilling, completing, and producing operations plus a reasonable profit.[165] Thus, production sufficient to cover operating and marketing costs – the traditional paying-quantities measure – will hold the federal lease on which the well is located but, absent profitable production, will not qualify as a unit well that will hold other federal leases within an exploratory unit.

If a qualifying unit well is drilled, then the operator must submit a plan of development that includes the number of wells to be drilled, their locations, and the sequence of drilling. This plan must be updated yearly.

---

[162] For more detailed information, see BLM Handbook 3180–1, Unitization (Exploratory) (Rel. 3–102, October 28, 1984).

[163] *Gordon H. Barrows*, 36 IBLA 160 (1978).

[164] *Chevron U.S.A., Inc.*, 111 IBLA 96 (1989), overruled in part on other grounds, *Orvin Froholm*, 111 IBLA 301 (1995).

[165] 43 C.F.R. 3186.1, Section 9 and BLM Handbook 3180–1, Unitization (Exploratory) 10 (Rel. 3–102, October 28, 1984). See also *Amoco Production Co.*, 41 IBLA 348 (1979) and *Yates Petroleum Corp.*, 67 IBLA 246 (1982).

Only approved development wells may be drilled except as may be necessary to protect drainage.[166]

Only those portions of the unit that are reasonably proven to be productive of unitized substances share in production from a unit well. Generally, separate participating areas are established for each productive reservoir. The participating area is supposed to reflect the area that the well would reasonably and effectively drain. Radial drainage is presumed. The participating area for a vertical well is determined by drawing a circle around the well that encompasses the area that would be efficiently and effectively drained by the well. As more unit wells are drilled the participating area is expanded. Productive (profitable) unit wells that are drilled within an established distance of a producing unit well may be included in an expanded participating area by the 'circle-tangent' method. A corresponding drainage circle is drawn around the new well and then straight tangent lines are drawn to create a corridor that connects with the circle of the prior well (creating a participating area that resembles the presumed drainage area for horizontal wells). While the primary allocation basis for sharing production is area (acreage), working-interest owners are generally free to negotiate their own allocation formula for costs and working-interest production. The fundamental point is that a lease included in an exploratory unit will not share in production except to the extent that it is part of the 'participating area' or is the situs of the well.

## 8. APPLICATION OF MARITIME LAW

Whenever oil and gas exploration or development occurs on navigable waters, including inland and offshore waters, the applicability of maritime (admiralty) law must be considered. Indeed, the general maritime law of the United States, if applicable, will preempt state law unless the parties contractually choose otherwise. The full body of maritime law consists of regulations, statutes, case law, and even international law. Although much of the body of maritime law has some impact on oil and gas activities on the offshore and inland navigable waters, a few maritime statutes and cases are especially important. The following discussion relates to offshore oil and gas operations specifically. While maritime law is also relevant to operations on inland navigable waters and on state

[166] 43 C.F.R. 3186.1, section 10.

coastal waters, the interplay of maritime law and OCSLA is not relevant to these waters.[167]

Federal courts are given express maritime jurisdiction under Article III, Section 2 of the United States Constitution. Under federal statutory law, 28 U.S.C. section 1333, federal district courts possess original jurisdiction over maritime matters; however, maritime claimants may also pursue certain *in personam* actions in state courts under the 'Savings to Suitors' clause. For maritime law to apply, there must be some nexus between the dispute and maritime commerce, which, under 46 U.S.C. section 30101, includes claims for damage or loss that occur on land resulting from wrongful conduct on navigable water. In addition, certain maritime statutes can trigger federal-question jurisdiction under 28 U.S.C. section 1331, and diversity of citizenship can provide an additional basis for federal jurisdiction over maritime claims. In general, if maritime law does not apply through its own force or by the parties' express choice of law, then state law applies, either through its own force or as surrogate federal law under OCSLA.

The general test for maritime jurisdiction is whether the matter involves or relates to the use and operation of a 'vessel' in navigation.[168] Generally, a vessel is any floating structure having a significant transportation use. In the oil and gas context, marine seismic ships, drilling barges, jack-up drilling rigs,[169] and semi-submersible rigs[170] are vessels, but fixed platforms,[171] and other more permanent production facilities

---

[167] For further discussion of maritime law in the context of offshore oil and gas operations, see W. M. Daniel Wellons, 'Maritime Law Impacts on Offshore Development,' 52 Inst. on Oil & Gas L. & Tax'n 8–1 (2001).

[168] See *Stewart v. Dutra Const. Co.*, 543 U.S. 481, 487 (2005) (defining a 'vessel' for purposes of the Longshore and Harbor Workers' Compensation Act (LHWCA)).

[169] *Offshore Co. v. Robison*, 266 F.2d 769 (5th Cir.1959). A jack-up rig is a mobile floating drilling rig with adjustable legs that may be placed on the ocean floor to hold the rig in place, and allow the rig to be jacked up out of the water, during drilling operations.

[170] A semi-submersible rig is a mobile offshore drilling rig that has large underwater compartments in the hull(s) that may be flooded with ballast to stabilize the rig for drilling operations.

[171] A fixed platform is a drilling and production structure built on the seabed and extending above the surface. Because they are not mobile, they are classified as artificial islands, not vessels. Moreover, a contract to build a fixed platform offshore is governed by state law, even though performance of the contract contemplates the hiring of vessels and seamen. *Laredo Offshore Constructors, Inc. v. Hunt Oil Co.*, 754 F.2d 1223, 1230–31 (5th Cir.1985).

(such as tension-leg platforms[172] and spars[173]) are not.[174] Thus, if a drilling contractor agrees to furnish a drilling barge, jack-up rig, or semi-submersible rig, the drilling contract will be considered a maritime contract, subject to maritime law; however, if the use of a vessel is merely incidental to the work performed, the contract will be considered non-maritime and will ordinarily be governed directly by state law or by state law adopted as surrogate federal law under OCSLA. Whether the use of a vessel is 'merely incidental' to the work performed is a difficult question.[175] In general, if the objective of the contract is to perform

---

172 *Overstreet v. Transocean Offshore, Inc.*, 1998 WL 709809 (E.D.La.1998). A tension leg platform or tension leg well platform is a floating platform used in deep waters that is held in place by steel tendons attached to the platform and to a foundation on the seabed. The tendons are under tension to pull the hull down into the water to minimize motion. Norman Hyne, *Dictionary of Petroleum Exploration & Drilling Production*, p. 519 (PennWell, Tulsa, OK, 1991). See also *McKay v. Offshore Specialty Fabricators, Inc.*, 1997 WL 289365 (E.D.La.1997). In a tax context, a tension leg platform has also been held to be a fixed platform. *Gulf Oil Corporation v. Commissioner*, 87 T.C. 324 (1986). But see *Jordan v. Shell Offshore, Inc.*, 2007 WL 128313 (S.D.Tex January 10, 2007) (slip copy) (stating, contrary to existing precedent, that a tension leg platform is a vessel).

173 *Fields v. Pool Offshore, Inc.*, 182 F.3d 353 (5th Cir.1999) (holding that a spar designed and used primarily as a stationary work platform is not a vessel). A spar is a floating structure that is tensioned by and attached to the seabed by cables so that it remains relatively stationary, much like a buoy. Like the tension leg platform, it does not have a significant transportation function once it has been placed in position offshore. When this book went to press, no court had considered whether a floating production, storage, and off-loading (FPSO) facility is a 'vessel'. Many FPSOs are converted oil tankers and many are flagged as vessels; however, they are used much like a spar. The difference is they can generally navigate from place to place under their own power, and they can also remain in a fixed location under their own power, often through the use of computer-controlled propellers at each corner of the facility (vessel). At least theoretically, FPSOs may be used in multiple locations, although their actual use is generally limited to one location for the productive life of a reservoir. They are also attached to the seabed by cable and production lines.

174 For a list of various factors used to determine whether a facility is a 'vessel', see *Burchett v. Cargill, Inc.*, 48 F.3d 173 (5th Cir.1995); see generally F. Maraist, *Admiralty in a Nutshell* 14–25 (3d ed. West, St. Paul, MN, 1996) [hereinafter Maraist].

175 Cf., *Thurmond v. Delta Well Surveyors*, 836 F.2d 952 (5th Cir.1988) (holding that a work order relating to work on a fixed platform pursuant to a contract to furnish a vessel, crew, and wireline services was primarily non-maritime in nature and thus governed by state law) and *Texaco Exploration and*

services on the vessel that contribute to the mission of the vessel, then the contract is maritime in nature.[176]

The frustration that judges experience when faced with determining whether a particular contract or work is maritime or non-maritime is evident in the following opening sentence of an opinion written by Judge John R. Brown for the Fifth Circuit Court of Appeals: 'Once more we embark on a voyage through the familiar marshland area of the law set aside for classifying the oil and gas exploration services as wet or dry.'[177] Nevertheless, classifying a contract or work as maritime or non-maritime is critical in many contract and tort cases dealing with offshore oil and gas operations, because the ultimate outcome often turns on whether the case is governed by state or maritime law.

In addition to classifying the contract as maritime or non-maritime, courts must determine the governing law in light of the situs of the events giving rise to the dispute, whether in tort or contract. Generally, if the contract is not maritime, state law governs of its own force within a state's territorial waters, and the law of an adjacent state[178] may govern beyond its territorial waters as surrogate federal law under OCSLA: (1) if the controversy arises on a situs covered by OCSLA, such as the subsoil, seabed, or artificial structures attached thereto (e.g., fixed platforms);[179]

---

*Production, Inc. v. AmClyde Engineered Products Co., Inc.*, 448 F.3d 760 (5th Cir. 2006) (holding that, where claims arose out of an offshore crane barge dropping a platform section into the water during installation, any connection to maritime law was eclipsed by the work's connection to the development of the Outer Continental Shelf).

[176] But see *Hoda v. Rowan Companies, Inc.*, 419 F.3d 379, 381 (5th Cir. 2005) (stating that '[t]he maritime or non-maritime status of the contract ultimately depends on its "nature and character", *not* on its place of execution or performance'). For a case illustrating both the importance and difficulty of determining the maritime or non-maritime nature of a contract and contract work, see *Davis & Sons, Inc. v. Gulf Oil Corp.*, 919 F.2d 313 (5th Cir. La. 1990).

[177] *Domingue v. Ocean Drilling & Exploration Co.*, 923 F.2d 393 (5th Cir.1991) (dubiously holding that a contract to perform wireline services on a jack-up drilling rig located 30 miles off the coast of Louisiana was non-maritime and thus governed by Louisiana law, even though a jack-up drilling rig is a 'vessel' governed by maritime law).

[178] Determining the 'adjacent state' is itself an important and potentially thorny issue. See, e.g., *Snyder Oil Corp. v. Samedan Oil Corp.*, 208 F.3d 521 (5th Cir.2000) and *Reeves v. B & S Welding, Inc.*, 897 F.2d 178 (5th Cir.1990). Although OCSLA gives the President authority to make this determination, 43 U.S.C. § 1333(2)(A), this determination has never been made.

[179] 43 U.S.C. § 1333(a)(1) and (2)(A).

(2) if federal law does not apply of its own force; and (3) if state law is not inconsistent with federal law.[180] Generally, if federal (e.g., maritime) law applies of its own force, state law will not apply except in some contract cases where the parties have made an express choice of state law having a substantial relationship to the parties.[181] If adjacent state law applies as surrogate law under OCSLA, that state's law will govern even if the parties have attempted to contract out of OCSLA or that state's law.[182]

A host of maritime doctrines and statutes govern maritime matters. The following brief discussion highlights those doctrines and statutes that are most relevant to offshore oil and gas operations, and discuss the potential application of state law as well.

When a worker is injured in connection with an offshore oil operation, the worker's remedies will depend in part on whether the injured worker is classified as a seaman, maritime worker, or regular (non-seaman, non-maritime) worker. If the employee is a regular worker, the employer would ordinarily be immune from suit under applicable state workers' compensation acts, but the injured employee will generally be able to sue third parties for negligence or other wrongdoing. While these third parties may have the benefit of contractual indemnities from the employee's employer, they might not be able to enforce their contractual indemnity rights against the employer due to state oilfield anti-indemnity statutes. Both Texas and Louisiana have such statutes, and both are important 'adjacent states' in the Gulf of Mexico.

On the other hand, if the employee is a 'seaman', the Jones Act of 1920 allows the employee to sue their employer for injuries resulting from negligence.[183] Together with the additional rights to sue for maintenance and cure[184] and for unseaworthiness[185] under general maritime law, injured seamen have far greater recovery rights than employees

---

[180] Id. § 1333(a)(2)(A).

[181] *Stoot v. Fluor Drilling Services, Inc.*, 851 F.2d 1514 (5th Cir.1988).

[182] *Matte v. Zapata Offshore Co.*, 784 F.2d 628 (5th Cir.1986).

[183] 46 U.S.C. app. § 688.

[184] Maintenance and cure refers to the duty of a vessel and vessel owner to care for a seaman who is injured or falls ill while in service to the vessel.

[185] A vessel and its owner are strictly liable for unseaworthiness. Unseaworthiness includes the failure of a vessel to withstand the perils of an ordinary voyage, a vessel that could not reasonably be expected to make a voyage, and manning a vessel with an incompetent crew.

who are limited to state workers' compensation benefits. Whether the injured employee is a seaman is of critical importance, but this question may be difficult to answer.[186] In general, to be classified as a seaman, one must be more or less permanently assigned to a vessel or group of identifiable vessels under common control and perform a job that contributes to the mission and function of the vessel.[187] In any event, a member of the crew of a jack-up or semi-submersible drilling rig is a seaman, even though an employee performing somewhat similar work on a fixed platform is not.

If the injured employee is not a seaman, he may nevertheless be a maritime worker (longshoreman or harbour worker). A maritime worker's rights vis-à-vis his employer are governed by the Longshore and Harbor Workers' Compensation Act of 1927, 33 U.S.C. sections 901–50 (LHWCA), a form of federal workers' compensation coverage that is automatically extended to offshore platform employees by OCSLA. The LHWCA permits a maritime worker to sue his employer for negligence if the employer is the owner or charterer of a vessel and the injury is caused by vessel-based negligence.[188] Vessel-based negligence generally refers to negligence of the vessel owner or charterer and the captain and crew as opposed to an unseaworthy condition of the vessel that was not created by or known by the vessel crew. The LHWCA also preserves the injured maritime worker's rights against third parties. OCSLA provides longshoremen with workers' compensation coverage under the LHWCA, 43 U.S.C. section 1333(b), if the claimant can show a 'substantial nexus' between the injury and the employer's extractive operations on the OCS.[189]

In case of death caused by tort on the high seas beyond a marine league from shore, a wrongful death claim arises against those responsible under the Death on the High Seas Act of 1920 (DOHSA), 46 U.S.C. sections 761–8. A DOHSA action may arise even though the accident does not have a maritime flavour.

---

[186] See, e.g., *Theriot v. Bay Drilling Corp.*, 783 F.2d 527, 538–9 (5th Cir.1986).

[187] See, e.g., *Harbor Tug and Barge Co. v. Papai*, 520 U.S. 548 (1997) and *Offshore Co. v. Robison*, 266 F.2d 769, 776 (5th Cir.1959).

[188] 33 U.S.C. § 905(b).

[189] *Pacific Operators Offshore, LLP v. Valladolid*, 132 S.Ct. 680 (2012).

Catastrophic losses can occur when cargo on board a vessel is lost or where a vessel collides with another vessel, offshore structure, or harbour facility. Where the potential claims exceed the value of the vessel and the cargo, the vessel owner (or demise charterer) may limit its liability to the value of the vessel and cargo after the casualty.[190] This limitation of liability may also extend to the vessel owner's insurers depending on the language of the insurance policy.[191] Limitation of liability, however, is not available unless the casualty results from a cause that is not within the privity or knowledge of the vessel owner or demise charterer. This limitation of liability is generally intended to apply to transient conditions or operational errors as opposed to problems of which the vessel owner was or should have been aware. Proving the value of both the culpable vessel and the damaged property, particularly unique and aging offshore oil installations, and then determining the appropriate measure of damages, can also be a difficult task.

Yet another limitation of liability in maritime law is court made. In *Robins Dry Dock & Repair Co. v. Flint*,[192] Justice Holmes held that a plaintiff may not recover for economic loss resulting from damage to property in which the plaintiff had no proprietary interest. Thus, an offshore operator who transported production from its platform through a third-party pipeline was barred from recovering damages for lost profits from a vessel that disabled the pipeline with its anchor.[193]

Despite the aforementioned limitations, however, the oil and gas industry is strictly liable for the consequences of oil spills. A major focus of the federal Oil Pollution Act (OPA) of 1990 was the prevention of and reaction to oil spills resulting from offshore production operations and the shipping of oil.[194] The federal OPA overrides a vessel owner's limited liability under the Limitation of Vessel Owners Liability Act and the limitations of *Robins Dry Dock*. The OPA limits monetary damages assessed to US $75 million for losses to private parties, such as property damage or lost profits from businesses affected by the spill, but the culpable parties are still liable for all cleanup costs

---

[190] Limitation of Vessel Owner's Liability Act, 46 U.S.C. App. §§ 181–95.

[191] *Crown Zellerbach Corp. v. Ingram Indus., Inc.*, 783 F.2d 1296 (5th Cir.1986).

[192] 275 U.S. 303 (1927).

[193] *Texas E. Transmission Corp. v. McMoRan Offshore Exploration*, 877 F.2d 1214 (5th Cir.1989).

[194] Pub. L. No. 101–380, 104 Stat. 486 (1990).

under the law. Following the Deepwater Horizon oil spill in the Gulf of Mexico, a number of bills were introduced that would raise the US $75 million cap limit to US $10 billion, retroactive to before the spill occurred. In addition, the OPA liability limits do not apply if the responsible party is grossly negligent, and the OPA does not preempt actions filed under state law.

A host of offshore oil and gas workers qualify as maritime workers[195] or seamen; however, oil and gas workers employed on a fixed platform in state territorial waters have often been classified as regular workers protected only by state workers' compensation laws;[196] although the *Valladolid* decision may affect future cases. Regardless of status, an employee who is injured on the job may generally sue any third party responsible for the injury. In anticipation of such tort claims, the parties to offshore oil and gas contracts commonly agree to indemnify each other against negligence and other claims made by their respective employees. Clearly expressed indemnity obligations are generally valid under maritime law[197] but may be invalid under certain state's laws.[198] Hence, whether maritime law governs the contract and events giving rise to the

---

[195] OCSLA provides that the provisions of the Longshore and Harbor Workers' Compensation Act apply to all oil and gas workers on the OCS who are not classified as seamen or public employees. 43 U.S.C. § 1333(b) (1994). Nevertheless, some oil and gas workers (e.g., an employee working on a fixed platform in state waters) remain protected only by state workers' compensation law. However, in light of *Valladolid, supra*, this may no longer be the case.

[196] For detailed discussion of the classification of offshore employees and remedies for injury, see Maraist, 171–293.

[197] Id. 298–303. Nevertheless, the Longshore and Harbor Workers' Compensation Act restricts indemnity by the employer of a longshoreman in favour of a 'vessel' (including the owner and a time charterer, such as the operator of a jack-up drilling rig), unless the indemnity is reciprocal. 33 U.S.C. §§ 905(b) and (c). However, insurance coverage is not prohibited even if the indemnity is invalid. *Price v. Zim Israel Navigation Co., Ltd.*, 616 F.2d 422 (9th Cir. 1980); *Voisin v. O.D.E.C.O. Drilling Co.*, 744 F.2d 1174, 1177–8 (5th Cir. 1984). In addition, a tower cannot get indemnity for its own negligence from its tow, *Bisso v. Inland Waterways Corporation*, 349 U.S. 85 (1955). Nevertheless, similar protection can often be obtained through an additional-insured provision or a waiver of subrogation.

[198] See discussion of state oilfield anti-indemnity acts in Lowe, *et al.*, Chapter 6, pp. 1060–62, nn. 2–3.

dispute, or if not, which state's law governs, is often critical to the enforceability of the indemnity provisions.[199]

[199] See e.g., *Hollier v. Union Texas Petroleum Corp.*, 765 F.Supp. 330 (W.D.La.1991), reversed 972 F.2d 662 (5th Cir.1992) (invalidating an indemnity under Louisiana law, as the law of the adjacent state); *Smith v. Penrod Drilling Corp.*, 960 F.2d 456 (5th Cir.1992) (validating an indemnity under maritime law even though the accident occurred on an OCSLA situs); *Hodgen v. Forest Oil Corp.*, 87 F.3d 1512 (5th Cir.1996) (holding that the contract was not maritime in nature, that the situs of the accident was governed by OCSLA, that Louisiana law applied as surrogate federal law, and that the indemnity provisions in the master service contract were thus unenforceable); *Grand Isle Shipyard, Inc. v. Seacor Marine*, LLC, 589 F.3d 778 (5th Cir. 2009) (en banc) (holding that despite previous holdings that indicated otherwise, a 'focus-of-the-contract' test should be used to determine the situs of the controversy in contract cases under OCSLA and that a 'situs of the tort' analysis was only intended for tort claims, overruling several cases including *Hollier* and *Hodgen* to the extent they held that a contract claim was governed by the law of the situs of the tort).

# 7. Offshore petroleum resource access and regulation in Canada

**Kylie Fletcher-Johnson***

## 1. INTRODUCTION

Canada is one of the world's leading petroleum producers.[1] It claims significant proven reserves of oil and natural gas. Canada's reserves are estimated to be in the order of 173 billion barrels of oil and 70 trillion cubic feet of natural gas.[2]

Canada's provinces, listed in order of entry into confederation, are Ontario (1867), Quebec (1867), Nova Scotia (1867), New Brunswick (1867), Manitoba (1870), British Columbia (1871), Prince Edward Island (1873), Saskatchewan (1905), Alberta (1905) and Newfoundland and Labrador (1949).[3] Its territories are the Northwest Territories, Yukon and Nunavut.

Canada has an extensive coastline, and lays claim to significant offshore resources. Petroleum resources can be found off the west coast

* The author would like to thank Andrew Stanger, who assisted the author in the early stages of research. The author would also like to thank Professor Rick Bigwood for his guidance and editorial comments. All mistakes are the author's own.

[1] Canadian Association of Petroleum Producers, 'Canada's Petroleum Resources', www.capp.ca/energySupply/canadaPetroleumResources/Pages/default.aspx, accessed 1 June 2014.

[2] BP, *Statistical Review of World Energy June 2013* (London: BP, 2013) 6, 20.

[3] Parliament of Canada, 'Our Country, Our Parliament: An Introduction to How Parliament Works', www.parl.gc.ca/about/parliament/education/ourcountry ourparliament/html_booklet/confederation-e.html, accessed 2 December 2014; see also Peter W Hogg, *Constitutional Law of Canada* (5th edn, Ontario: Thomson Carswell, 2007) 40–46.

and off the north coast (including in the Arctic).[4] However, most of Canada's offshore development and production is occurring off the east coast, in the Atlantic.[5] The area off the coast of British Columbia (Canada's west coast) is subject to a policy-based moratorium on oil and gas activities.[6] Consequently, there are no current petroleum operations in this area.[7]

For the most part, the resources located in Canada's north and in the offshore areas (except those within the stated jurisdiction of separate arrangements between governments) are within the jurisdiction of the federal government.[8] However, responsibility for onshore oil and gas (and other resources) located in Yukon and the Northwest Territories has been devolved to the relevant territory.[9] At this stage, the offshore resources in these areas remain under the control of the federal government.[10] The resources located off the east coast (in the Atlantic) are regulated through separate arrangements between the federal government

---

4 Senate of Canada, Standing Senate Committee on Energy, the Environment and Natural Resources, *Facts Do Not Justify Banning Canada's Current Offshore Drilling Operations: A Senate Review in the Wake of BP's Deepwater Horizon Incident* (2010) 9–10.

5 Robert Bott and David M Carson, *Canada's Evolving Offshore Oil and Gas Industry* (2nd edn, Calgary: Canadian Centre for Energy Information, 2007) 2 and 7–9; Senate of Canada, Standing Senate Committee on Energy, the Environment and Natural Resources (n 4) 10.

6 Katia Opalka, 'Oil and Gas: The View from Canada' (2011) 26 Nat Resources & Env't 37, 39; Natural Resources Canada, 'Offshore British Columbia: Review of the Federal Moratorium on Oil and Gas Activities Offshore British Columbia', www.nrcan.gc.ca/energy/offshore-oil-gas/5843, accessed 1 June 2014; Bott and Carson (n 5) 10.

7 Senate of Canada, Standing Senate Committee on Energy, the Environment and Natural Resources (n 4) 9–10; Natural Resources Canada (n 6).

8 Alastair R Lucas and Constance D Hunt, *Oil and Gas Law in Canada* (Toronto: Carswell, 1990) 39 and 51.

9 Yukon Northern Affairs Program Devolution Transfer Agreement (2001); Yukon Act, SC 2002, c 7; Northwest Territories Lands and Resources Devolution Agreement (2013); Northwest Territories Devolution Act SC 2014, c 2.

10 Yukon Government, Department of Energy, Mines and Resources, 'Roles and Responsibilities of Government', www.emr.gov.yk.ca/oilandgas/roles_responsibilities.html, accessed 14 June 2014; National Energy Board, 'NEB Regulatory Oversight in NWT Post-Devolution', www.neb-one.gc.ca/clf-nsi/rthnb/nrthffshr/nb-nwt-dvltn/nb-nwt-dvltn-eng.html, accessed 12 August 2014.

and each of the provincial governments of Newfoundland and Labrador and Nova Scotia.[11]

Canada's onshore petroleum reserves are typically owned by the province in which they are located,[12] and each province has constitutional jurisdiction over most petroleum activities conducted within the province.[13] Consequently, Canada's onshore petroleum activities are, for the most part, regulated at a provincial level, with the laws varying among provinces.[14] This chapter does not explore the regulation of onshore petroleum activities in each of Canada's provinces; rather, it focuses on the regulation of Canada's offshore petroleum activities. It does so primarily by discussing the federal government's regulation of those activities while making, where relevant, reference to the separate arrangements between the federal government and the provincial governments of each of Newfoundland and Labrador and Nova Scotia.

Aboriginal and treaty rights are protected, including under the Constitution Act, 1982.[15] This important aspect of international and Canadian law is beyond the scope of this chapter and is not discussed, except to say that the aboriginal peoples of Canada have certain rights and title that may need to be considered by parties intending to acquire tenure in petroleum and/or conduct petroleum activities.[16] Notably, government may have an obligation to consult with aboriginal peoples (which obligation may extend to a responsibility to make appropriate accommodations) before allowing or approving the acquisition of petroleum rights or petroleum activities.[17] There are various areas that are covered by specific settlement agreements with aboriginal peoples. These agreements

---

11 Senate of Canada, Standing Senate Committee on Energy, the Environment and Natural Resources (n 4) 65 (Appendix C).

12 Alexander Black, 'Jurisdiction over Petroleum Operations in Canada' (1986) 35 Int'l & Comp LQ 446, 446–447; Nigel Bankes, 'Canada' in Eduardo G Pereira and Kim Talus (eds), *Upstream Law and Regulation: A Global Guide* (London: Globe, 2013) 224.

13 Guy Régimbald and Dwight Newman, *The Law of the Canadian Constitution* (1st edn, Ontario: LexisNexis Canada, 2013) 460–2; Jakub Handrlica, 'Oil and Gas Law in Canada' (2004) 6 Common L Review 59, 59; Black (n 12) 447–9.

14 Bankes (n 12) 224.

15 Constitution Act, 1982, being Schedule B to the Canada Act 1982 (UK), 1982, c 11, s 35.

16 MA Stephenson, 'Resource Development on Aboriginal Lands in Canada and Australia' (2002–2003) 9 JCULR 21, 23–4 and 64–72; Opalka (n 6) 37.

17 Opalka (n 6) 37.

are given effect in legislation.[18] Referring to several of these enactments, Quesnel states:

> The legislation addresses the right of Aboriginal groups to have a major role in the development of resources in their settlement areas and to benefit economically from such development. As well, the legislation provides Aboriginal groups a major role in the protection of the environment and in the preservation of their way of life and heritage.[19]

There are differences among the various settlement agreements.[20] Where a settlement agreement is in place, petroleum rights may only be granted, and petroleum activities may only be conducted, in accordance with the relevant conditions, if any, of that agreement.

---

[18] See, for example, the Inuvialuit Final Agreement (1984) (effectuated under the Western Arctic (Inuvialuit) Claims Settlement Act, SC 1984, c 24), the Gwich'in Comprehensive Land Claim Agreement (1992) (effectuated under the Gwitch'in Land Claim Settlement Act, SC 1992, c 53), the Inuit of the Nunavut Settlement Agreement (1993) (effectuated under Nunavut Land Claims Agreement Act, SC 1993, c 29), the Yukon Umbrella Final Agreement (1993) (effectuated under the Yukon First Nations Land Claims Settlement Act, SC 1994, c 34), the Sahtu Dene and Métis Comprehensive Land Claim Agreement (1993) (effectuated under the Sahtu Dene and Métis Land Claim Settlement Act, SC 1994, c 27) and the Tlicho Land Claims and Self-Government Agreement (2003) (effectuated under the Tlicho Land Claims and Self-Government Act, SC 2005, c 1). For a synopsis of these and further settlement agreements, and a discussion of the right to self-government, refer to Jennifer E Dalton, 'Aboriginal Title and Self-Government in Canada: What is the True Scope of Comprehensive Land Claims Agreements?' (2006) 22 Windsor Rev Legal & Social Issues 29. See also the Mackenzie Valley Resource Management Act, SC 1998, c 25, which, for example, gives effect to various requirements (such as the establishment of land use planning boards, land and water boards, and an environmental impact review board) set out in the Gwich'in Comprehensive Land Claim Agreement and the Sahtu Dene and Métis Comprehensive Land Claim Agreement.

[19] Raymond Quesnel, 'A Review of the Canada Petroleum Resources Act and the Canada Oil and Gas Operations Act as the Legal Framework for Future Development in the Northwest Territories' (2002) 40 Alta L Rev 83, 84 and 85.

[20] Dwight Newman, Michelle Biddulph and Lorelle Binnion, 'Arctic Energy Development and Best Practices on Consultation with Indigenous Peoples' (2014) 32 BU Int'l LJ 449, 464–72.

## 2. OVERVIEW OF OFFSHORE PETROLEUM ACTIVITIES

Petroleum discoveries have been made off the north coast of Canada (in the Canadian Arctic), for example in the Mackenzie Delta-Beaufort Sea region.[21] However, a lack of infrastructure and other difficulties in the Canadian Arctic have meant that the petroleum resources in that area remain, for the most part, undeveloped.[22] Yet, there is increasing interest in the resources located in the Canadian Arctic, and it appears that potential exists for further production in this area.[23]

As mentioned above, most of Canada's offshore development and production has occurred and is occurring off the east coast.[24] The Cohasset oil field was discovered off the coast of Nova Scotia in the 1970s, with the Panuke oil field being discovered much later in the 1980s.[25] These two discoveries were developed together as part of Canada's first offshore project, the Cohasset-Panuke.[26] The Cohasset-Panuke project commenced production in the early 1990s, continuing through to the late 1990s.[27] Two major discoveries, the Venture natural gas field off Sable Island and the Hibernia oil field on the Grand Banks, were made in the late 1970s, with both progressing to commercial production in the 1990s.[28] The Hibernia oil field, off Newfoundland and Labrador, continues to produce today.[29] The other large offshore oil fields that are currently producing are the Terra Nova, the White Rose and the

---

[21] Robert D Bott, *Evolution of Canada's Oil and Gas Industry* (Calgary: Canadian Centre for Energy Information, 2012) 22; Senate of Canada, Standing Senate Committee on Energy, the Environment and Natural Resources (n 4) 10.

[22] Bott and Carson (n 5) 10; Senate of Canada, Standing Senate Committee on Energy, the Environment and Natural Resources (n 4) 10; Bott (n 21) 22.

[23] Bott and Carson (n 5) 30–31; Bott (n 21) 23; Michael T Geiselhart, 'The Course Forward for Arctic Governance' (2014) 13 Wash U Global Stud L Rev 155, 163.

[24] Bott and Carson (n 5) 2 and 7–9; Senate of Canada, Standing Senate Committee on Energy, the Environment and Natural Resources (n 4) 10.

[25] Bott and Carson (n 5) 7 and 26.

[26] Ibid; Senate of Canada, Standing Senate Committee on Energy, the Environment and Natural Resources (n 4) 19.

[27] Bott and Carson (n 5) 7 and 26; Senate of Canada, Standing Senate Committee on Energy, the Environment and Natural Resources (n 4) 20.

[28] Bott and Carson (n 5) 7 and 27; Bott (n 21) 23.

[29] Senate of Canada, Standing Senate Committee on Energy, the Environment and Natural Resources (n 4) 19.

North Amethyst fields off Newfoundland and Labrador.[30] The Hebron oil field, also off Newfoundland and Labrador, is projected to commence production in 2017.[31] Natural gas production near Sable Island off Nova Scotia continues today through Sable Offshore Energy from the Thebaud, Venture, North Triumph, Alma and South Venture fields.[32] The Deep Panuke gas field, off Nova Scotia, has also now commenced production.[33]

## 3. JURISDICTION OVER AND OWNERSHIP OF OFFSHORE RESOURCES

Like Australia, Canada has a federal system of government.[34] The Constitution Act, 1867 grants federal parliament exclusive legislative jurisdiction over stated matters, such as making laws in relation to '[t]he Public Debt and Property', '[t]he Regulation of Trade and Commerce', 'Sea Coast and Inland Fisheries' and 'Indians, and Lands reserved for the Indians'.[35] It also grants to the provincial legislatures exclusive jurisdiction over other stated matters, such as to make laws in relation to '[t]he Management and Sale of the Public Lands belonging to the Province and of the Timber and Wood thereon', 'Property and Civil Rights in the Province', 'Generally all Matters of a merely local or private Nature in the Province' and, subject to certain listed exceptions, 'Local Works and Undertakings'.[36] Further, certain amendments made by the Constitution Act, 1982 (known as the Natural Resources Amendments) ensure that the exclusive legislative jurisdiction of each provincial government extends to making laws in relation to 'exploration for non-renewable natural resources in the province' and 'development,

---

30 Bott and Carson (n 5) 26 and 28–9; Senate of Canada, Standing Senate Committee on Energy, the Environment and Natural Resources (n 4) 18–19.

31 Senate of Canada, Standing Senate Committee on Energy, the Environment and Natural Resources (n 4) 19.

32 Ibid; Bott and Carson (n 5) 27.

33 Bott and Carson (n 5) 29; Senate of Canada, Standing Senate Committee on Energy, the Environment and Natural Resources (n 4) 19.

34 Hogg (n 3) 103; Handrlica (n 13) 59.

35 Constitution Act, 1867 (UK), 30 & 31 Vict, c 3, ss 91(1A), 91(2), 91(12) and 91(24), reprinted in RSC 1985, App II, No 5; for a detailed discussion of Canadian constitutional jurisdiction over petroleum activities and ownership of petroleum resources, refer to Black (n 12) 446–449 and Régimbald and Newman (n 13) 460–70.

36 Ibid ss 92(5), 92(13), 92(16) and 92(10).

conservation and management of non-renewable natural resources … in the province, including laws in relation to the rate of primary production therefrom'.[37] Further, the provincial governments may also make laws in relation to:

> the export from the province to another part of Canada of the primary production from non-renewable natural resources … in the province … but such laws may not authorize or provide for discrimination in prices or in supplies exported to another part of Canada.[38]

While the federal parliament has exclusive legislative jurisdiction under the Constitution Act, 1867 to make laws in relation to '[t]he raising of Money by any Mode or System of Taxation',[39] the provincial legislatures also have exclusive legislative jurisdiction to make laws in relation to 'Direct Taxation within the Province in order to the raising of a Revenue for Provincial Purposes'.[40] Moreover, the Natural Resources Amendments ensure that the provinces 'may make laws in relation to the raising of money by any mode or system of taxation in respect of … non-renewable natural resources … in the province and the primary production therefrom …'.[41] This power is subject to a proviso that disallows discrimination in taxes between exported and non-exported production.[42]

The federal parliament has exclusive jurisdiction to make laws relevant to the territories.[43] However, it should be noted that certain powers have been devolved to the territories.[44]

All residual matters are, by virtue of the 'Peace, Order and good Government' powers found in section 91 of the Constitution Act, 1867, within the jurisdiction of the federal parliament.[45] Despite the distribution of powers outlined above, it is not unusual for complicated matters, such as petroleum activities, to be regulated at both levels of government

---

37 Constitution Act, 1982 (n 15) ss 92A(1)(a) and 92A(1)(b).

38 Ibid s 92A(2).

39 Constitution Act, 1867 (n 35) s 91(3).

40 Ibid s 92(2).

41 Constitution Act, 1982 (n 15) s 92A(4)(a).

42 Ibid.

43 Marie-Claude Gaudreault, Legislative Services Branch, Department of Justice Canada, 'Federal Legislation and the Private Law of the Canadian Territories: An Argument for Complementarity' (2010), www.justice.gc.ca/eng/rp-pr/csj-sjc/harmonization/gaudr/territories/index.html, accessed 31 July 2014.

44 Ibid.

45 Black (n 12) 447.

exercising separate legislative jurisdiction. To the extent of any inconsistency, the federal legislation prevails.[46]

The original provinces – Ontario, Quebec, Nova Scotia and New Brunswick – retained ownership of their resources after entering confederation by virtue of sections 108, 109 and 117 of the Constitution Act, 1867.[47] The federal government transferred mineral rights to the provinces of Manitoba, Saskatchewan and Alberta under the Natural Resource Transfer Agreements,[48] while at the same time transferring certain lands, originally reserved to the federal government, to the province of British Columbia.[49] British Columbia, Prince Edward Island and Newfoundland (which later changed its name to Newfoundland and Labrador) each retained ownership of its resources by virtue of each of its Terms of Union with Canada.[50] Whatever the legal source of each of the provinces' resource ownership, there was a lack of clarity surrounding the status of resources located offshore.[51] Consequently, ownership of and jurisdiction over Canada's offshore resources has been the focus of various disputes between the provincial and federal governments.

Canada is a signatory to and has ratified the 1982 United Nations Convention on the Law of the Sea (UNCLOS).[52] Relevantly, UNCLOS

---

46 Hogg (n 3) 424 and 425; Eric Colvin, 'Legal Theory and the Paramountcy Rule' (1979) 25 McGill LJ 82, 82.

47 Black (n 12) 446–7.

48 Alberta Natural Resources Act, SC 1930, c 3; Saskatchewan Natural Resources Act, SC 1930, c 41; Manitoba Natural Resources Act, SC 1930, c 29; Constitution Act, 1930, 20 & 21 Geo V, c 26 (UK); Black (n 12) 447; Michael A Thackray, 'Oil and Gas' in *Halsbury's Laws of Canada 2013 Oil and Gas Reissue* (Ontario: LexisNexis Canada, 2013) 108.

49 The Railway Belt and Peace River Block Act, SC 1930, c 37; Constitution Act, 1930, 20 & 21 Geo V, c 26 (UK); Black (n 12) 447.

50 British Columbia Terms of Union, 1871; Prince Edward Island Terms of Union, 1873; Newfoundland Act, 12–13 Geo V1, c 22 (UK).

51 For a discussion of the issues and disputes, refer to Lucas and Hunt (n 8) 69–86, Ian Townsend Gault, 'Recent Developments in the Federal-Provincial Dispute Concerning Jurisdiction Over Offshore Petroleum Resources' (1983) 21 Alta L Rev 97, Van Penick, 'Legal Framework in the Canadian Offshore' (2001) 24 Dal LJ 1, 11–15 and Black (n 12) 452–5.

52 1982 United National Law of the Sea (adopted 10 December 1982, entered into force 16 November 1994) 1833 UNTS 3; for a discussion of Canada's offshore jurisdiction, refer to Keith Miller, 'Implications of UNCLOS for Canada's Regulatory Jurisdiction in the Offshore: The 200-Mile Limit and the Continental Shelf' (2007) 30 Dal LJ 341.

divides maritime zones into the 'territorial sea', the 'contiguous zone', the 'exclusive economic zone' and the 'continental shelf'.[53]

Subject to certain exceptions, UNCLOS recognizes a coastal state's right to establish its territorial sea from baselines seaward not exceeding 12 nautical miles.[54] The 'internal waters' are the waters landward from baselines.[55] UNCLOS provides that the sovereignty of a coastal state extends beyond its internal waters to the territorial sea,[56] including its 'bed and subsoil',[57] and that such sovereignty 'is exercised subject to [UNCLOS] and to the other rules of international law'.[58] A coastal state may exercise control in its contiguous zone, from baselines seaward not exceeding 24 nautical miles, as is necessary to prevent, and punish, breach of certain laws and regulations.[59]

UNCLOS also recognizes that a coastal state has certain rights in the exclusive economic zone, from baselines seaward not exceeding 200 nautical miles.[60] Relevantly, these rights include 'sovereign rights for the purpose of exploring and exploiting, conserving and managing the natural resources … of the waters superjacent to the seabed and of the seabed and its subsoil …'.[61]

Additionally, UNCLOS provides that 'the coastal State exercises over the continental shelf sovereign rights for the purpose of exploring it and exploiting its natural resources'.[62] Such rights are recognized from baselines seaward to the greater of 200 nautical miles or, subject to certain limitations, the outside edge of the continental margin.[63] The continental shelf beyond the 200-nautical-mile limit is often referred to as an 'extended continental shelf'.[64]

---

53 Tina Hunter and John Chandler, *Petroleum Law in Australia* (Chatswood: LexisNexis Butterworths, 2013) 68. It should be noted that there are specific provisions, not discussed in this chapter, that relate to archipelagic states. These provisions are set out in 1982 United Nations Law of the Sea (n 52) Part IV.

54 1982 United Nations Law of the Sea (n 52) art 3.

55 Ibid art 8.

56 Ibid art 2.

57 Ibid.

58 Ibid.

59 Ibid art 33.

60 Ibid art 57.

61 Ibid art 56.

62 Ibid art 77.

63 Ibid art 76.

64 Foreign Affairs, Trade and Development Canada, 'Canada's Extended Continental Shelf', www.international.gc.ca/arctic-arctique/continental/index.aspx?lang=eng, accessed 1 May 2014. The 1982 United Nations Law of the Sea

Canada has taken various steps to assert jurisdiction in respect of its offshore, including, more recently, by enacting the Oceans Act in the late 1990s.[65] That Act legislates in respect of the maritime zones dealt with in UNCLOS and defines them in a manner largely consistent with UNCLOS.[66] The Act provides that both 'the internal waters of Canada and the territorial sea of Canada form part of Canada'.[67] It also states that, except within a province, the seabed and subsoil below are 'vested in Her Majesty in right of Canada'.[68] The Act defines the continental shelf and provides that it extends, subject to certain exceptions, to the outer edge of the continental margin which is to be 'determined in the manner under international law that results in the maximum extent of the continental shelf of Canada'.[69] Canada asserts its sovereign rights over the exclusive economic zone and the continental shelf of Canada using, for the most part, the same terms as those used in UNCLOS.[70] It stipulates that its rights in 'the seabed and subsoil of the exclusive economic zone … and their resources' and its rights in the continental shelf are 'vested in Her Majesty in right of Canada'.[71]

Canada is still in the process of exploring and making extended continental shelf submissions to the Commission on the Limits of the Continental Shelf. It made a partial continental shelf submission to the Commission in late 2013, which submission related to areas of its continental shelf in the Atlantic Ocean.[72] The partial submission will be

(n 52) art 82 provides that an international royalty must be paid on specified production in this 'extended continental shelf' area.

[65] Oceans Act, SC 1996, c 31; for a discussion of Canada's assertion of continental shelf rights, refer to Miller (n 52) 359–61.

[66] Miller (n 52) 360; Aldo Chircop and Bruce A Marchand, 'Oceans Act: Uncharted Seas for Offshore Development in Atlantic Canada' (2001) 24 Dal LJ 23, 26 and 27.

[67] Oceans Act (n 65) s 7.

[68] Ibid s 8.

[69] Ibid s 17.

[70] Ibid ss 14(a) and 18.

[71] Ibid ss 15(1) and 19(1).

[72] United Nations, Division for Ocean Affairs and the Law of the Sea, 'Circular Communications from the Division for Ocean Affairs and the Law of the Sea Office of Legal Affairs' (9 December 2013), www.un.org/depts/los/clcs_new/submissions_files/can70_13/clcs70_2013.pdf, accessed 1 June 2014; Government of Canada, 'Partial Submission of Canada to the Commission on the Limits of the Continental Shelf Regarding its Continental Shelf in the Atlantic Ocean: Executive Summary' (2013), www.un.org/depts/los/clcs_new/submissions_files/can70_13/es_can_en.pdf, accessed 1 June 2014.

considered by the Commission.[73] It is stated that 'Canada intends to submit information on the limits of the continental shelf beyond 200 nautical miles in the Arctic Ocean at a later date'.[74]

Canada's maritime neighbours are the United States, France (so far as Canada shares maritime boundaries with the territorial collectivity of Saint-Pierre and Miquelon) and Denmark (so far as Canada shares maritime boundaries with the island nation of Greenland).[75] The issues over maritime boundaries as between coastal states, including in the Arctic, are beyond the scope of this chapter, except that it is useful to point out that the maritime boundary dispute with France, in relation to France's exclusive economic zone off Saint-Pierre and Miquelon, and with Denmark, in relation to the delimitation of the continental shelf off Greenland (excluding Hans Island), have been resolved or determined.[76] However, unresolved issues remain regarding various other maritime boundaries.

As noted above, the ownership and jurisdiction of Canada's offshore resources has been the focus of various disputes between provincial and federal governments. A number of offshore reference cases have ensued, each of which represents an important milestone on the issue of offshore ownership and jurisdiction.[77] In each case the relevant court was asked to consider what sovereign rights the relevant province had exercised or could lay claim to prior to or at the time of entry into confederation, the position being that certain rights continued to be held by the province after such a time. The relevant court was also asked to consider the related issue of jurisdiction. These cases are commonly known as the *BC Reference*,[78] the *Newfoundland Reference*,[79] the *Georgia Strait*

---

[73] United Nations, Division for Ocean Affairs and the Law of the Sea (n 72).

[74] United Nations, Division for Ocean Affairs and the Law of the Sea (n 72).

[75] Miller (n 52) 343.

[76] Ibid 364–70.

[77] For a discussion of the offshore reference cases, refer to Lucas and Hunt (n 8) 70–74. It should be noted that a reference case is an opinion of the court rather than a binding decision of the court. For a discussion of the court's reference opinion capacity, refer to Ivan L Head, 'The Canadian Offshore Minerals Reference: The Application of International Law to a Federal Constitution' (1968) 18 UTLJ 131, 131.

[78] *Reference re Off-Shore Mineral Rights of British Columbia* [1967] SCR 792.

[79] *Reference re Mineral and Other Resources of Continental Shelf* (1983) 145 DLR (3rd) 9.

*Reference*,[80] and the *Hibernia Reference*.[81] Lucas and Hunt state that these cases:

> have laid to rest doubts about federal authority over the continental shelf [see the *Hibernia Reference*], established federal authority over the British Columbia territorial sea [see the *BC Reference*], and make British Columbia rights clear in regard to the passage between the mainland and Vancouver Island [see the *Georgia Strait Reference*]. Doubts have been expressed as to whether (if an appeal is ever pursued on the point) the Newfoundland Court of Appeal decision [see the *Newfoundland Reference*] on the territorial sea [that Newfoundland did hold the requisite rights with respect to the three-nautical-mile territorial sea, such rights having vested prior to 1949 and being retained under the Terms of the Union] will be upheld. It has also been argued that no other offshore province could establish so persuasive a case as Newfoundland concerning the territorial sea; thus the federal position in this regard is very solid.[82]

The federal government grants rights to petroleum found on 'frontier lands'.[83] Frontier lands are:

> lands that belong to Her Majesty in right of Canada, or in respect of which Her Majesty in right of Canada has the right to dispose of or exploit the natural resources, and that are situated in
>
> (a) that part of the onshore that is under the administration of a federal minister,
> (b) Nunavut,
> (c) Sable Island,
> (d) the submarine areas in that part – of the internal waters of Canada or the territorial sea of Canada – that is not situated
> (i) in a province other than the Northwest Territories, or
> (ii) in that part of the onshore that is not under the administration of a federal minister, or
> (e) the continental shelf of Canada,
>
> but does not include the adjoining area, as defined in section 2 of the *Yukon Act*.[84]

---

80 *Reference re Ownership of the Bed of the Strait of Georgia and Related Areas* [1984] 1 SCR 388.

81 *Reference re Continental Shelf Offshore Newfoundland* [1984] 1 SCR 86.

82 Lucas and Hunt (n 8) 73.

83 Canada Petroleum Resources Act, RSC 1985, c 36 (2nd Supp).

84 Ibid s 2.

Petroleum rights and activities off Newfoundland and Labrador and Nova Scotia are now, for all intents and purposes, treated as rights and activities under shared control by virtue of the Atlantic Accord (a memorandum of agreement between the federal government and the government of Newfoundland and Labrador) and the Canada-Nova Scotia Offshore Petroleum Resources Accord (a similar agreement between the federal government and the government of Nova Scotia) (together referred to in this chapter as 'the Accords'), respectively.[85] The Canada-Nova Scotia Agreement on Offshore Oil and Gas Resource Management and Revenue Sharing was signed on behalf of the federal government and the government of Nova Scotia in the early 1980s.[86] A similar arrangement was resisted by Newfoundland and Labrador at the time.[87] In the mid-1980s, following the *Hibernia Reference* (above), the Atlantic Accord was signed on behalf of the federal government and the government of Newfoundland and Labrador.[88] This Accord was implemented through reciprocal legislation, that is, the Canada-Newfoundland and Labrador Atlantic Accord Implementation Act[89] and the Canada-Newfoundland and Labrador Atlantic Accord Implementation Newfoundland and Labrador Act[90] (together referred to in this chapter as the 'Newfoundland and Labrador Accord Implementation Acts').[91] Nova Scotia then renegotiated its arrangement with the federal government and entered into the Canada-Nova Scotia Offshore Petroleum Resources Accord.[92] This Accord was also implemented through reciprocal legislation, that is, the Canada-Nova Scotia Offshore Petroleum Resources Accord Implementation Act[93] and the Canada-Nova Scotia Offshore

---

85 Angus Taylor and Jim Dickey, 'Regulatory Regime: Canada-Newfoundland/Nova Scotia Offshore Petroleum Board Issues' (2001) 24 Dal LJ 51, 55–6; see also J Marshall Burgess, 'Effective and Efficient Regulation in Nova Scotia (2003) 26 Dal LJ 303 and Shawn Denstedt and RJ (Jack) Thrasher, 'The Accord Act Twenty Years Later' (2007) 30 Dal LJ 287.

86 Taylor and Dickey (n 85) 55.

87 Thackray (n 48) 114; Taylor and Dickey (n 85) 54.

88 Taylor and Dickey (n 85) 54.

89 Canada-Newfoundland and Labrador Atlantic Accord Implementation Act, SC 1987, c 3.

90 Canada-Newfoundland and Labrador Atlantic Accord Implementation Newfoundland and Labrador Act RSNL 1990, c C-2.

91 Taylor and Dickey (n 85) 55.

92 Thackray (n 48) 114; Taylor and Dickey (n 85) 55.

93 Canada-Nova Scotia Offshore Petroleum Resources Accord Implementation Act, SC 1988, c 28.

Petroleum Resources Accord Implementation (Nova Scotia) Act[94] (together referred to in this chapter as the 'Nova Scotia Accord Implementation Acts', with the Newfoundland and Labrador Accord Implementation Acts and the Nova Scotia Accord Implementation Acts together referred to in this chapter as the 'Accord Implementation Acts').

## 4. THE GRANTING OF OFFSHORE PETROLEUM RIGHTS

Petroleum rights in respect of frontier lands (including offshore areas) that are not within the stated jurisdiction of separate arrangements between governments are granted under and in accordance with the Canada Petroleum Resources Act,[95] and the relevant regulations enacted under that Act: the Frontier Lands Petroleum Royalty Regulations[96] and the Frontier Lands Registration Regulations.[97]

The Minister may grant interests in petroleum 'in respect of any frontier lands in accordance with [the] Act and the regulations'.[98] Depending on who has the relevant administrative responsibility over the resources which are the subject of the grant, the Minister is either the Minister of Indian Affairs and Northern Development or the Minister of Natural Resources.[99] The Department of Indian Affairs and Northern Development carries out responsibilities on behalf of the Minister of Indian Affairs and Northern Development,[100] whereas the Department of Natural Resources carries out responsibilities on behalf of the Minister of Natural Resources.

The Canada-Newfoundland and Labrador Offshore Petroleum Board, an agency jointly founded by the federal government and the government of Newfoundland and Labrador under the Newfoundland and Labrador Accord Implementation Acts, administers and regulates oil and gas

---

[94] Canada-Nova Scotia Offshore Petroleum Resources Accord Implementation (Nova Scotia) Act, SNS 1987, c 3.

[95] Canada Petroleum Resources Act (n 83); Lucas and Hunt (n 8).

[96] Frontier Lands Petroleum Royalty Regulations 1992, SOR/92-26.

[97] Frontier Lands Registration Regulations 1988, SOR/88-230.

[98] Canada Petroleum Resources Act (n 83) s 13(1).

[99] Ibid s 2 definition of 'Minister'.

[100] Handrlica (n 13) 59; Aboriginal Affairs and Northern Development Canada, 'Rights Management' (2010), www.aadnc-aandc.gc.ca/eng/1100100036364/1100100036369, accessed 1 May 2014.

activities in its own defined 'offshore area'.[101] The Canada-Nova Scotia Offshore Petroleum Board, an agency jointly founded by the federal government and the government of Nova Scotia under the Nova Scotia Accord Implementation Acts, administers and regulates oil and gas activities in its own defined 'offshore area'.[102] Tenure to petroleum is granted by each Board under the relevant Accord Implementation Act in much the same way as outlined below.[103] The various provisions set out in the Accord Implementation Acts relating to the various types of tenure are also substantially the same as those in the Canada Petroleum Resources Act, which are outlined below.[104]

The Canada Petroleum Resources Act defines 'petroleum' to mean 'oil or gas'.[105] '[O]il' is defined to mean:

> (a) crude oil, regardless of gravity, produced at a well head in liquid form, and
> (b) any other hydrocarbons, except coal and gas, and, without limiting the generality of the foregoing, hydrocarbons that may be extracted or recovered from deposits of oil sand, bitumen, bituminous sand, oil shale or from any other types of deposits on the surface or subsurface or the seabed or its subsoil of any frontier lands.[106]

---

[101] '[O]ffshore area' is defined in s 2 of the Canada-Newfoundland and Labrador Atlantic Accord Implementation Act (n 89) and s 2 of the Canada-Newfoundland and Labrador Atlantic Accord Implementation Newfoundland and Labrador Act (n 90).

[102] '[O]ffshore area' is defined in s 2 and Schedule 1 of the Canada-Nova Scotia Offshore Petroleum Resources Accord Implementation Act (n 93) and s 2 and Schedule 1 of the Canada-Nova Scotia Offshore Petroleum Resources Accord Implementation (Nova Scotia) Act (n 94).

[103] See Part II of the Canada-Newfoundland and Labrador Atlantic Accord Implementation Act (n 89) and Part II of the Canada-Newfoundland and Labrador Atlantic Accord Implementation Newfoundland and Labrador Act (n 90); further, see Part II of the Canada-Nova Scotia Offshore Petroleum Resources Accord Implementation Act (n 93) and Part II of the Canada-Nova Scotia Offshore Petroleum Resources Accord Implementation (Nova Scotia) Act (n 94).

[104] See Part II of the Canada-Newfoundland and Labrador Atlantic Accord Implementation Act (n 89) and Part II of the Canada-Newfoundland and Labrador Atlantic Accord Implementation Newfoundland and Labrador Act (n 90); further, see Part II of the Canada-Nova Scotia Offshore Petroleum Resources Accord Implementation Act (n 93) and Part II of the Canada-Nova Scotia Offshore Petroleum Resources Accord Implementation (Nova Scotia) Act (n 94).

[105] Canada Petroleum Resources Act (n 83) s 2.

[106] Ibid s 2.

'[G]as' is defined to mean 'natural gas and includes all substances, other than oil, that are produced in association with natural gas'.[107]

The licenses that may be granted in relation to petroleum rights found on frontier lands are exploration licenses, significant discovery licenses, and production licenses.

An interested party may typically nominate frontier lands for inclusion in a call for bids.[108] The Minister will consider that nomination.[109] If a decision is made to grant petroleum rights in relation to those frontier lands, the Minister will, subject to certain exceptions, call for bids by publishing a notice.[110] Nothing appears to preclude the Minister from doing so where no nomination is forthcoming. Bids must be assessed on a sole criterion, which criterion must be set out in the relevant call for bids.[111] The sole-criterion requirement removes ministerial discretion and, as one commentator points out, 'is intended to make the rights issuance process fair and objective'.[112] Bids, it would appear, are commonly called on a work-expenditure basis,[113] and this is certainly the case for those called for exploration licenses. In Newfoundland and Labrador, bids for significant discovery and production licenses would normally be called on a cash-bonus basis.[114] A bid may not be selected if, among other things, it does not satisfy the terms and conditions,[115] or the selection does not accord with the sole selection criterion set out in the call.[116]

A successful bidder who is granted an exploration license over a parcel will typically be required to pay certain fees and a deposit as surety over the work-expenditure bid amount.[117] The Canada Petroleum Resources Act states that an exploration license grants:

---

107 Ibid s 2.

108 Aboriginal Affairs and Northern Development Canada, *Northern Oil and Gas Annual Report 2013* (Ottawa: Aboriginal Affairs and Northern Development Canada, 2014) 11; Bankes (n 12) 226.

109 Canada Petroleum Resources Act (n 83) s 14(2).

110 Ibid ss 14(1)(a) and 19.

111 Ibid s 14(3)(g).

112 Quesnel (n 19) 89.

113 Bankes (n 12) 226; Quesnel (n 19) 89.

114 Canada-Newfoundland and Labrador Offshore Petroleum Board, 'Legal and Land: Land Rights Issuance Process', www.cnlopb.nl.ca/land_issuance.shtml, accessed 2 May 2014.

115 Canada Petroleum Resources Act (n 83) s 15(1)(a).

116 Ibid s 15(1)(b).

117 Aboriginal Affairs and Northern Development Canada (n 108) 21.

(a) the right to explore for, and the exclusive right to drill and test for, petroleum;
(b) the exclusive right to develop [the relevant] frontier lands in order to produce petroleum; and
(c) the exclusive right, subject to compliance with the other provisions of [the] Act, to obtain a production license.[118]

Subject to certain exceptions, the term of an exploration license is limited to a nine-year period.[119] However, it appears that that term is, under the terms and conditions of the license, further divided into two periods.[120] The intention is that the exploration license interest owner spends the work expenditure bid amount during the first period.[121] They must drill an exploration well during the first period as a qualification to continuing the license into the second period.[122] An exploration license will automatically expire upon the granting of a significant discovery license or production license in respect of the same frontier lands.[123]

Where a significant discovery has been made, the National Energy Board shall, upon application from the exploration license interest owner, or may of its own volition, declare a significant discovery 'in relation to those frontier lands in respect of which there are reasonable grounds to believe that a significant discovery may extend'.[124] The rights held under a significant discovery license mirror those held under an exploration license.[125] Where a significant discovery is declared and continues, the Minister shall, upon application from the exploration license interest owner, or may, under the bid process outlined above where there is no relevant interest owner, grant a significant discovery license.[126] The Canada Petroleum Resources Act does not stipulate a maximum license term for a significant discovery license. However, such a license may only continue for the life of the significant discovery declaration or until

---

118 Canada Petroleum Resources Act (n 83) s 22.
119 Ibid s 26(2).
120 Aboriginal Affairs and Northern Development Canada (n 108) 12.
121 Ibid.
122 Ibid; Bankes (n 12) 228.
123 Canada Petroleum Resources Act (n 83) s 32(1) and 42(1).
124 Ibid ss 28(1) and (2); 'Significant discovery' is defined in s 2 as 'a discovery indicated by the first well on a geological feature that demonstrates by flow testing the existence of hydrocarbons in that feature and, having regard to geological and engineering factors, suggests the existence of an accumulation of hydrocarbons that has potential for sustained production'.
125 Ibid s 29.
126 Ibid ss 30(1) and (2).

a production license is granted.[127] The absence of a maximum term provides the exploration license interest owner adequate time to assess and progress their discovery,[128] and is said to 'reward successful exploration by providing indefinite tenure to the discovery'.[129]

Where the National Energy Board has declared a significant discovery, and provided that the relevant interest owner has not completed a well on the relevant frontier lands within six months, and that more than three years have elapsed since the well-termination date of the well indicating the relevant significant discovery, the Minister may, under the provisions relating to drilling orders, require any relevant interest owner 'to drill a well … and to commence drilling' on any portion of the significant discovery to which their interest extends within a year, at least, of the relevant order.[130] Such an order cannot require the significant discovery license interest owner to drill more than one well at any one time.[131] The issuing of the abovementioned drilling order is reviewable, at the request of the relevant interest owner, by a hearing of the Oil and Gas Committee.[132]

The National Energy Board shall upon application from the exploration license or significant discovery license interest owner, or may of its own volition, declare a commercial discovery, where such a discovery is made, 'in relation to those frontier lands in respect of which there are reasonable grounds to believe that the commercial discovery may extend'.[133] Where such a declaration has been made, the Minister may, under the provisions relating to development orders, give the relevant interest owner of any portion where commercial production has not commenced, at least six months' notice of a prospective reduction to the term of any relevant interest.[134] The consequence of such a notice is that, unless after receiving submissions from the relevant interest owner the Minister determines differently, the term of the relevant license may by order be reduced to at least three years where the Minister is of the

---

127 Ibid ss 32(3) and 42(1).

128 Quesnel (n 19) 89.

129 Handrlica (n 13) 60.

130 Canada Petroleum Resources Act (n 83) ss 33(1), (2) and (3).

131 Ibid s 33(4).

132 Ibid ss 33(1) and 106.

133 Ibid ss 35(1) and (2). 'Commercial discovery' is defined in s 2 as 'a discovery of petroleum that has been demonstrated to contain petroleum reserves that justify the investment of capital and effort to bring the discovery to production'.

134 Ibid s 36(1).

opinion that it is in the public interest.[135] The relevant interest may then of course expire in accordance with that reduced term.[136] The relevant interest owner can avoid such consequences if they commence commercial production before the expiry of the reduced term.[137] The issuing of the abovementioned development order is reviewable, at the request of the relevant interest owner, by a hearing of the Oil and Gas Committee.[138]

A production license effectively grants the same rights as the two licenses described above.[139] However, a production license also grants 'an exclusive right to produce petroleum' and 'title to the petroleum so produced'.[140] Upon application, a production license will be granted in relation to a commercial discovery area to either an exploration license interest owner or a significant discovery license interest owner of the relevant area.[141] If there is no relevant interest owner, then a production license may be granted under the bid process outlined above.[142] A production license may only be held by a corporation incorporated in Canada.[143] It is issued for a term of 25 years,[144] but is subject always to the commercial discovery declaration continuing in relation to a commercial discovery area.[145] Upon the expiry of the 25-year term, a production license can be continued in certain circumstances, including where commercial production continues.[146]

The terms of any of the licenses discussed above may require the holder to pay rent.[147] The holder(s) of a production license must pay royalties 'on petroleum that is transported from the project lands by a transportation facility or [subject to certain exceptions] consumed, lost or wasted'.[148] The details of those royalties are set out in the Frontier Lands

---

135 Ibid ss 36(2) and (3).
136 Ibid s 36(4).
137 Ibid s 36(5).
138 Ibid ss 36(3) and 106.
139 Ibid ss 37(1)(a) and (b).
140 Ibid ss 37(1)(c) and (d).
141 Ibid s 38(1).
142 Ibid s 38(2).
143 Ibid s 44.
144 Ibid s 41(1).
145 Ibid s 41(2).
146 Ibid s 41(3).
147 Ibid s 14(3)(c).
148 Ibid s 55; Frontier Lands Petroleum Royalty Regulations (n 96) s 3(4).

Petroleum Royalty Regulations.[149] A royalty is payable for the production months preceding the 'month of payout'.[150] This royalty starts at one per cent of gross revenues and increases at the rates and times stipulated until the royalty payable reaches five per cent, where it remains until the month of payout.[151]

A royalty is also payable for the month of payout and each month onwards.[152] The royalty payable in this period is the greater of 30 per cent of the net revenues and 5 per cent of the gross revenues.[153] The relevant interest owner may take advantage of limited rights to defer the payment of part of a royalty for the relevant month payable under the first of these calculations, but not so that it results in the payment of a royalty for the relevant month below the second of these calculations.[154] These limited rights of deferral exist where the relevant interest owner has made a contribution to an abandonment and restoration royalty trust.[155] The relevant interest owner may also take advantage of an investment royalty credit of 25 per cent that relates to qualified frontier exploration expenses.[156]

As one would instinctively appreciate, the royalties payable in respect of petroleum found in the Newfoundland and Labrador and Nova Scotia offshore areas differ. Each of the Accord Implementation Acts incorporates the relevant provincial royalty arrangements,[157] with the relevant royalty being paid to the relevant province.[158]

---

149 Frontier Lands Petroleum Royalty Regulations (n 96).

150 Ibid s 3(1)(a); '[M]onth of payout' of 'an interest holder of a production license in relation to a project' is defined in s 2(1) to mean 'the first month in respect of which the interest holder's cumulative gross revenues in relation to the project are equal to or greater than the interest holder's cumulative costs in relation to the project'; '[C]umulative gross revenues' and 'cumulative costs' are also defined in s 2(1), as too are various terms used within those defined terms.

151 Ibid s 3(1)(a); '[G]ross revenues' is defined in s 2(1).

152 Ibid s 3(1)(b).

153 Ibid ss 3(1)(b)(i) and (ii); '[N]et revenues' is defined in s 2(1).

154 Ibid ss 3(1)(2) and 6.

155 Ibid s 6.

156 Ibid s 3(3).

157 Michael Harrington, QC, Colm Seviour, Mark MacDonald and James Dickson, 'Emerging Issues in East Coast Oil and Gas Development' (1997) 35 Alta L Rev 269, 294–5; Matthew Clarke, 'Primary Beneficiaries: Newfoundland and Nova Scotia's Struggle to Achieve the Promise of Petroleum Wealth' (2004) 13 Dal J Legal Stud 1, 11 and 12; see also Alan T Pettie, 'Are Royalty Agreements Required for Canada's East Coast Offshore Oil and Gas?' (2001) 24 Dal LJ 151, 154 and 156–163.

158 Harrington, Seviour, MacDonald and Dickson (n 157) 294–5; RJ (Jack) Thrasher, 'Newfoundland Generic Royalty Regime' (2003) 26 Dal LJ 365, 368.

## 5. REGULATION OF OFFSHORE PETROLEUM ACTIVITIES

The National Energy Board is an independent federal agency established under the National Energy Board Act.[159] The Board regulates, among other things, petroleum activities on 'frontier lands', including in the offshore areas, not within the stated jurisdiction of separate arrangements between governments, under and in accordance with the Canada Oil and Gas Operations Act[160] and the following regulations enacted under that Act: the Canada Oil and Gas Drilling and Production Regulations,[161] the Canada Oil and Gas Certificate of Fitness Regulations,[162] the Canada Oil and Gas Installations Regulations,[163] the Canada Oil and Gas Diving Regulations,[164] the Canada Oil and Gas Geophysical Operations Regulations,[165] the Oil and Gas Spills and Debris Liability Regulations[166] and the Canada Oil and Gas Operations Regulations.[167] The Canada-Newfoundland and Labrador Offshore Petroleum Board and the Canada-Nova Scotia Offshore Petroleum Board administer a similar arrangement under each of the relevant Accord Implementation Acts.[168]

The Canada Oil and Gas Operations Act and its regulations regulate many aspects of petroleum operations.[169] The basic structure of the regulatory scheme is one where there is a wide prohibition that operates to prohibit any person from carrying on oil-and-gas-related work or activity without the requisite license, authorizations and approvals. Applicants for licenses, authorizations and approvals are required to comply

---

[159] National Energy Board Act RSC 1985, c N-7.

[160] Canada Oil and Gas Operations Act RSC 1985, c O-7; Lucas and Hunt (n 8).

[161] Canada Oil and Gas Drilling and Production Regulations SOR/2009-315.

[162] Canada Oil and Gas Certificate of Fitness Regulations SOR/96-114.

[163] Canada Oil and Gas Installations Regulations SOR/96-118.

[164] Canada Oil and Gas Diving Regulations SOR/88-600.

[165] Canada Oil and Gas Geophysical Operations Regulations SOR/96-117.

[166] Oil and Gas Spills and Debris Liability Regulations SOR/87-331.

[167] Canada Oil and Gas Operations Regulations SOR/83-149.

[168] See Part III of the Canada-Newfoundland and Labrador Atlantic Accord Implementation Act (n 89) and the Canada-Newfoundland and Labrador Atlantic Accord Implementation Newfoundland and Labrador Act (n 90); further, see Part III of the Canada-Nova Scotia Offshore Petroleum Resources Accord Implementation Act (n 93) and the Canada-Nova Scotia Offshore Petroleum Resources Accord Implementation (Nova Scotia) Act (n 94).

[169] It should be noted that various of these regulations are supported by various guidelines.

with certain requirements, and applications for licenses, authorizations and approvals are required to be supported by certain documents and information. The National Energy Board is required to take certain matters into account before it confers such licenses, authorizations and approvals.

The Canada Oil and Gas Operations Act and its regulations do more than establish a licensing, authorization and approval regime. The regulatory instruments also establish a regime that deals with, among other things, the regulation of petroleum operations (including by establishing operator's duties), spill regulation and spill liability and the regulation of production arrangements. This regime, at least insofar as it relates to offshore drilling, has been described as follows:

> The National Energy Board has adopted a hybrid approach that combines the use of prescriptive and performance-based requirements depending on which is considered to be the most appropriate. Prescription is used when compulsory means of compliance are desired. Goals are used when circumstances can differ greatly among the regulated companies or where superior outcomes are likely to be achieved through innovation or new technology.[170] … The *Canada Oil and Gas Drilling and Production Regulations* … contain mostly performance-based regulations, while other regulations (in particular, the *Canada Oil and Gas Installations Regulations*, the *Canada Oil and Gas Geophysical Operations Regulations* and the *Canada Oil and Gas Diving Regulations*) are mainly prescriptive.[171]

It is impossible in this chapter to do justice to the full suite of regulatory provisions. What follows instead is a discussion of a few relevant aspects of the regime aimed at giving the reader an insight into the expansive nature of the regime. Before progressing, it should be noted that spill regulation and spill liability have become the subjects of much discussion and scrutiny following the Deepwater Horizon incident in the Gulf of Mexico. Given this focus and the topical nature of this issue, it seems pertinent to discuss this aspect of regulation in more detail. For this reason, the spill regulation and spill liability regime is discussed in a separate section below. There are, however, many overlaps between the provisions discussed in this section and those in the section on spill regulation and spill liability that follows.

---

[170] Jennifer Dagg, Peggy Holroyd, Nathan Lemphers, Randy Lucas and Benjamin Thibault, *Comparing the Offshore Drilling Regulatory Regimes of the Canadian Arctic, the US, the UK, Greenland and Norway* (Drayton Valley: The Pembina Institute, 2011) 20.

[171] Ibid 27.

As discussed above, certain licenses, authorizations and approvals must be held before petroleum activities are carried out on frontier lands. The Canada Oil and Gas Operations Act provides:

> No person shall carry on any work or activity related to the exploration or drilling for or the production, conservation, processing or transportation of oil or gas in any area to which this Act applies unless
>
> (*a*) that person is the holder of an operating license issued under paragraph 5(1)(*a*);
> (*b*) that person is the holder of an authorization issued, before the commencement of operations, under paragraph 5(1)(*b*) for each such work or activity; and
> (*c*) where it is required, that person is authorized or entitled to carry on business in the place where that person proposes to carry on the work or activity.[172]

Under section 5(1)(a), the National Energy Board may on application grant an operating license. Under section 5(1)(b), the National Energy Board may on application grant 'an authorization with respect to each work or activity proposed to be carried on' (referred to in this chapter as 'authorizations'). An operating license expires on a specified date each year but may be renewed.[173] It acts as a precondition for carrying on any oil-and-gas-related work or activity,[174] while an authorization acts as a precondition for carrying on a certain (more specific) oil-and-gas-related work or activity. Importantly, before issuing an authorization, the National Energy Board:

> shall … consider the safety of the work or activity by reviewing, in consultation with the Chief Safety Officer, the system as a whole and its components, including its installations, equipment, operating procedures and personnel,[175] [and] … ensure that the applicant has complied with the requirements of subsection 27(1) in respect of that work or activity.[176]

---

172 Canada Oil and Gas Operations Act (n 160) s 4.
173 Ibid s 5(2).
174 National Energy Board, 'Filing Requirements for Offshore Drilling in the Canadian Arctic' (2011), www.neb-one.gc.ca/clf-nsi/rthnb/pplctnsbfrthnb/rctcff shrdrllngrvw/rctcrvwflngrqrmnt/rctcrvwflngrqrmnt-eng.html, accessed 26 April 2014.
175 Canada Oil and Gas Operations Act (n 160) s 5.02.
176 Ibid s 5.03.

Section 27(1) requires an applicant to provide proof of financial responsibility.[177]

Operating licenses and authorizations are subject to National Energy Board and prescribed requirements,[178] and authorizations are subject to certain approvals, including National Energy Board approvals.[179] Proceeding without an authorization is an offence,[180] as is failing to comply with an approval or requirement of an authorization.[181] An operating license or any authorization may be suspended or revoked in certain circumstances, including for non-compliance.[182]

For the avoidance of doubt and subject to certain exceptions,[183] approval must be obtained before an operator may, among other things, 'drill, re-enter, work over, complete or recomplete a well or suspend or abandon a well or part of a well' (such approval is referred to as a 'well approval').[184] Notably, a well approval may be suspended in specified circumstances related to non-compliance, safety concerns, waste or pollution, and be revoked if those circumstances are not remedied within 120 days.[185]

Certain approvals, including a well approval that relates to a production project and an approval that relates to an authorization to carry on work or activity related to the development of a pool or field, may not be granted without development plan approval.[186] Generally, a pool or field development plan is required to include certain information.[187] Firstly, it must include a description of the 'general approach of developing the pool or field' and include information regarding:

(i) the scope, purpose, location, timing and nature of the proposed development,

---

[177] Proof of financial responsibility is discussed in further detail below in the section on spill regulation and spill liability.

[178] Canada Oil and Gas Operations Act (n 160) ss 5(3) and (4).

[179] Ibid s 5(4).

[180] Ibid s 60(1)(c).

[181] Ibid s 60(1)(c).

[182] Ibid s 5(5).

[183] Canada Oil and Gas Drilling and Production Regulations (n 161) s 10(2).

[184] Ibid s 10(1).

[185] Ibid ss 14(1) and (2).

[186] Canada Oil and Gas Operations Act (n 160) s 5.1(1); Canada Oil and Gas Drilling and Production Regulations (n 161) s 15.

[187] Canada Oil and Gas Operations Act (n 160) s 5.1(3).

> (ii) the production rate, evaluations of the pool or field, estimated amounts of oil or gas proposed to be recovered, reserves, recovery methods, production monitoring procedures, costs and environmental factors in connection with the proposed development, and
> (iii) the production system and any alternative production systems that could be used for the development of the pool or field ...[188]

Secondly, it must set out the details, including the technical details.[189] The Canada Oil and Gas Drilling and Production Regulations require the inclusion of a resource management plan in certain circumstances.[190]

An authorization may not be issued, and a development plan may not be approved, unless and until the benefits plan is approved by the Minister.[191] A benefits plan, which is required to include many prescribed details, is defined to mean:

> [a] plan for the employment of Canadians and for providing Canadian manufacturers, consultants, contractors and service companies with a full and fair opportunity to participate on a competitive basis in the supply of goods and services used in any proposed work or activity referred to in the benefits plan.[192]

Interestingly, the requirement for ministerial approval of a benefits plan is also specifically referred to in the Canada Petroleum Resources Act.[193] It

---

188 Ibid s 5.1(3)(a).

189 Ibid s 5.1(3)(b).

190 Canada Oil and Gas Drilling and Production Regulations (n 161) s 16.

191 Canada Oil and Gas Operations Act (n 160) s 5.2(2); however, such a requirement may be waived by the Minister.

192 Ibid s 5.2(1). A Canada-Newfoundland and Labrador benefits plan is required under s 45(2) of the Canada-Newfoundland and Labrador Atlantic Accord Implementation Act (n 89) and s 45(2) of the Canada-Newfoundland and Labrador Atlantic Accord Implementation Newfoundland and Labrador Act (n 90). A Canada-Nova Scotia benefits plan is required under s 45(2) of the Canada-Nova Scotia Offshore Petroleum Resources Accord Implementation Act (n 93) and s 45(2) of the Canada-Nova Scotia Offshore Petroleum Resources Accord Implementation (Nova Scotia) Act (n 94). Under s 45(1) of these Acts, the relevant 'benefits plan' is defined to mean 'a plan for the employment of Canadians and, in particular, members of the labour force of the Province and, subject to [certain exceptions], for providing manufacturers, consultants, contractors and service companies in the Province and other parts of Canada with a full and fair opportunity to participate on a competitive basis in the supply of goods and services used in any proposed work or activity referred to in the benefits plan'.

193 Canada Petroleum Resources Act (n 83) s 21.

prohibits, without such approval, any work or activity on the relevant frontier lands. It is also the case that, subject to certain exceptions, no authorization may be issued unless the National Energy Board is in receipt of a declaration that relates to (and ensures into the relevant future): (i) the fitness for purpose of; (ii) the appropriateness of operating procedures relating to; and (iii) the qualifications and competency of employees in connection with, equipment and installations.[194] Further, no authorization relating to prescribed equipment or installations (whether prescribed by class or more specifically) may be issued unless the National Energy Board is in receipt of a certificate relating to fitness for purpose and safety and compliance from the relevant certifying authority.[195] The Canada Oil and Gas Certificate of Fitness Regulations set out further requirements on this point.[196]

The Canada Oil and Gas Drilling and Production Regulations require that parties applying for an authorization 'develop an effective management system that integrates operations and technical systems with the management of financial and human resources to ensure compliance with the Act and these Regulations'.[197] The Canada Oil and Gas Drilling and Production Regulations provide that a management system is to include, among other things:

> (a) the policies on which the system is based;
> (b) the processes for setting goals for the improvement of safety, environmental protection and waste prevention;
> (c) the processes for identifying hazards and for evaluating and managing the associated risks;
> (d) the processes for ensuring that personnel are trained and competent to perform their duties;
> (e) the processes for ensuring and maintaining the integrity of all facilities, structures, installations, support craft and equipment necessary to ensure safety, environmental protection and waste prevention;
> (f) the processes for the internal reporting and analysis of hazards, minor injuries, incidents and near-misses and for taking corrective actions to prevent their recurrence; ...[198]

The Canada Oil and Gas Drilling and Production Regulations also require that applications for authorizations include, among other things,

---

[194] Canada Oil and Gas Operations Act (n 160) s 5.11(1).
[195] Ibid s 5.12(1).
[196] Canada Oil and Gas Certificate of Fitness Regulations (n 162).
[197] Canada Oil and Gas Drilling and Production Regulations (n 161) s 5(1).
[198] Ibid s 5(2).

an execution plan (with a schedule),[199] a safety plan,[200] an environmental protection plan,[201] flaring and venting information,[202] oil burning information,[203] contingency plans (including procedures for emergency response)[204] and decommissioning and abandonment information.[205] Applications for certain authorizations must also include additional information.[206] The information required will depend on the authorization. For example, an application for an authorization for a production project must include, among other information, a field data acquisition program that will provide certain information.[207] An application for a well approval to drill a well must include, among other information, the drilling program and a well data acquisition program.[208]

In relation to petroleum operations, the Canada Oil and Gas Operations Act provides that the Chief Conservation Officer may, subject to there being a capability to do so, order a commencement, continuation or increase in production in order to stop waste.[209] The Chief Conservation Officer may also give an order to decrease, suspend or cease production for the same reason.[210] A person who does not comply with such an order is guilty of an offence,[211] and, for the most part, so too is a person who commits 'waste'.[212] Waste is defined to have its ordinary meaning and the meaning 'as understood in the oil and gas industry'.[213] Waste is also defined by reference to a non-exhaustive list of inclusions.[214] The Canada Oil and Gas Operations Act also includes provisions enabling the Governor in Council to make certain regulations for 'the purposes of safety and the protection of the environment as well as for the production

---

199 Ibid s 6(b).
200 Ibid s 6(c).
201 Ibid s 6(d).
202 Ibid s 6(e).
203 Ibid s 6(f).
204 Ibid s 6(j).
205 Ibid s 6(k).
206 For example, Canada Oil and Gas Drilling and Production Regulations (n 161) ss 6(g),(h) and (i).
207 Ibid s 6(i).
208 Ibid s 11.
209 Canada Oil and Gas Operations Act (n 160) s 17(1).
210 Ibid s 17(2).
211 Ibid s 60(d).
212 Ibid s 18(1).
213 Ibid s 18(2).
214 Ibid s 18(2).

and conservation of oil and gas resources'.[215] The Canada Oil and Gas Drilling and Production Regulations are enacted under those provisions. Consequently, most of the 105 sections of the Canada Oil and Gas Drilling and Production Regulations set out operator duties aimed at achieving these goals. Such duties include, to mention but a few: (i) ensuring compliance with management systems;[216] (ii) taking reasonable precautions to ensure safety and environmental protection (and listing specific duties in that regard);[217] (iii) duties as to design, construction, testing, maintenance and operation of wells, installations, equipment and facilities;[218] (iv) duties as to drilling practices;[219] (v) duties as to well control;[220] and (vi) duties as to training and competency of personnel.[221]

In relation to production arrangements, the Canada Oil and Gas Operations Act allows, and mandates, pooling and unitization in certain circumstances.[222] The National Energy Board has certain powers under the Canada Oil and Gas Drilling and Production Regulations to make orders determining, among other things, the size of spacing units and well production rates.[223] The Canada Oil and Gas Drilling and Production Regulations also deal with production, conservation and, among other things, require that an operator ensures that 'maximum recovery from a pool or zone is achieved in accordance with good oilfield practice'[224] and that 'wells are located and operated to provide for maximum recovery from a pool'.[225]

## 6. OFFSHORE SPILL REGULATION AND SPILL LIABILITY

As noted above, offshore spill regulation and spill liability have become the subjects of much discussion and scrutiny following the Deepwater Horizon incident in the Gulf of Mexico. This section discusses the key aspects of the offshore spill regulation and spill liability regime set out in

---

215 Ibid s 14(1).
216 Canada Oil and Gas Drilling and Production Regulations (n 161) s 18.
217 Ibid s 19.
218 Ibid s 25(a).
219 Ibid s 30.
220 Ibid ss 35–8.
221 Ibid s 72.
222 Canada Oil and Gas Operations Act (n 160) ss 30–48.
223 Canada Oil and Gas Drilling and Production Regulations (n 161) s 2.
224 Ibid s 65(a).
225 Ibid s 65(b).

the Canada Oil and Gas Operations Act, which applies to frontier lands, but is also, for the most part, similar to the regime adopted under each of the Accord Implementation Acts.[226]

The Canada Oil and Gas Operations Act defines 'spill' to mean:

> [a] discharge, emission or escape of petroleum, other than one that is authorized under the regulations or any other federal law or that constitutes a discharge from a vessel to which Part 8 or 9 of the *Canada Shipping Act, 2001* applies or a ship to which Part 6 of the *Marine Liability Act* applies.[227]

An applicant for an authorization must provide proof of financial responsibility (for example, by way of letter of credit or guarantee) in an amount satisfactory to the National Energy Board and, upon becoming an authorization holder, must ensure that the relevant instrument remains in force.[228] The amount necessary is typically set taking into account the risks associated with the particular project.[229]

The Canada Oil and Gas Operations Act provides that '[n]o person shall cause or permit a spill on or from any area to which [the] Act applies'.[230] This prohibition is followed by obligations to report a spill.[231] Such obligations fall to a person carrying on any exploration, development or production work or activity in the area.[232] The circumstances that give rise to an obligation to report a spill also trigger an obligation to take all reasonable measures to repair or remedy, prevent further spill and reduce or mitigate dangers (including to the environment).[233] The Chief

---

[226] For a detailed discussion of Canada's offshore spill regulation and spill liability regimes, refer to William Amos and Ian Miron, 'Protecting Taxpayers and the Environment Through Reform of Canada's Offshore Liability Regime' (2013) 9 JSDLP 3. In relation to the offshore spill regulation and spill liability regime in Newfoundland and Labrador offshore area, refer to ss 160–65 of the Canada-Newfoundland and Labrador Atlantic Accord Implementation Act (n 89) and ss 155–61 of the Canada-Newfoundland and Labrador Atlantic Accord Implementation Newfoundland and Labrador Act (n 90). In relation to the offshore spill regulation and spill liability regime in Nova Scotia, refer to ss 165–70 of the Canada-Nova Scotia Offshore Petroleum Resources Accord Implementation Act (n 93) and ss 157–62 of the Canada-Nova Scotia Offshore Petroleum Resources Accord Implementation (Nova Scotia) Act (n 94).

[227] Canada Oil and Gas Operations Act (n 160) s 24(1).

[228] Ibid ss 27(1) and (1.1).

[229] Amos and Miron (n 226) 18.

[230] Canada Oil and Gas Operations Act (n 160) s 25(1).

[231] Ibid s 25(2).

[232] Ibid s 25(2).

[233] Ibid s 25(3).

Conservation Officer may take, or direct another to take, the measures set out above where such measures are not being taken and there are reasonable grounds to believe that immediate action is necessary.[234] A person who incurs any costs in taking the reasonable measures referred to above may recover those costs against the federal government.[235] The federal government, however, may be able to recover such costs from the relevant authorization holder (ie the holder of the authorization for the work or activity from where the spill emanated).[236] The Chief Conservation Officer may direct a third party to 'take over the management and control of any work or activity' at the spill site, whereby such a party is required to do so and also to take the reasonable measures referred to above.[237] The relevant authorization holder is liable for costs incurred by this third party, which can be recovered as a debt due to the federal government.[238]

Section 26 applies where 'any discharge, emission or escape of oil or gas that is authorised by regulation, or any spill, occurs in any area to which [the] Act applies'. Where such a spill or discharge, emission or escape occurs, the relevant authorization holder (which when referred to in this paragraph, includes any person who was required to hold the relevant authorization) is liable, without proof of fault or negligence, and so too is any person who is at fault or negligent, for all actual loss and damage incurred by a person as a result of the spill or discharge, emission or escape.[239] Where the relevant authorization holder is not at fault or negligent, that authorization holder is also expressly required to pay the costs reasonably incurred in taking action.[240] Notwithstanding the preceding, the total liability of the relevant authorization holder who is not at fault or negligent is limited to a prescribed limit (which is set out in the Oil and Gas Spills and Debris Liability Regulations[241] and, depending on location, ranges between $30 million and $40 million).[242]

---

234 Ibid s 25(4).

235 Ibid s 25(7.1).

236 Amos and Miron (n 226) 24.

237 Canada Oil and Gas Operations Act (n 160) ss 25(5) and (6).

238 Ibid s 25(7).

239 Ibid ss 26(1)(a) and (b); the term 'actual loss or damage' includes the losses set out in s 24(3).

240 Ibid 26(1)(a)(ii).

241 Oil and Gas Spills and Debris Liability Regulations (n 166).

242 Ibid s 3; unless the spill is located in Yukon Territory or the Northwest Territories.

Obviously, the prescribed limit does not serve to protect a party (including a relevant authorization holder) who is at fault or is negligent.[243] Such a person will be liable, to the extent proved against them, for all actual loss and damage incurred by a person as a result of the spill or discharge, emission or escape.[244] The section goes on to provide that no person is liable beyond the greater of either the prescribed limit or the amount that they would be liable for under any other law.[245] Claims under section 26 are to be brought within three years after the occurrence of the relevant loss, damage or costs, and must be brought within six years after the spill or the discharge, emission or escape.[246] Section 26 also sets out substantially similar statutory arrangements in relation to 'debris' as defined.[247]

Upon the requirement of the National Energy Board, a claim, provided that it is one for which proceedings may be instituted, can be paid out (up to a prescribed or determined amount) under an instrument provided as proof of financial responsibility (above).[248] This may be done irrespective of whether proceedings have been instituted.[249] As would be expected, any such amount is set off against any subsequent award made under section 26.[250]

A contravention of the Canada Oil and Gas Operations Act and/or its subordinate legislation is also an offence that could result in, if the proceedings are on summary conviction, the offender being fined a maximum of $100,000 and/or being imprisoned for a maximum of one year, or, if the proceedings are on indictment, the offender being fined a maximum of $1 million and/or being imprisoned for a maximum of five years.[251]

It should be noted that to the extent that any spill occurs in Arctic waters, the Arctic Waters Pollution Prevention Act will apply.[252] Amos and Miron state that 'in most of the Arctic offshore, the liability provisions of the [Arctic Waters Pollution Prevention Act] operate in

---

243 Canada Oil and Gas Operations Act (n 160) s 26(1)(b).
244 Ibid.
245 Ibid s 26(2.1).
246 Ibid s 26(5).
247 Ibid ss 26(2)(a) and (b).
248 Ibid s 27(2).
249 Ibid s 27(2).
250 Ibid s 27(4).
251 Canada Oil and Gas Operations Act (n 160) s 60(1) and (2).
252 Arctic Waters Pollution Prevention Act, RSC 1985, c A-12. 'Arctic waters' are defined in s 2 of the Act.

conjunction with, and may in some circumstances supersede those of the [Canada Oil and Gas Operations Act]'.[253] The Inuvialuit Final Agreement, effectuated under the Western Arctic (Inuvialuit) Claim Settlement Act,[254] and the Fisheries Act[255] also contain provisions relating to spill liability.[256] Liability may also arise under either or both of these statutes in certain circumstances.[257]

Following the Deepwater Horizon incident, the Standing Senate Committee on Energy, the Environment and Natural Resources reviewed Canada's offshore oil and gas industry and its regulation. It released a report in August 2010. The Committee did not recommend a permanent or temporary ban of offshore drilling but did make a number of recommendations, being:

> Recommendation 2 … exploring in greater detail the structure and role of the offshore petroleum Boards to determine whether there may be in fact a material conflict between regulatory roles. Recommendation 3 … a thorough discussion by regulators and industry respecting whether and under what circumstances relief wells should be prescribed. … Recommendation 4 … that there be greater collaboration between all those responsible for responding to an oil spill in developing, preparing and practicing in advance of an event. Recommendation 5 … that all offshore operators be required to organize Tier Three spill response tabletop drills at regular intervals. Recommendation 6 … a comprehensive review of the issue of liability, including whether the thresholds should be adjusted to reflect current economic realities.[258]

The National Energy Board also conducted a safety and environmental regulatory review known as the Arctic Offshore Drilling Review.[259] The report, titled 'Review of Offshore Drilling in the Canadian Arctic' and dated December 2011, sets out a number of findings. As a result of these findings, the National Energy Board has developed the National Energy Board Filing Requirements for Offshore Drilling in the Arctic, which

---

253 Amos and Miron (n 226) 27.
254 Western Arctic (Inuvialuit) Claim Settlement Act (n 18).
255 Fisheries Act, RSC 1985, c F-14.
256 Amos and Miron (n 226) 16 and 27–30.
257 Ibid.
258 Senate of Canada, Standing Senate Committee on Energy, the Environment and Natural Resources (n 4) 5.
259 National Energy Board, *Review of Offshore Drilling in the Canadian Arctic* (Calgary: National Energy Board, 2011).

clarify the National Energy Board's requirements regarding the information to be filed in relation to an application for an authorization.[260] The National Energy Board has also more recently released, in draft for consultation, its Financial Viability and Financial Responsibility Guidelines, which contain, among other information, details as to how an applicant will be required to demonstrate financial responsibility, including in relation to responding to a spill.[261]

It should also be noted that, at the time of writing, the Energy Safety and Security Act[262] was in Bill form. Among other things, this Bill proposes a number of important changes to the spill regulation and spill liability regime set out in the Canada Oil and Gas Operations Act (above) and in the Accord Implementation Acts. The changes include, among others, the inclusion of an express reference to the 'polluter-pays principle' and an increase (to $1 billion dollars) to the prescribed amount that limits the total liability of an authorization holder not at fault or negligent. The changes, if enacted, will certainly strengthen the spill regulation and spill liability regime in those Acts.

## 7. ENVIRONMENTAL REGULATION

The Constitution Act does not grant exclusive legislative jurisdiction over the environment per se.[263] Instead, one must look to the other legislative powers found within the Constitution Act in order to determine which government has the relevant jurisdiction.[264] Consequently, certain issues relating to the environment are regulated at a provincial level, and certain other issues relating to the environment are regulated at the federal level. While offshore petroleum activities may also be regulated at a provincial level, this chapter will focus on regulation at the federal level.

---

[260] National Energy Board, *Filing Requirements for Offshore Drilling in the Canadian Arctic* (Calgary: National Energy Board, 2011).

[261] National Energy Board, *Draft Financial Viability and Financial Responsibility Guidelines* (Calgary: National Energy Board, 2013).

[262] Bill C-22, An Act Respecting Canada's Offshore Oil and Gas Operations, Enacting the Nuclear Liability and Compensation Act, Repealing the Nuclear Liability Act and Making Consequential Amendments to Other Acts, 2nd Sess, 41st Parl, 2014 (second reading and referral to committee in the House of Commons, 29 May 2014).

[263] Régimbald and Newman (n 13) 452.

[264] Opalka (n 6) 37.

A proponent of a designated project is prohibited under section 6 of the Canadian Environmental Assessment Act, 2012[265] from acting in connection with the carrying out of the whole or a part of a designated project if it may cause an environmental effect set out in section 5(1), unless there is compliance with certain requirements (set out below).[266] Section 5(1) lists various environmental effects that are linked to certain aspects of the environment over which the federal government has constitutional jurisdiction, being aboriginal peoples, federal lands, inter-provincial and international matters, fish and fish habitat, aquatic species and migratory birds.

Section 7 provides that a federal authority must not 'exercise any power or perform any duty or function conferred on it' under a statute that could permit the carrying out of the whole or a part of a designated project unless there is compliance with certain requirements (set out below).[267]

A 'designated project' is any physical activity (and any incidental physical activity) that is carried out in Canada or on federal lands and that is designated by regulation (or ministerial order) and linked to a responsible authority specified in such a designation.[268] The Regulation Designating Physical Activities[269] designates a list of activities (including petroleum-related activities) and links them to the Canadian Environmental Assessment Agency.[270] The regulations also designate a list of activities (including petroleum-related activities) and links them to the National Energy Board.[271] There is some duplication in the various lists of activities. In relation to those activities linked to the Canadian Environmental Assessment Agency, that agency is the responsible authority for such activities where they are not regulated by, relevantly, the National Energy Board Act or the Canada Oil and Gas Operations Act.[272] In relation to those activities linked to the National Energy Board, that

---

265 Canadian Environmental Assessment Act, SC 2012, c 19. A summary of the Canadian Environmental Assessment Act 2012 can be found in Jamie Benidickson, *Environmental Law* (4th edn, Toronto: Irwin Law, 2013) 262–5.

266 Canadian Environmental Assessment Act (n 265) ss 6(a) and (b).

267 Ibid ss 7(a) and (b).

268 Ibid s 2(1). Federal lands are defined in s 2(1) to include the internal waters (not within a province), the territorial sea, the exclusive economic zone and the continental shelf of Canada.

269 Regulations Designating Physical Activities SOR/2012-147.

270 Ibid s 4(1) and items 1–30 of the Schedule.

271 Ibid s 4(3) and items 39–48 of the Schedule.

272 Ibid s 4(1).

Board is the responsible authority for those activities that are regulated by the National Energy Board Act or the Canada Oil and Gas Operations Act.[273]

A proponent of a designated project will be permitted to act in connection with the carrying out of the whole or a part of a designated project, even though it may cause an environmental effect, in the following circumstances. In the event that an environmental assessment is required, a proponent will be permitted to act after the relevant environmental assessment and in compliance with a resulting decision statement.[274] An environmental assessment will be required, for example, where the designated project is linked to the National Electricity Board (above).[275] Otherwise, a proponent will be permitted to act if, after screening, the Canadian Environmental Assessment Agency makes a decision that no environmental assessment is required.[276]

A federal authority will be permitted to exercise a power or perform a duty or function conferred on it under a statute, even though it could permit the carrying out of the whole or a part of a designated project, in the following circumstances. In the event that an environmental assessment is required, the federal authority will be permitted to exercise such a power or perform such a duty or function after the environmental assessment, and only if the resulting decision statement 'indicates that the designated project is not likely to cause significant adverse environmental effects or that the significant adverse environmental effects that it is likely to cause are justified in the circumstances'.[277] Otherwise, a federal authority will be permitted to exercise a power or perform such a duty or function if, after screening, the Canadian Environmental Assessment Agency makes a decision that no environmental assessment is required.[278]

The Canadian Environmental Assessment Act, 2012 details the requirements for an environmental assessment conducted under it no matter which authority is the relevant responsible authority.[279] These requirements obviously apply to the environmental assessments carried out by the National Energy Board acting as a responsible authority under the Act. The Minister may refer an environmental assessment to a review

---

273 Ibid s 4(3).
274 Canadian Environmental Assessment Act (n 265), s 6(b).
275 Ibid ss 13 and 15(b).
276 Ibid ss 6(a) and 10(b).
277 Ibid s 7(b).
278 Ibid s 7(a).
279 Ibid ss 17–27.

panel in certain circumstances.[280] However, this cannot occur, for example, where the National Energy Board is the responsible authority for the designated project.[281]

When carrying out an environmental assessment, the factors set out in section 19(1), including comments from the public, must be taken into account. The 'environmental effects' to be taken into account are set out in section 5(1) (above). However, the additional 'environmental effects' set out in section 5(2) also need to be taken into account 'if the carrying out of the physical activity, the designated project or the project requires a federal authority to exercise a power or perform a duty or function conferred on it' under a statute.

The National Energy Board also conducts environmental assessments as part of its authorization processes.[282] Environmental assessments may also be required under settlement agreements with aboriginal peoples and the legislation effectuating those settlement agreements.

Petroleum activities in the offshore area of Newfoundland and Labrador may trigger the environmental assessment requirements of the Canada-Newfoundland and Labrador Offshore Petroleum Board.[283] Likewise, petroleum activities in the offshore area of Nova Scotia may trigger the environmental assessment requirements of the Canada-Nova Scotia Offshore Petroleum Board.[284] It is understood that after the passage and commencement of the proposed Energy Safety and Security Act,[285] each of these Boards will be prescribed by regulation to be 'responsible authorities' under the Canadian Environmental Assessment Act, 2012 in relation to the activities that will be linked to them.[286] That being the

---

280 Ibid s 38(1).

281 Ibid ss 38(6) and 15(b).

282 National Energy Board, 'National Energy Board Fact Sheet', www.neb-one.gc.ca/clf-nsi/rthnb/whwrndrgvrnnc/nbfctsht-eng.html, accessed 1 June 2014.

283 Canada-Newfoundland and Labrador Offshore Petroleum Board, 'Environment: Project Based Environmental Assessments', www.cnlopb.nl.ca/env_project.shtml, accessed 1 June 2014; Canada-Nova Scotia Offshore Petroleum Board, 'Environmental Assessment', www.cnsopb.ns.ca/environment/environmental-assessments, accessed 1 June 2014.

284 CA Parker, Canada-Nova Scotia Offshore Petroleum Board, 'Environmental Considerations Related to Petroleum Activities Offshore Nova Scotia', prepared for presentation at the 2005 Offshore Technology Conference held in Houston TX, USA, 2–5 May 2005.

285 Bill C-22 (n 262).

286 Canadian Environmental Assessment Agency, 'Responsible Authorities Regulations', www.ceaa.gc.ca/default.asp?lang=en&n=C0E0137C-1, accessed 4 August 2014.

case, the Boards will each also conduct environmental assessments as required under the Canadian Environmental Assessment Act, 2012.[287]

Subject to certain exceptions, an authority (including a federal authority and each of the Boards) is prohibited from, among other things, carrying out a project (defined to include 'a physical activity that is carried out on federal lands … and is not a "designated project"') on federal lands, or 'exercis[ing] any power or perform[ing] any duty or function conferred on it' under a statute that could permit the whole or a part of a project to be carried out on federal lands.[288] The authority may proceed where it is determined by the authority that 'the carrying out of the project is not likely to cause significant adverse environmental effects'.[289] Where the authority determines that such effects are likely, the authority may proceed provided that the Governor in Council determines that those effects are justified in the circumstances.[290]

As discussed above, the Canada Oil and Gas Operations Act and various regulations enacted under that Act (and, in the same manner, the Accord Implementation Acts and various regulations enacted under the Accord Implementation Acts) set out various requirements aimed at protecting the environment. Several of these are discussed above. Among other things, the Canada Oil and Gas Drilling and Production Regulations require the development of an environmental protection plan as part of the authorizations process.[291] Numerous other duties are imposed on the operator under these regulations that relate to environmental protection. However, other statutes and regulations also impose environmental protection obligations on parties involved in offshore petroleum activities. The Canada Environmental Protection Act[292] is one source of environmental protection obligations banning, for example, unauthorized disposal of substances at sea.[293] Obviously, other statutes also exist that impose environmental protection obligations, the Fisheries Act[294] being just one example.[295]

---

287 Bill C-22 (n 262).

288 Canadian Environmental Assessment Act (n 265) ss 66 and 67.

289 Ibid s 67(a).

290 Ibid s 67(b) and s 69(3).

291 Canada Oil and Gas Drilling and Production Regulations (n 161) s 6(k).

292 Canadian Environmental Protection Act, SC 1999, c 33.

293 Ibid; Douglas Black and FVW Penick, 'Survey of Legal Issues: Canadian Offshore Oil and Gas Developments' (1992) 30 Alta L Rev 178, 215.

294 Fisheries Act, RSC 1985, c F-14.

295 Black and Penick (n 293) 215.

## 8. SAFETY FRAMEWORK

Many statutory instruments make up the offshore petroleum safety framework, such as the Canada Oil and Gas Operations Act, and various regulations enacted under that Act (including the Canada Oil and Gas Drilling and Production Regulations). The Canada Oil and Gas Operations Act, and various regulations enacted under that Act (and in the same manner the Accord Implementation Acts and various regulations enacted under the Accord Implementation Acts), set out various occupational health-and-safety related requirements. Several of these are discussed above. Among other things, the Canada Oil and Gas Drilling and Production Regulations require the development of a management system and a safety plan as part of the authorizations process.[296] Many other duties are imposed on the operator under these regulations that relate to occupational health and safety. Additionally, the Oil and Gas Occupational Safety and Health Regulations[297] enacted under Part II of the Canada Labour Code[298] deal with the 'occupational safety and health of employees employed on or in connection with exploration or drilling for, or the production, conservation, processing or transportation of, oil or gas in Canada Lands'.[299] National Energy Board inspectors have been appointed to administer this aspect of the Code in relation to facilities and activities that fall under the regulatory jurisdiction of the National Energy Board.[300]

The Accord Implementation Acts have recently been amended so as to establish a new occupational health and safety regime in the relevant offshore areas.[301] The amending legislation has received royal assent but, at the time of writing, is not yet in force.

---

[296] Canada Oil and Gas Drilling and Production Regulations (n 161) ss 5(1) and 6(c).

[297] Oil and Gas Occupational Safety and Health Regulations SOR/87-612.

[298] Canada Labour Code, RSC 1985, c L-2.

[299] Oil and Gas Occupational Safety and Health Regulations (n 297).

[300] National Energy Board, 'National Energy Board 2013–14 Report on Plans and Priorities', www.neb-one.gc.ca/clf nsi/rpblctn /rprt/plnprrt/2013_2014/rpp_2013-2014-eng.pdf, accessed 1 June 2014.

[301] Offshore Health and Safety Act, SC 2014, c 13; An Act to Amend the Canada-Newfoundland and Labrador Atlantic Accord Implementation Newfoundland and Labrador Act, RSNL 2013, c 3; An Act to Amend Chapter 3 of the Acts of 1987, the Canada-Nova Scotia Offshore Petroleum Resources Accord Implementation (Nova Scotia) Act, SNS 2013, c 16.

## 9. CONCLUSION

Petroleum resources can be found in many areas off Canada's extensive coastline. The federal government controls those resources, except where they are within the stated jurisdiction of a separate arrangement between that government and another government. For example, by agreement, petroleum resources discovered off the coastline of the maritime provinces of Newfoundland and Labrador and Nova Scotia each fall under the shared control of the federal government and the relevant provincial government. Much of Canada's offshore petroleum development and production is occurring in those two areas of shared control. However, the legislation applicable in those areas – the Accord Implementation Acts – is partly based on, hence in many respects similar to, the legislation applicable in the broader offshore areas falling under federal jurisdiction, namely, the Canada Petroleum Resources Act and the Canada Oil and Gas Operations Act. For that reason this chapter has focused on the latter two Acts and their supporting regulations, although the Accord Implementation Acts were noted where relevant.

It has been impossible in a chapter of this nature to do justice to the full suite of regulatory provisions in the subject field. Accordingly, no exhaustive summary of all of the relevant provisions of each of the relevant Acts and regulations was provided. Rather, the chapter has attempted to provide the reader with an insight into the expansive nature of the offshore petroleum regulatory regime in Canada. It did so by parsing various aspects of the regulation of offshore petroleum activities in that country, particularly those most relevant to the publication title *Regulation of the Upstream Petroleum Sector: A Comparative Study of Licensing and Concession Systems.*

The recent parliamentary and regulatory reviews of the offshore spill regulation and spill liability regime certainly warrant special noting. Into the future, the reader is encouraged to 'watch this [regulatory] space', as little doubt exists that Canada will soon witness the introduction of legislation making significant changes to, and strengthening, the current regime.

# PART III

# Comparative petroleum regulation in developing petroleum provinces

# 8. The Brazilian concession system for petroleum extraction in Brazil

**Eduardo G. Pereira**

## 1. INTRODUCTION AND BACKGROUND

The oil and gas history of Brazil goes back even further than the well-known 21st century and the pre-salt discoveries. The first permissions were not granted for oil but rather to explore for and to supply gas in the city of Rio de Janeiro as early as the 1820s and 1830s.[1]

The 19th century was probably the most relevant century for the petroleum history of Brazil, with several individuals proving the existence of hydrocarbons in Brazil. These individuals received 'permissions' from the emperor throughout the 19th century and these continued to be granted in the 20th century by the relevant presidents, as the regency/imperial period finished with the first republic in 1890.[2] Thus, without the persistency of these individuals our petroleum history could be very different or even never exist as the first onshore commercial discovery was achieved by 1939 and its first offshore discovery later on by 1968.[3]

Although the Brazilian upstream sector began as an open market it changed after the creation of the Brazilian state petroleum company, Petrobras, in 1953. Since 1953 no private entity has been authorized to participate in the upstream sector in Brazil. Petrobras was the only company authorized to explore for and produce petroleum as it represented the state's interests. However, this was altered as a result of the economic turbulence that prevailed in the 1990s, and was stabilized through economic policies to prevent inflation and social gaps (i.e. poverty) and the privatization of certain industry sectors. In order to support the growth of the economy more energy was required for the

---

[1] See: Planalto, *O petroleo no Brasil* http://blog.planalto.gov.br/o-petroleo-no-brasil/, accessed 18 August 2014.

[2] Ibid.

[3] Ibid.

petroleum industry. This meant encouraging more exploration and production activities, resulting in the Brazilian government opening up the upstream sector to enable the direct participation of private parties.

Private access to Brazilian offshore petroleum resources was effected in 1995 by amendments to the Federal Constitution, thus ending Petrobras' monopoly on petroleum exploration and production and ushering in a new era for the Brazilian petroleum industry by allowing private investors to participate in the upstream sector.

The constitutional amendment was implemented by Federal Law 9478/97 ('Petroleum Law'). This law created a licensing and concession system (LCS) in Brazil, which remains the main guidance and framework for oil and gas companies in Brazil. However, the Petroleum Law was modified in 2010 when the Brazilian government adopted a production sharing agreement regime (PSAs) for selected areas with higher potential in the country (known as the pre-salt and strategic areas). The introduction of PSAs alongside the existing LCS created a hybrid regime that remains dominated by the LCS, which governs nearly all agreements between private parties and the government and the vast majority of petroleum activities in Brazil. Therefore, this chapter will address only the process and procedures concerning the LCS in Brazil.

## Legal Framework

Up until 1995 upstream activities were restricted to private investors, as the legal regime did not authorize direct private participation. Only Petrobras had the right to explore for and produce hydrocarbons. In 1995 the Brazilian Congress approved the amendment of Article 177 of the Federal Constitution (amendment number 9).

As a consequence of the amendment, Article 177 reads as follows:

> Paragraph I – The Union may contract with state-owned or with private enterprises for the execution of the activities provided for in items I through IV of this article, with due regard for the conditions set forth by law. (…)
>
> I. Prospecting and exploitation of deposits of petroleum and natural gas and of other fluid hydrocarbons;
> II. Refining of domestic or foreign petroleum;
> III. Import and export of the products and basic by-products resulting from the activities set forth in the preceding items;
> IV. Ocean transportation of crude petroleum of domestic origin or of basic petroleum by-products produced in the country, as well as pipeline transportation of crude petroleum, its by-products and natural gas of any origin;

V. Prospecting, mining, enrichment, reprocessing, industrialization and trading of nuclear mineral ores and minerals and their by-products.

This amendment did not remove the state monopoly over the natural resources below the ground, but removed restriction against private parties, as it allowed the participation of private parties in the upstream sector if the government decided to do so. This was a historical change as Brazil went back to its roots of allowing private parties to explore for and produce hydrocarbons, similar to the 19th and early 20th centuries.

However, the Brazilian congress took another two years to implement the constitutional change and pass the Petroleum Law. The Petroleum Law regulates the upstream sector in Brazil as analyzed below.

## 2. REGULATORY FRAMEWORK

The Petroleum Law implemented a new system of regulation for upstream activities. Firstly, it set the general principles and objectives for the national energy policy. Secondly, it created two new regulatory entities: the National Council for Energy Policy and the National Petroleum Authority (NPA). These two entities regulate and manage upstream activities. The Petroleum Law established the NPA. This entity is responsible for all the relevant attributions concerning the petroleum industry (including upstream and downstream) and biofuels. In terms of the upstream sector, the NPA is responsible for regulating the sector, awarding concession agreements and monitoring the performance of the concession agreements. Some of the relevant duties of the NPA include the following:

- Implementing the National Council for Energy Policy resolutions
- Encouraging a wider participation of players
- Defining exploration blocks
- Setting the minimum requirements for the bid rounds
- Drafting the concession agreements
- Managing the concession agreements

The NPA is the ultimate body managing and controlling upstream activities. However, it is not the sole authority in the petroleum sector, as it has to cooperate with the Ministry of Energy and the National Council for Energy Policy, neither of which are involved in the regulation of day-to-day activities. Rather, their role is that of policy, only intervening when national interests are threatened.

The Petroleum Law also implemented a number of other significant changes. It modified the petroleum regime to implement a concession regime. In this system, the concessionaire has the right to explore for and produce petroleum from the concession area in the same way as the US, Canada, UK and Australia. Thirdly, interested parties could obtain a concession agreement via licence rounds. Such licence rounds were required to be transparent and open to a wider audience through competitive bids. The NPA has adopted a standard form concession agreement in order to provide a fair and transparent agreement and to avoid direct negotiations.

## Licensing Regime

The licensing regime implemented under the Petroleum Law is applied equally to private entities and the Brazilian national oil company (NOC) Petrobras. Thus, when Petrobras seeks to acquire new blocks for exploration and production of petroleum, then it has to compete with any other national or international companies in accordance with the criteria defined in the relevant bidding process.[4]

In order to participate in a bid round the applicant must fulfil the basic requirements of qualifications. These requirements include technical expertise, financial resources and legal elements, thereby enabling the applicant to qualify to participate in the respective licence round.[5] The bidding process works with clear and pre-determined methods of qualification for the award of a concession agreement. This is a highly positive aspect of the bid round, as it adopts an objective method of evaluation; there is no bias in the award process or through government interference. Normally, Brazilian tender protocols divide the blocks to be offered into three different categories, as shown in Table 8.1.[6]

There are several dispositions regarding the procedures for punctuations in order to be qualified within each category, such as exploration or production experience, personnel experience, and environmental qualifications, among others.[7] The more points any company obtains, then the more likely it will be able to participate in any area offered in the country. The fewer points any participant obtains in this pre-qualification stage, then it is less likely to participate in more risky and complex areas.

---

4 Law No. 9478 of 6 August 1997 (Brazil) (Petroleum Law), Art. 61.

5 Ibid Arts 37–41.

6 www.brasil-rounds.gov.br/arquivos/editais/FT%20Protocol%20R10%20Final%20Version.pdf, accessed 18 August 2014.

7 Law No. 9478 of 6 August 1997 (Brazil) (Petroleum Law), Arts 37–41.

*Table 8.1 Operatorship categories and point allocations*

| Operator category | A | B | C |
|---|---|---|---|
| Block location(s) | Onshore or shallow waters or deep waters | Onshore or shallow waters | Onshore |
| Points needed | > or = to 81 | From 30 to 80 | From 1 to 29 |

Consequently, more experienced players are allowed to participate in offshore operations and less experienced players in onshore.

The common criteria adopted in the tender to award the concession are divided into the following:[8]

1. Signatory Bonus – the amount that the investor is willing to pay in order to acquire the concession agreement – 40 per cent;
2. Minimum Exploratory Program – the minimum work that the investor will perform in the licensed area – 40 per cent; and
3. Local Content – the amount of services and equipment that will be acquired within the Brazilian territory – 20 per cent.

The criteria are transparent and apply equally to all participants.[9] It is important to note that there is no intervention from the governmental authorities on the result of the bidding process, thereby establishing a highly democratic and transparent procedure to regulate the entrance of new players in the upstream sector.

**Concession agreements**

The licence or concession agreement is an instrument issued by the host government, which authorizes an oil and gas company to explore, develop and produce petroleum under certain terms.[10] The concession agreement gives the oil and gas company full ownership (proprietary

---

[8] www.brasil-rounds.gov.br/arquivos/editais/FT%20Protocol%20R10%20Final%20Version.pdf, accessed 18 August 2014.

[9] Law No. 9478 of August 6, 1997 (Brazil) (Petroleum Law), Art. 17.

[10] It is important to note here that it might be necessary to obtain consent from third parties such as land owners or salmon fishers (Greg Gordon, 'Petroleum Licensing' in John Paterson and Greg Gordon (eds.), *Oil and Gas Law: Current Practice and Emerging Trends* (DUP, Dundee 2007) 31.

rights) of the petroleum produced.[11] This is the major difference between a licence and a production sharing contract (PSC).

The Petroleum Law approved the concession regime as the new legal regime regulating the Brazilian upstream sector. However, the Petroleum Law did not provide the details of the concession agreement. Therefore, in 1999 the NPA approved the first model form Brazilian concession agreement (BCA), which has not been substantially altered since this time.

**Delimitation of the area**

The rights referred to any licence will have geographical limitations.[12] Although Brazil is the fifth largest country in the world, the size of each block offered in the bid rounds are relatively small, averaging 2,000km$^2$. From a European perspective 2,000km$^2$ for a concession area is fairly common, but for other countries, such as Central Africa, blocks of 100,000km$^2$ are not uncommon, even though the size of the country is comparable to Brazil.

**Duration of a licence**[13]

The duration of a licence is directly related to the phase of activity. Usually, a petroleum agreement is divided into exploration, development and production phases. Exploration primarily involves seismic and exploratory wells and typically does not take longer than 10 years. In Brazil, the exploratory phase is commonly divided into two periods, lasting between 4 and to 8 years. The development and production phases involve appraisal and production activities, which generally involve a period between 20 and 30 years. In Brazil these phases are commonly fixed at 27 years at the relevant tender protocols.

The BCA provides more assurances and comfort for investors, as the same agreement covers everything from the exploration to the production phase. As such, there is no need to acquire one agreement for exploration and another for development or production. Therefore, the duration of a licence is directly connected to the duration of the activity. If a licensee does not fulfil the requirements necessary to continue its activities after the exploration phase, the licence terminates sooner. However, if the licensee hits a commercial discovery, then the licence is valid until the end of the production phase for that relevant area.

[11] Ibid 31.

[12] Concession Agreement for the Exploration and Production of Oil and Natural Gas (2008) (10th Licensing Round), Annex I.

[13] Ibid clause 4.

**Relinquishment**

The major principle of a licence is to maximize the economic recovery of its natural resources.[14] If there is an area that is not being explored or produced it shall be returned (relinquished) to the government so that it can be used in another licensing round.[15] It is also possible to make an early relinquishment. However, the respective government must be notified, and the relinquishment cannot affect the minimal work program already agreed between the licensee and the government.[16]

The relinquishment terms under the BCA are fairly standard in the petroleum industry. However, the main difference with the BCA is the absence of an early relinquishment to move between different exploratory phases as long as the minimal work program is fully executed. In other words, in the BCA the licensee can maintain the whole licensed area until the end of the exploratory period. However, at the end of the exploratory period the licensee must move to the development stage or terminate the concession agreement. But in the case of a development stage, the licensee can only retain the area related to the relevant development plan and relinquish the remaining areas to the government so that is available for exploration again.

**Work program**

The minimum work provision stipulates the minimum work program the licensee promises to perform in the licensed area.[17] This clause is extremely relevant for the government because the property of any information obtained within the work executed will be transferred to the government. Therefore, the greater amount of exploration that occurs provides the government with greater amounts of data, thereby increasing the government's knowledge and understanding of the petroleum provinces, thereby increasing the government's chances of awarding future acreage at a higher price.

The BCA includes the minimum work program in several parts of the agreement.[18] The most relevant is attachment number II that requires the licensee to undertake the same amount of work as outlined in the bid for the minimum work program. However, if the licensee does not fully observe the minimum work requirement imposed in the licence, then the

---

14 Ibid clauses 1.2.24 and 13.12.

15 Ibid clause 5.

16 Ibid clause 18.9.

17 Ibid Annex II.

18 Ibid clauses 5, 10, 13, 15, 16, 22 and 27.

government can terminate the licence and execute any financial guarantees available.

In a concession agreement the government does not participate actively in the petroleum activities. Therefore, whenever a licensee makes a discovery of oil or gas it is necessary to notify the government within 72 hours, and then to evaluate the discovery in order to assess whether the discovery is economic.[19]

If the evaluation of the discovery proves it to be economic, the licensee is required to prepare and submit a plan to develop these resources to the government.[20] The government will verify whether or not the proposed plan adheres to the best oil and gas practices, and whether it can maximize the economic recovery of the relevant field.[21] After the relevant governmental approval, the licensee shall commence its development and then move into the production phase.[22]

**Ownership of petroleum**

Although the licensing regime provides proprietary rights to the licensee, it does not include the reserves *in situ*.[23] Until the delivery point, the host government maintains ownership of its natural resources. But after the delivery point (known as the measure point as per Petroleum Law), the licensee shall have the proprietary rights of the product and will be able to decide what it wants to do with the production.[24]

The BCA provides freedom for the licensee, as it does not establish any restriction to export production, nor determine preferential supply to the local market. However, there is an exception due to the event of a national emergency declared by the President of the Republic or the Congress.[25] Therefore, the general rule is that the licensee shall have the full ownership of its production upon reaching delivery point, but in specific situations such as a national emergency, there will be restrictions concerning the disposal of such a production.

---

19 Ibid clause 5.

20 Ibid clause 9.

21 Ibid clause 9.

22 Ibid clauses 2–11.

23 Law No. 9478 of 6 August 1997 (Brazil) (Petroleum Law), Art. 26.

24 Concession Agreement for the Exploration and Production of Oil and Natural Gas (2008) (10th Licensing Round), clauses 2–11.

25 Ibid clause 11.4.

**Unitization**

A reservoir could extend beyond the delimitation of a licensed area. Therefore, the idea behind the unitization provision is to avoid over-exploitation of the natural resources and to protect the longevity of the reservoir. This is a clear lesson learnt from the excessive drilling that happened throughout the 20th century in the USA. If the parties cannot reach a unitization agreement, the government will enforce the joint operations of all the parties involved (unitization) in order to avoid harmful practices of capture, that is, excessive drilling activities.[26] Before the enactment of the production sharing regime the licensees involved in affected areas were encouraged to reach an agreement and to cooperate and develop this area jointly, otherwise the NPA will find a solution on their behalf. This was the provision from the Petroleum Law.[27] The introduction of law no. 12.351 from 22 December 2012 (the Production Sharing Contract Law) revoked the unitization provision to avoid a conflict of laws.[28] This is an exception to the BCA where a licensee is subject to the production sharing law even though the activity is governed by a BCA. The main reason for this approach is to avoid the uncertainty to which rules will apply to unitization between a concession area and that of a production area. In doing so, even if the field straddles both concessions and PSC areas, all parties apply the Production Sharing Contract Law.

**Guarantee**

The government needs additional guarantees from the licensee in case of failure to comply with the relevant obligations established in the licence, such as the minimal exploration work, environmental and decommissioning liabilities.[29] Therefore, the licensee is required to issue: (i) a bank guarantee in accordance with the models provided in the BCA in order to secure the performance of the minimal work program; and (ii) insurance against the performance of the obligations of the concession agreement, such as damage to property, human life and the environment, among others.[30] The aim of this provision is to secure the government's interest that the licensee will comply with the obligations established in the BCA.

---

26 Law No. 9478 of 6 August 1997 (Brazil) (Petroleum Law), Art. 27.

27 Ibid.

28 Production Sharing Contract Law, Arts 33 to 41.

29 Ibid clauses 15 and 22.

30 Ibid Arts 15.1 and 22.1.

**Data**

Any government is keen to better understand the geological structure and basins of its licence area. Therefore, under the BCA the government, as a general rule is the owner of all of the relevant data and information acquired in the licensed area, such as seismic information and drilling information.[31] This information applies directly to all licensing rounds, as it is important to distinguish a frontier from a mature area, and to attract more players and better fiscal terms in the new licence rounds.[32] The BCA expressly requires the licensee to keep the NPA fully informed about the progress and results of its operations, and confirms that the ownership of any data acquired from the licensed area belongs to the governmental authorities.[33] This obligation might seem obvious, but several companies throughout the 20th century explored in less developed countries and regions, and took all the data back with them once the operations finished, thereby leaving the host government with little or no information about the activities in its country. Therefore, the data acquired in respect of the licence is a matter of national interest and for this reason shall be detained and controlled by the government.

**Local content requirements**[34]

This clause is more sensitive in developing countries than in developed countries, as it aims to protect national industries relating to services, suppliers and personnel required for the performance of the petroleum activities within the scope of the licence.[35] The BCA determines the minimum percentage of equipment and services that should be acquired from local companies, but it does not require the training and developing of local communities, which is fairly common in production sharing agreements. Therefore, the local content provisions aim to protect and benefit the national interest, as a large variety of sectors can benefit from the petroleum activities.

**Government take**[36]

The government take is the monetary consideration that the government receives from the activities carried out by the licensee. In a licensing

---

[31] Law No. 9478 of 6 August 1997 (Brazil) (Petroleum Law), Art. 22.

[32] Concession Agreement for the Exploration and Production of Oil and Natural Gas (2008) (10th Licensing Round), clause 17.

[33] Ibid clauses 17.1 and 17.2.

[34] Ibid clause 20.

[35] Ibid clause 20.

[36] Ibid clause 23.

regime, this is the most important issue since the ownership of the oil is fully transferred to the licensee after the measured point.[37] In other words, this provision is the compensation that the oil and gas companies have to pay to the relevant host government in order to obtain the rights to explore and produce oil and gas. The Petroleum Law and the BCA recognize the following government take:

(i) Bonus of signature;
(ii) Royalties;
(iii) Special participation;
(iv) Payment for the occupation and withholding of areas.

Bonus of signature is the cash consideration that the licensee pays in order to obtain the rights in the BCA. The precise amount depends on the bid offer during the licensing round, but the minimum will be fixed at the relevant tender protocol.

Royalties is another payment that the licensee shall pay the host government. This amount is a fraction of the relevant production. In the Brazilian case, it is usually between 5–10 per cent, with 10 per cent the norm.

Special participation is also known as windfall profits, but this designation can vary in different regions. The main idea of this provision is to prevent excessive profits by charging additional tax on significant production fields.[38] In the context of the Brazilian oil industry, only fields with a determined amount of production are governed by the scope of this provision, and the percentage of this special participation varies from 0–40 per cent of the production.

Payment for the occupation and reservation of licence areas is known as 'rental fees'. This provision is an additional annual income the government charges the licensee, with the fee usually calculated per square kilometre of the licensed area.[39] However, it is important to note that this specific fee is more expensive in development and production phases than in exploration phases, in order to encourage the best use of the given acreage.[40]

From a governmental perspective, royalties and special participation are the 'big numbers' as most of their compensation will come from

[37] Law No. 9478 of 6 August 1997 (Brazil) (Petroleum Law), Art. 26.
[38] Ibid Arts 45 to 52.
[39] Ibid Art. 45.
[40] Concession Agreement for the Exploration and Production of Oil and Natural Gas (2008) (10th Licensing Round), Annex V.

these routes. But they might not occur as royalty and special participation are based upon production. In this sense, bonus of signature and rental fees tends to be smaller amounts charged, regardless of the results of exploration in the licence area.

Although the Brazilian legal system makes a distinction between the ownership of the surface and the natural resources below the ground, the surface owner has a right to compensation for activities carried out within their land.[41] This compensation is based upon production, and might vary from 0.5 to 1 per cent of the production, according to the NPA determination. However, this amount might represent an excessive compensation in the case of a huge reservoir and may prove challenging in the development of onshore petroleum fields.

Therefore, the government take is the most sensitive issue in a licence regime because it will determine the level of profitability of a field. Companies need to make appropriate calculations that consider the fees, royalties and government take before bidding for the blocks during the licensing round, as there is little opportunity to negotiate given that the royalty, special participation and rental fees are fixed.

**Assignment[42]**

Assignment is the formal procedure of transfer of interest from the licensee to another party. There could be several restrictions within this clause because the government wants to maintain the overall control of parties that are ultimately responsible for the performance of the obligations established in the licence. Thus, it is common for a licence to require government approval for assignment of the interest, including assignment to an affiliate company.

The BCA is clear and rigid about the governmental approval of assignment under the Petroleum Law:

> 28.1 With the prior agreement of the ANP [Agencia Nacional do Petroleum or National Petroleum Agency], in accordance with Article 29 of the Petroleum Law, the Blocks within the Concession Area may be assigned, according to the provisions of this Clause Twenty-Eight, which defines the conditions to be observed by the assignor and its assignees.
>
> 28.2 All and any transfer of ownership under this contract, even in the event of merger, spin-off, and incorporation of a company belonging to the Concessionaire shall be considered an Assignment.

---

41 Law No. 9478 of August 6, 1997 (Brazil) (Petroleum Law), Art. 52.

42 Concession Agreement for the Exploration and Production of Oil and Natural Gas (2008) (10th Licensing Round) clause 28.

This provision is deliberately vague as the wording here might include any kind of assignment or corporate restructure, but it is not clear if it applies to financing the operations through an asset valuation. Consequently, a licensee must take precautionary action before executing any sort of assignment, as government approval might be required.

**Force majeure**[43]

Force majeure is an event that it is not possible for any of the parties to predict and/or which is out of their control (e.g. hurricane, flood, civil war). Whenever there is a case for force majeure, there shall be no penalty or prejudice to the parties for the lack of execution of a contractual obligation.[44] However, payments are usually excluded from any allegation of force majeure. The rationale from the government is that fiscal obligations are fully described since day one of the contract and cannot be postponed under any circumstances. The BCA expressly provides that 'Under no circumstances shall the situation described in paragraph 32.1 exempt the Concessionaire from paying the Governmental Participations'.[45]

It was not necessary to provide for a force majeure clause in the BCA as the Brazilian Civil Code (BCC) already deals with this matter. However, it is important to establish the details involving the force majeure (e.g. notification, best efforts to continue the operations as soon as possible), as the BCC only provides general regulation. Therefore, force majeure is essential for petroleum companies, since unforeseeable events might occur frequently throughout the work, either because the weather does not allow the completion of the activities in time (e.g. the Hurricane Katrina) or because civil unrest might create difficulties for continuing work, like in Somalia or Libya.

**Confidentiality**[46]

This clause is directly connected to the data that is submitted in accordance with the conditions of the BCA. All the relevant information and data acquired in the licence shall not be disposed of by the licensee to third parties without governmental approval, except for: (i) affiliated companies; (ii) consultants; (iii) compliance with laws and court orders; (iv) prospective assignees; and (v) financial assignees. In turn, the

43 Ibid clause 32.
44 Ibid clause 32.
45 Ibid clause 32.
46 Ibid clause 33.

government is required not to disclose confidential information, except as required by law.

**Applicable law and dispute resolution**[47]
In the case of any conflict involving the licence, it shall be resolved according to the provisions of the licence. The applicable law and dispute resolution law is commonly the same law of the host government since the government is not keen to delegate sovereignty over decisions concerning its natural resources to a foreign authority. In Brazil, the BCA secures the national interest by stipulating that Brazilian law governs the concession agreement. However, the government also allows for international investors by allowing disputes to be resolved through arbitration in Rio de Janeiro in accordance with the International Chamber of Commerce rules.[48]

Therefore, it is possible to suggest that the Brazilian licence regime is fairly standard. It is easily comparable with the rules governing the United Kingdom, Australia, and Norway, among others. This is probably another reason for the success of the licence regime, as the legal framework was very clear, transparent and stable.

**Evaluation of the licensing regime**
There is little doubt about the success of the Brazilian licence regime. Several international and national companies are presently engaged in concession agreements in order to explore for, and produce, hydrocarbons in Brazil. The bid rounds of the BCA play a key role in the effectiveness of the licensing regime, as the best bid shall be the winner regardless of the nationality and the size of the company. Independent companies (e.g. Petra Energia and OGX) have competed and won licences against giant companies such as Petrobras and Shell. Thus, the bid round provides for transparent and fair procedures to encourage a large variety of investors in the upstream sector. Although the licence regime represented a successful scheme, it is not free from flaws. The most relevant flaw relates to the encouragement of exploration in new frontiers, since the vast majority of the Brazilian territory is still unexplored.

Brazil is a large continental country and the majority of its territory is waiting to be explored. Flexible and fair regulations are very important in

---

47 Ibid clause 31.

48 www.brasil-rounds.gov.br/round10/ingles_r10/edital.asp, accessed 18 August 2014.

order to attract new investors. Thus, it is reasonable to expect that such a large country would have differing regional geology requiring different approaches to exploration and production. However, there is only one model concession agreement, applicable for all areas, be it onshore, offshore, in mature or frontier areas. There is only one exception to this rule.[49] The marginal fields (i.e. mature areas) received a distinct treatment from the Brazilian government in order to adjust to the particularities required for such mature areas even though it is not very different from the model BCA. In addition, the Brazilian government can issue an extraordinary bid round for mature areas. However, this is not a common procedure and has only occurred twice during the course of the ten bid rounds.[50]

However, there are no specific procedures or treatment for frontier areas. This is probably the most significant problem of the current licensing regime. A frontier area is by definition an area which has extremely adverse conditions for the potential licensee, either because there is little or no data available, or because the location of the area is complex and difficult to operate, such as in the Amazon jungle or in ultra-deep water.

Commonly host governments provide better terms for investors to come and explore frontier areas. The most critical issue for a frontier area is the amount of time required to complete the minimal work program. Usually frontier areas cover huge areas, requiring more time for exploration and analysis of the data collected. Another critical issue is the fiscal term. Investors would like to receive better fiscal terms in order to justify the higher risks involved.

The United Kingdom is a good example. Their licence regime makes distinction between onshore, offshore, mature and frontier areas. The most relevant distinction between them is the duration of the exploration period and the fiscal terms. Generally speaking, a Brazilian exploration period lasts between 4 to 8 years. But for frontier areas the UK regimes can go up to 15 years and the fiscal terms are reduced to attract more investors. The Brazilian licence regime does not make such a distinction. If a company has a block in the middle of the jungle, next to the developed areas or close to international waters, they will be allocated a similar exploration period and fiscal terms. This is a significant problem

---

49 It is important to note that it is extremely rare for such models to exist. So far such provisions were made only on the 7th (2005) and 9th rounds (2008).

50 For further information see: www.brasil-rounds.gov.br/round1am/index.asp and www.brasil-rounds.gov.br/round2am/index.asp, both accessed 18th August 2014.

for operators in Brazil, as it might be extremely difficult to fulfil the work commitment in a short period of time and there might not be a fair correlation between risk and reward.

The Brazilian government should be keen to encourage more activities by allowing more time to fulfil the work commitment. This is based on international best practices, as certain areas require a particular approach in order to face its challenges. On the other hand, the Brazilian government should require guarantees that the incremental period will be compensated with further work and timely report of progress; otherwise these areas should be relinquished. In the end, it is in the best interests of the government to help the relevant players to perform their activities successfully, rather than being unreasonable and unsupportive.

More recently the Brazilian government offered areas for unconventional and more frontier areas. Minor adjustments were made in the tender protocol and concession agreement (e.g. a slightly longer exploration period), but still fail to fully recognize the challenges required in an unconventional or frontier area.

## 3. REGULATION OF PETROLEUM ACTIVITIES: SAFETY, ENVIRONMENT AND WELL INTEGRITY

### Regulation of the Environment

The petroleum licence usually does not involve or include environmental authorizations and permissions to explore or produce oil. For this reason, the licensee shall obtain, by its own cost and risk, all the necessary authorizations and permissions. This is the main environmental control in day-to-day operations as any activity with potential hazard to the environment can only be conducted with an adequate study and an environmental licence for such activity. In Brazil this is likely to be far more complicated onshore operations, as these permissions involve federal, regional and municipal authorities.[51] Offshore operation should be more straightforward as it involves only the federal authorities.

[51] 'The Brazilian Institute of the Environment (Instituto Brasileiro do Meio Ambiente or IBAMA) is the agency responsible for implementing environmental policy at federal level. IBAMA was created by Federal Law No. 7,735, of February 22 1989, which established among its functions exercising environmental police power; executing actions of national policies for the environment, referring to federal attributions related to environmental licensing, control of

Environmental issues are extremely important. The petroleum industry works under pressure and in hostile environments. Although higher standards and the best practices are broadly required by the BCA[52] and the Petroleum Law,[53] it is not possible to ignore the existence of environmental risks and the liabilities attached. The tendency is to increase environmental standards as the licensee's reputation is at stake (e.g. the BP spill in the Gulf of Mexico), but the BCA guarantees the Brazilian governmental position as it establishes that the licensee shall be liable for all the environmental damage and harm related to its activities regardless of the cause or guilt for such damage.[54]

Under the BCA the licensee must take all the necessary efforts to avoid any damage to the environment.[55] But if any environmental damage happens, then the licensee must notify the governmental authorities and be fully and jointly liable for all and any compensation, regardless of the cause of the damage.[56]

## Regulation of Safety and Well Integrity

Safety is broadly regulated in the Petroleum Law and the BCA. The Petroleum Law does not provide details concerning safety regulations but only general guidance expecting higher standards on safety practices by

---

environmental quality, authorisation for the use of natural resources and environmental inspection, monitoring and control, observing the guidelines issued by the Ministry of the Environment; and executive supplementary actions under the competency of the Union, according to the current environmental legislation. Brazilian states have their own environmental agencies and have the power to analyze proposed activities and issue environmental licences, and also to impose supervision, restrictions and penalties, except in areas where IBAMA enjoys exclusive jurisdiction. Each Brazilian municipality may also legislate on local environmental rules, provided the issue is not under federal or state legislative competency.' Luis Pacheco, 'Environmental Permission and Regulation' in Eduardo G. Pereira, *Brazilian Midstream and Downstream Oil and Gas Law and Regulation* (Globe Law and Business, London 2014).

[52] Concession Agreement for the Exploration and Production of Oil and Natural Gas (2008) (10th Licensing Round), clauses 13, 19, 21 and 31.

[53] Law No. 9478 of August 6, 1997 (Brazil) (Petroleum Law), Arts 1 (IV), 8 (IX) and 44 (I).

[54] Concession Agreement for the Exploration and Production of Oil and Natural Gas (2008) (10th Licensing Round), clause 21.

[55] Ibid clause 21.

[56] Ibid clause 21.5.

any licensee.[57] Additional regulation may be issued by the compentent authorities (especially the NPA). The BCA provides further details. Firstly the licensee is expected to conduct the operations diligently and safely.[58] Secondly the licensee shall observe the labour laws regarding safety at work.[59] Thirdly the licensee shall adopt the best practices in the industry to avoid any damage to the reservoir, other natural resources and the environment. In short, the licensee shall adopt industry best practices and meet legislative requirements regarding safety and environmental standards.

## 4. CONCLUSION

Petrobras has done an excellent job with deep-water exploration throughout the 20th century. However, onshore basins were almost forgotten, and the level of production was not enough to avoid importation of petroleum during that period. In order to increase production in a manner to reach self-sufficiency, or even to export oil and gas, it was necessary to attract more participants. The Brazilian government understood the necessity of opening its upstream industry before the end of the 1990s.

Constitutional amendment number 9 and the Petroleum Law provided the necessary changes to open the upstream sector to private parties. Although some adjustments might be required to increase exploration throughout the country, it is possible to affirm that the licence regime has been very successful from its creation until the present. This regime allowed private and public companies to compete in the licence rounds on equitable rules and procedures. Transparency and fairness of the procedures involved are highly positive in terms of attracting investors.

The BCA has maintained most of the same terms over the last ten years, which provides consistency and assurance for investors. However, some economic changes have occurred in the local content requirement, as illustrated in Figure 8.1.

Higher and higher rates of local content have been required in the tender protocols in the past decade. This represents a real challenge for operations in Brazil, since the supply chain is not completely ready to cope with such demand. Several international companies and services

---

[57] Law No. 9478 of August 6, 1997 (Brazil) (Petroleum Law), Arts 31, 44 (I), 53, 56, 68-A.

[58] Concession Agreement for the Exploration and Production of Oil and Natural Gas (2008) (10th Licensing Round), clause 13.5.

[59] Ibid clauses 19.2.

**Aggregated Results**

| Local Content average | Round 1 | Round 2 | Round 3 | Round 4 | Round 5 | Round 6 | Round 7 | Round 9 |
|---|---|---|---|---|---|---|---|---|
| | 1999 | 2000 | 2001 | 2002 | 2003 | 2004 | 2005 | 2007 |
| *Exploration Stage* | 25% | 42% | 28% | 39% | 79% | 86% | 74% | 69% |
| *Development Stage* | 27% | 48% | 40% | 54% | 86% | 89% | 81% | 77% |

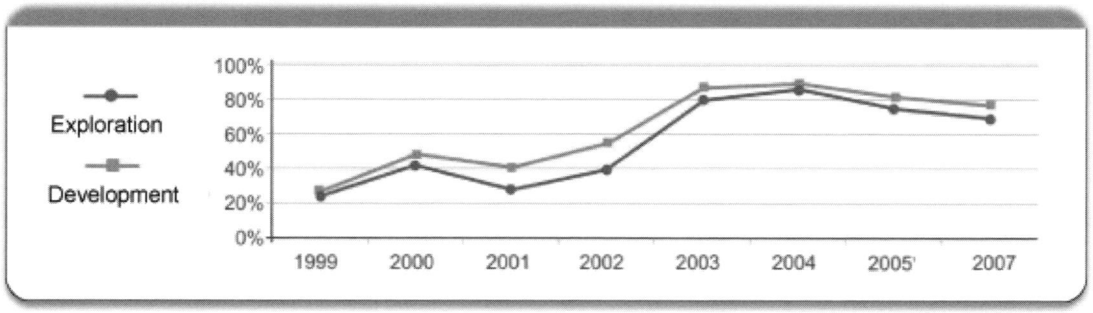


*Note*: The average Local Content during the exploration stage was of 25% in Round 1. In Round 9, it increased to 69%. During the development stage, Local Content moved from 27% to 77% in the respective Rounds.

*Source*: http://www.brasil-rounds.gov.br/portuges_topo/resumo_geral.asp

*Figure 8.1 Changes to local content requirements in Brazil*

providers are based in Brazil (e.g. Halliburton, SLB) but still a lot of work is yet to be done to provide the necessary manpower and related products and machinery to attend this demand.

Although some changes have occurred in the last ten years, it is possible to state that the regime in the upstream sector in Brazil has been stable. No substantial changes occurred (especially in the fiscal terms and the type of agreements in place). The production sharing agreement will be applicable only in specific areas defined in the law, and the remaining parts of the country continue to be governed by the concession agreement.

# 9. The legal regime for petroleum activities in Nigeria

**Simon Warikiyei Amaduobogha**

## 1. INTRODUCTION

Nigeria is the largest producer of oil in Africa, with reserves of 36–37.2 billion barrels of proven crude oil as at January 2013. Daily production varied between 2.1 million and 2.3 million barrels per day (bpd) at the close of the third quarter of 2013.[1] There are about 500 oilfields in the Niger Delta of Nigeria. Over 55 per cent of these are onshore, while the remaining are in the shallow waters (less than 500 metres). Of these fields, 193 are currently producing, while 23 have either been abandoned or closed down. About 5,284 wells (approximately 10 per cent discovery wells) have been drilled, mainly in the Niger Delta region of Nigeria. The average production cost per barrel of crude is $3.50 onshore and $5.00 offshore.[2] Nigeria's crudes are of the finest quality, being sweet light crude, and therefore in high demand.[3]

Petroleum accounts for 40 per cent of the country's gross domestic product (GDP), 80 per cent of government revenue and 95 per cent of

[1] Dayo Oketola, 'Missing Oil Revenue, Crude Theft Threaten Nigerian Economy', *Punch*, 31 December 2013, available at www.punchng.com/special-feature/missing-oil-revenue-crude-theft-threaten-nigerian-economy/, last visited 11 March 2014.

[2] NAPIMS, 'Crude Oil Reserves/Production', available at http://dev.nnpc group.com/napims/Crude.aspx, last visited 3 March 2014.

[3] Nigeria produces only high value, low sulphur content, light crude oils such as Antan Blend, Bonny Light, Bonny Medium, Brass Blend, Escravos Light, Forcados Blend, IMA, Odudu Blend, Pennington Light, Qua-Iboe Light and Ukpokiti; see NNPC, Oil Production, available at http://nnpcgroup.com/NNPCBusiness/UpstreamVentures/OilProduction.aspx, last visited 14 March 2014.

foreign exchange earnings.[4] Concessions, until recently, were discretionarily allocated, with the industry gaining a reputation of being less than open and transparent. The industry, although slow to change, has moved from the granting of large concessions to the award of licenses for defined acreage in licensing rounds. This chapter examines licensing in the petroleum industry in Nigeria from inception to date, and also considers the future direction of the regulatory regime.

## 2. HISTORY AND OVERVIEW OF THE NIGERIAN PETROLEUM INDUSTRY

The first piece of petroleum legislation was the Petroleum Ordinance of 1889, which was followed by the Mineral Regulation (Oil) Ordinance of 1907 (Mineral Oil Ordinance), both of which laid down a basic framework for the development of petroleum and other natural resources. The Mineral Oil Ordinance stipulated, inter alia, that only British subjects or companies controlled by British subjects would be entitled to explore for petroleum resources in Nigeria. Ironically, the first company ever to engage in petroleum exploration in Nigeria was the German Bitumen Company, in 1908 around Okitipupa in present day Ondo State. This exploration effort was unsuccessful and the company terminated its operations following the outbreak of the First World War.[5]

In 1914, after the amalgamation of the Northern and Southern Protectorates into one administrative unit called Nigeria, the Mineral Oils Ordinance was enacted by the colonial administration. In 1938, Shell D'Arcy Petroleum Development Company was granted a concession[6] covering the entire mainland of Nigeria. In its monopoly status, Shell was able to explore at its leisure until 1962, when 15,000 square miles of the original concession area was obtained by Shell. Although exploration activities were interrupted in Nigeria during the Second World War, Shell D'Arcy (later Shell-BP Petroleum Development Company of Nigeria

---

[4] Alike, Ejiofor, 'Nigeria Produces 2.5mbpd of Crude Oil', Thisday live, 23 July 2013, available at www.thisdaylive.com/articles/nigeria-produces-2-5mbpd-of-crude-oil/154098/, last visited on 13 March 2014.

[5] For detailed early history of oil production in Nigeria, see Schatzl, Ludwig, *Petroleum in Nigeria* (Nigerian Institute of Social and Economic Research (NISER), lbadan 1969), especially pp. 1–4.

[6] NNPC, 'History of the Nigerian Petroleum Industry', available at www.nnpcgroup.com/NNPCBusiness/BusinessInformation/OilGasinNigeria/IndustryHistory.aspx, last visited 21 February 2014.

Limited) subsequently made the historic discovery of petroleum in commercial quantities in 1956 at Oloibiri town in present day Ogbia Local Government Area of Bayelsa State (Niger Delta). Production was originally maintained at 5,100 bpd, peaking at 2.44 million bpd in 1979.

In 1959, the nationwide concession right given to Shell was revised and the Nigerian government granted concessions to other companies of different nationalities such as Mobil, Gulf, Agip, Elf, Tenneco and Amoseas (Texaco/Chevron) in both offshore and onshore areas of Nigeria. In 1961, Texaco Overseas started operations in Nigeria while Elf and Agip commenced their operations in 1962. Elf discovered the Obagi field and Gulf started production at the Ubata gas field in 1963. Phillips Oil Company started operations in Bendel State in 1965 and Agip had another find at Ebocha. In 1966 Elf started production in Rivers State with a 12,000 bpd output while Phillips drilled its first well (dry) at Osari–I in 1967, before its first oil discovery at Gilli-Gilli–I. The year 1968 saw the commissioning of Gulf's terminal at Escravos and the formation of Mobil Producing Nigeria Limited. Then 1970 and 1971 were very eventful years in the Nigerian oil industry as Mobil started production from four wells at Idoho Field and the Department of Petroleum Resources (DPR) establishment in 1970 while Shell's Forcados Terminal and Mobil's Qua Iboe terminal was commissioned in 1971.

Since the first laws a host of legislation has been passed, regulating the activities of the oil industry.[7] In 1969, the Petroleum Decree No. 51[8] was promulgated, which repealed the Mineral Oils Ordinance of 1914 and reduced the term of oil mining leases from 30 years (onshore) and 40 years (offshore) to 20 years. Oil mining leases granted prior to 1969 were, however, not affected. The Petroleum (Drilling and Production) Regulations 1969 were also enacted.

In 1971, Nigeria joined OPEC, although OPEC resolutions and policy statements had previously influenced Nigerian oil policies. For instance, in 1969 a report of the Federal Ministry of Finance[9] proposed that

---

[7] One of such laws is the Petroleum Profits Tax Act of 1959 (as amended), imposes tax and royalty rates of 85 per cent and 20 per cent respectively.

[8] Now Petroleum Act, Chapter P10 Laws of the Federation of Nigeria (LFN) 2004.

[9] Federal Ministry of Finance, *Report of the Fact-Finding Mission on Petroleum Taxation: Problems Affecting Petroleum Revenue and Miscellaneous Matters on the Petroleum Industry* (Federal Ministry of Finance, Lagos) (1969).

government participate in exploration and production companies. The report emphasized the need for an efficient renegotiation of agreements with oil companies to give the state an undivided interest, and the need for a state oil corporation to administer this interest. The report recommended an organized, systematic strategy for participation, which was approved by the federal government in referring to the 'active participation in mining operations'.[10] The Nigerian National Oil Corporation (NNOC) was thus established in 1971[11] to engage in prospecting for, mining and marketing petroleum and participate in all other activities in the petroleum industry. NNOC was an offshoot of the Federal Ministry of Mines and Power. There were conflicts between NNOC and the Ministry, which seemed reluctant to allow the NNOC to develop the capabilities necessary to accomplish its duties. In the tussle, the NNOC's chief executive post remained vacant while the Ministry officials retained ultimate control and 'prevented the pursuit of active government-led technology transfer in the area of exploitation and production'.[12] This power tussle rendered NNOC ineffective in practice, until 1977 when the Oil Ministry was merged with NNOC to form the Nigerian National Petroleum Corporation (NNPC).[13]

## Existing Joint Ventures in Nigeria[14]

A number of participation agreements through joint ventures (JVs) were secured in the 1970s and 1980s, and there are now six existing JVs in Nigeria in which the NNPC is a participant. One, a JV operated by Shell, accounts for more than 40 per cent of Nigeria's total oil production (899,000 bpd in 1997) from more than 80 oilfields. In the JV, NNPC has a 55 per cent holding, Shell holds 30 per cent, Elf has 10 per cent and

---

See also Turner, Terisa, 'Nigeria: Imperialism, Oil Technology and the Comparador State', in Petter Nore-Turner, Petter and Terisa Turner, *Oil and Class Struggle* (Zed Press, London 1980) 208, at p. 221.

[10] Turner, T., supra note 9, at p. 209.

[11] By Decree No. 18 of April 1971.

[12] Omoregbe, Yinka, 'The Legal Framework for the Production of Petroleum in Nigeria', 5 J. Energy & Nat. Resources L. 273, 276 (1987).

[13] Through the instrumentality of the Nigerian National Petroleum Corporation Decree No. 33 of 1977. For a detailed analysis on the establishment and organization of NNPC, see Olisa, M.M., *Nigerian Petroleum Law and Practice* (Fountain Books, Ibadan 1987) p. 181.

[14] This is based on online information from the NNPC website under the head 'Joint Venture Operations' available at www.nnpcgroup.com/NNPC Business/UpstreamVentures.aspx, last visited on 27 February 2014.

Agip has 5 per cent. This JV operates largely onshore on dry land and in the mangrove swamp. Second is a JV between NNPC with a 60 per cent interest and Chevron with a 40 per cent interest that has in the past been the second largest producer (approximately 400,000 bpd). Chevron is the operator of the JV with fields located in the Warri region west of the Niger River and offshore in shallow water. Its target is to increase production to 600,000 bpd. Third is a JV between NNPC (60 per cent) and Mobil Producing Nigeria Unlimited (40 per cent) which operates in the shallow water off Akwa Ibom State of Nigeria in the south eastern delta with an average production of 632,000 bpd in 1997, making it the second largest producer, as against 543,000 bpd in 1996. Mobil as operator also holds a 50 per cent interest in a production sharing contract for a deep water block further offshore with a production target of 900,000 bpd by 2000. The fourth JV is operated by Agip with a 20 per cent interest, the NNPC (60 per cent) and Phillips Petroleum (20 per cent). This JV produces 150,000 bpd mostly from small onshore fields. A JV operated by Texaco is fifth with production of about 60,000 bpd from five offshore fields. The NNPC has a 60 per cent interest holding, Texaco Overseas Petroleum Company holds a 20 per cent interest and Chevron's interest is 20 per cent. The last JV is between NNPC with a 60 per cent interest and Elf with a 40 per cent interest. It produced approximately 125,000 bpd in 1997 from fields both onshore and offshore.

The 21st century saw the beginning of a new dawn in the oil and gas industry in Nigeria with the introduction of the open bidding process for the grant of petroleum licenses and leases. Bid rounds have so far been conducted in 2000, 2005 and 2007[15] and the DPR has commenced the 2013 marginal field licensing round.

## 3. REGULATION OF PETROLEUM ACTIVITIES IN NIGERIA

Nigeria has a robust regulatory regime for the petroleum industry. There are statutes, subsidiary legislation and an array of institutions undertaking various responsibilities within the petroleum industry. This section deals with the extant law and regulatory institutions.

---

[15] DPR, 'Road-Show on Marginal Fields Flags off in Lagos', online at http://dprnigeria.org.ng/road-show-on-marginal-fields-flags-off-in-lagos/, last visited on 7 March 2014; E&P, 'Nigeria Sets Strict Guidelines in Licensing Round', online at www.epmag.com/Technology-Regulations/Nigeria-Sets-Strict-Guidelines-Licensing-Round_126916 last visited on 7 March 2014.

## Legislation

The main legislation guiding the petroleum industry in Nigeria, particularly the oil exploration and production activities, is: the Petroleum Act 1969 and the Petroleum (Amendment) Decree 1996; the Petroleum (Drilling and Production) Regulations 1969, with amendments in 1973, 1979, 1995 and 1996; the Mineral Oils (Safety) Regulations 1963; the Petroleum Profits Tax Act 1959, with amendments in 1967, 1970, 1973 and 1979; the Oil Pipeline Act 1956, as amended in 1965; the Nigerian National Petroleum Corporation Decree 1977; the Associated Gas Re-Injection Decree 1979, as amended in 1985; the Production Sharing Decree 1999; the Deep Water Block Allocations to Companies (Back-in-Rights) Regulations 2003; and the Oil Prospecting Licenses (Conversion to Oil Mining Leases etc.) Regulations 2003.

Chief among the array of legislation implemented is the Petroleum Act of 1969, which vests ownership and control of all petroleum in, under or upon any land in Nigeria in the federal government. It provides for Nigerian citizens or companies incorporated in Nigeria to be granted a license (oil exploration license – OEL) to explore for petroleum,[16] a license to prospect for petroleum (oil prospecting license – OPL)[17] and a petroleum lease (oil mining lease – OML) to search for, win, work, carry away and dispose of petroleum.[18]

The duration and maximum areas of each license/lease are specified within the Act and the regulations. The OML, which is the largest of the rights available under this Act, may be granted only to the holder of an OPL who has satisfied all the conditions imposed on the license or otherwise imposed on them by the Act and discovered oil in commercial quantities.[19] Oil is deemed to have been discovered in commercial quantities if the DPR 'upon evidence adduced by the licensee is satisfied that the licensee is capable of producing at least 10,000 bbl/d of crude oil from the licensed area'.[20] The Petroleum Act also provides for assignment, farm-out, termination, surrender/relinquishment and revocation of licenses and leases.[21]

The prevailing government policy permits the involvement of private and public interests, whether indigenous or foreign, in the exploration,

---

16 Paras 1–4 of first schedule of the Petroleum Act 1969.
17 Id., paras 5–7.
18 Id., paras 8–13.
19 Id., para 8.
20 Id., para 9.
21 Id., paras14–18 and 24–25.

appraisal and development of petroleum resources. Recent modifications to the criteria for the acquisition of concessions in Nigeria now allow for the grant of OPL and OML, where appropriate, to interested parties seeking sole operations.[22] The essence of sole operations is to: expand the scope of participation in Nigeria's oil industry and diversify sources of investment and inflow of funds to grow the Nigerian economy; increase the oil and gas reserves base through aggressive exploration; promote indigenous participation in the oil industry to foster technological transfer, local goods and services utilization; and ensure indigene employees' skills and competencies acquisition. Also, the policy serves to institutionalize the conduct of periodic, open competitive bidding for acreages and eliminate hawking of these blocks by licensees to third parties in the open market.[23]

## Administration of the Nigerian Petroleum Industry

The Ministry of Petroleum Resources (MPR) is responsible for all petroleum matters, including the coordination of all related policies, manages the petroleum resources and administers the relevant laws and regulations. In addition, the MPR supervizes the activities of the executive organs of the public sector of the oil industry, the government's interests and investments in the industry, as well as co-operating with other relevant government ministries, departments and agencies in an advisory capacity. The Department of Petroleum Resources (DPR) in the MPR is responsible for regulatory control of the industry, revenues, royalties and fiscal provisions and the licensing of operations under the Petroleum Act and technical issues of a policy nature.

Generally, the DPR initiates petroleum policies and advises on policy matters affecting the management of petroleum resources, ensures compliance with petroleum laws and regulations, and regulates and monitors the activities of companies operating within the oil industry. The existing government policy permits the involvement of private and public interests in the exploration and development of petroleum resources, whether indigenous or foreign.

The National Petroleum Investment Management Services (NAPIMS) in the Exploration and Production (E&P) Directorate is the upstream arm of NNPC that oversees federal investment in the joint venture companies

---

[22] The Petroleum (Amendment) Decree No. 23 of 1996 was enacted to encourage such sole risk operations.

[23] Ministry of Petroleum Resources, *Guidance Information Notes for Bidders in the 2007 Licensing Round* (February 2007).

(JVCs), production sharing companies (PSCs) and services contract companies (SCs). NAPIMS was set up to earn margins arising from investments in the JVCs, PSCs, SCs, with the international oil companies (IOCs). It also protected the nation's strategic interests in the JVCs and engages in frontier exploration services in basins where the multinationals hesitate to venture, such as the Chad Basin. This is with a view to optimizing the benefits accruing to the federal government from investments in the upstream sector of the petroleum industry. NAPIMS also plays a role in the promotion of local content input in engineering and construction supplies and materials utilization through in-country technological capability, as well as enhancing the 'Nigerianization' of the industry and facilitating technology transfer. In the interests of all players in the industry, NAPIMS promotes co-operation in communities of petroleum producing areas and also ensures that environmental protection standards are strictly maintained.[24] However, the enforcement of Nigerian content regulations in the upstream petroleum industry is now the function of the Nigerian Content Monitoring Board, established under section 69 of the Nigerian Oil and Gas Industry Content Development Act 2010 (NOGICD Act).[25]

## 4. GRANT OF ACCESS TO PETROLEUM FOR PETROLEUM ACTIVITIES – PRODUCTION SHARING CONTRACT (PSC)

The PSC represents one of the modes by which an IOC can, through agreement rather than formal licensing rounds, acquire rights for exploration and production of oil. This form of contract, first pioneered in Indonesia in 1967, is a legal instrument for defining the relationship between host countries (HCs) and IOCs.[26] PSCs are the result of efforts by HCs to modify the modes of acquisition of oil rights by the IOCs and, above all, to find an alternative to the traditional concession agreement which would enable the HC to exercise more control over both petroleum

[24] Information on NAPIMS available online at http://dev.nnpcgroup.com/napims/AboutUs/NAPIMSProfiles.aspx, last visited 3 March 2014.

[25] See sections 69 and 70 of the NOGICD Act.

[26] Blinn, K.W., Duval, C., Le Leuch, H. and Pertuzio, A., *International Petroleum Exploration and Exploitation Agreements: Legal, Economic and Policy Aspects* (Barrows, New York 1986) p. 69; see also Atsegbua, Lawrence, 'Acquisition of Oil Rights under Contractual Joint Ventures in Nigeria', 37 J. Afr. L. 10, 13 (1993).

operations and the ownership of production.[27] Cost recovery provision, which is also a feature of the standard PSC, is particularly attractive to IOCs as it enables them to obtain a quick pay-out of 40 per cent of the oil towards reimbursement of the exploration, development and other operating costs they have incurred.[28] On the other hand, an HC might regard the 40 per cent cost recovery as excessive. While it has evolved to take into account the particular circumstances of the ventures concerned and to keep pace with the development of relationships between HCs and IOCs, it has nevertheless retained its basic features.

## Production Sharing Contract in Nigeria

The PSC used in Nigeria differs in some aspects from the standard PSC pioneered in Indonesia.[29] An early example of a PSC in Nigeria is that between the NNPC and Ashland Oil (Nigeria) Company (AON), the Nigerian subsidiary of an American company. Although the contract was signed on 12 June 1973, its effective date was 1979 and it extends for a duration of 20 years. No explanation was given as to why the effective date was 1979 instead of 1973.[30]

The Nigerian experience with PSCs has not been successful in enabling the Nigerian state to achieve its objectives. The NNOC/Ashland Petroleum Corporation PSC contract was for 20 years, and extended over two oil prospecting licenses. Under the terms of the contract, Ashland was the designated operator and title to petroleum passes to each party at the wellhead. Ashland paid for all equipment necessary for conducting operations in Nigeria and this equipment becomes the property of NNPC upon arrival in Nigeria save for those leased or said to belong to third parties. This is regarded as an operating cost, as are the production bonuses payable by Ashland to NNPC. Ashland may also include 2 per cent of the actual operating costs as overhead charges for the calculation of the total operating costs.[31] All operating costs, including rents and royalties paid by Ashland and interest on funds borrowed to conduct operations were completely recoverable out of the proceeds of sale of a maximum of the first 40 per cent of available crude oil. Then 55 per cent

---

27 Id., p. 3.

28 Hossain, Kamal, *Law and Policy in Petroleum Development* (Frances Printer, London 1979) p. 148.

29 Atsegbua, L., supra note 26, p. 14.

30 Id.; see also Basic Oil Laws and Concession Contracts South and Central Africa (Original Texts), Supplement Nos. 51–55 (1978–1979).

31 Omoregbe, Y., supra note 12, p. 280.

of the remaining crude oil was allocated to Ashland and applied towards Petroleum Profits Tax. Any additional amounts payable were to be paid by the respective parties in proportion to their 'participating interest shares'. The rest of the crude oil was shared between Ashland and NNPC according to their participating interests which are 35/65 per cent until production exceeds 50,000 barrels a day in which case the shares are 30/70 per cent in favour of NNPC.[32] Ashland undertook to prepare and carry out a programme for the industrial training and recruitment of Nigerians for all types of job classification.

This contract attracted a lot of adverse comment. However, the contract is casually worded and subject to different interpretations, unlike several other PSCs from other countries which are extremely detailed and precise, especially as regards fiscal provisions.[33] Further, there are some progressive clauses found in other PSCs. Indonesia, for example, required a contractor to provide 8.52 per cent of total production at a highly subsidized price, to cover domestic demand.[34] The Ashland PSC type of contract has been seriously condemned for many reasons including the feeling that Ashland has been allowed to earn 'windfall profits'. According to Omoregbe, the problem lies not so much in the group of contracts known as PSCs, as in the variant which is presently in force in Nigeria.[35]

It is the contention of commentators that although the PSC may have secured for the government its objective of maximizing its revenue from the oil industry, in terms of achieving the government's objectives of de facto control of the oil industry and the development of national capabilities in the field of petroleum operations, it failed. In the absence of real factual control of the oil industry by Nigerians, using the PSC as a means to guarantee permanent sovereignty over natural resources is in doubt.[36]

---

[32] Id.

[33] Id.; see also Sasegbon, Fola, 'Current Developments in Oil and Gas Law Nigeria: With Comparative Analysis with Other African Oil Producing Countries', Vol. 1, Energy Law, Proceedings of a Seminar Organized by the Committee and Energy and Natural Resources, Section on Business Law, International Bar Association, p. 361, 371 (1981).

[34] Asante, S.K.B., 'Restructuring Transnational Mineral Agreements', Vol. 73, A. J. I. L. 335, 364 (1979); also Mlotok, Paul, *The Oil Industry in the Far East: Indonesian Oil and Gas* (Salomon Brothers, New York 1984) p. 4.

[35] Omoregbe, Y., supra note 12, pp. 280–81.

[36] Atsegbua, L. supra note 26, pp. 19–20.

The NNPC issued a model PSC in September 1990 that is different from the NNPC/AON PSC. In respect of the recovery of operation costs and sharing of profits, it provides that the 'cost oil' recoverable by the IOC shall be 30 per cent, 'tax oil' is fixed at 40 per cent and the balance of 30 per cent shall be shared with 35 per cent to the NNPC and 65 per cent to the IOC. Taxes due on the IOC's share shall be paid by the IOC in accordance with the Companies Income Tax Act 1990, and shall not under any circumstances be reimbursable to the IOC. This provision differs from clause 6 of the NNPC/AON PSC, which was variously criticized.[37] Clause 6 gave AON the right to sell on behalf of the joint venture the quantity of production made up of 'cost oil', 'tax oil', its 'participating interest' and, on request, NNPC's 'participating interest'. In March 1991, the Nigerian government produced a confidential model PSC. The 1991 model contains some substantial differences from the 1990 model. The major changes are in respect of the recovery of operating costs and of crude oil allocation. Article 8(i) provides:

(a) Tax oil: to offset actual tax and royalty due, will be deducted in full in the relevant year. Government Agency shall take in kind, lift and dispose of tax oil and from the proceeds pay royalty, petroleum profit tax and concession rentals.
(b) Cost oil: for cost recovery purposes, operating costs will be recovered in the year of expenditure, while capital cost will be recovered in a minimum of 20 quarterly instalments. Contractor shall take in kind, lift and dispose of cost oil and from the proceeds pay all operating cost.
(c) Profit oil: crude oil after deduction of royalty, petroleum profit tax, concession rental and cost recovery.

The PSC shares the usual features of duration, work obligation, and so on, with the service contracts. The main differences between them lies in the mode for recovery of costs and the remuneration of the IOC. The distinctive feature of the SC is that the IOC is remunerated in cash and not in crude oil (although there may be provisions that permit the IOC to buy back the amount of crude oil at international prices from established production).[38]

Ultimately, however, the PSC was seen as an unfavourable method of allocating access to petroleum in Nigeria, and the use of a licensing regime was introduced.

---

[37] Id.
[38] Id., p. 20–21.

## 5. GRANT OF ACCESS TO PETROLEUM FOR PETROLEUM ACTIVITIES – LICENSES

Nigeria is a Federation of units (called states) and all petroleum in Nigeria is vested in the federal government, which controls the resources and only permits their exploitation under license, in accordance with the Petroleum Act 1969. The federal government grants licenses and allocates acreages to operators in areas deemed to have potential for petroleum reservoirs at the discretion of the Minister of Petroleum Resources (the Minister).[39] All grants and allocation of acreages (oil blocks) are at the sole discretion of the Minister, supposedly with the approval of the Head of State or the President.

Concessions and licenses over oil blocks in Nigeria were granted through an arbitrary exercise of discretion before and after independence until the 2000s when some level of open bidding process was introduced. Most grants and allocations of oil blocks are now based on a set of criteria made known to qualified and deserving companies at the time such blocks are open for bidding. The government has indicated its commitment to only allocate concessions or licenses to operators based on open competitive bidding rounds. Concessions or licenses may be granted only to a company incorporated in Nigeria under the Petroleum Act or any corresponding law without prejudice, and the company must be registered solely for activity in the petroleum business.[40]

The Nigerian government reserves the right to participate in the operations of any block and to determine the form of contractual relationship between the licensee or lessee and the government in accordance with the provisions of the Petroleum Act as amended and the Deep Water Block Allocations to Companies (Back-in-Rights) Regulations 2003. Presently, in line with the local content policy aspirations of Nigeria, the state also reserves the right to recommend local content vehicles (LCVs) to fulfil the provision of local content.[41] However, in view of the government's policy to increase both its oil and gas reserves base and daily production there is an aggressive push to explore its potential sedimentary basins. The Nigerian government is willing to make attractive offers to investors capable and willing to participate in

---

39 S 2 of the Petroleum Act.

40 DPR, Guidance Information for Prospective Bidders of the Year 2005 Licencing Round; see also DPR, Guidance Information for Prospective Bidders of the Year 2007 Licencing Round (Ministry of Petroleum Resources) (2007).

41 Id.

exploration and production activities in the inland basins (Chad, Benue, Bida, Sokoto and Anambra).[42]

The bid process had four stages. Firstly, the zoning of acreage, where available acreage is zoned by terrain and maturation, targeting the different spectrum of investors such as majors, independents, local operators, downstream operators and local content providers. Second is the registration of bidders, where companies provide all required technical, financial and legal information. This is followed by technical evaluation that qualifies and assigns participation to strategic partners and makes these public in interactive sessions with the bidders. At this stage the operators are analyzed to determine operators who will participate in the commercial bidding conference. The final stage is the commercial bidding conference which is conducted with the objective of obtaining a fair value. Only operators participate on the basis of established parameters via an automated process, and the blocks are awarded on the bidding day.[43]

Licensing of marginal fields is also being done to encourage indigenous participation and increase daily production of petroleum. Pursuant to section 1 of the Petroleum (Amendment) Decree No. 23, 1996, the Nigerian government carries out in collaboration with operating companies the farm-out of un-produced, unapprized, abandoned or producing fields on existing OMLs to independent indigenous companies on a periodic basis. Marginal fields licensing has become a practice in the Nigerian licensing system in the past decade with the first licenses granted in 2001.[44] The second marginal field licensing rounds commenced in 2013 with 31 fields offered, including 16 offshore.[45]

The broad objectives of the 2013 marginal fields awards included:

---

42 Id.; Orient Petroleum is currently engaged in exploration activities in the Anambra basin.

43 See generally, Guidance Information for Prospective Bidders of the Year 2005 Licencing Round, supra note 40.

44 A marginal field is defined as 'any field that has (oil and gas) reserves booked and reported annually to the Department of Petroleum Resources (DPR) and has remained un-produced for a period of over 10 years.' See DPR, *Guidelines for Farm Out and Operations of Marginal Fields – 2013* (Ministry of Petroleum Resources, Lagos).

45 Obafemi, Oredein, E&P, 'Nigeria Sets Strict Guidelines in Licensing Round', available at www.epmag.com/Technology-Regulations/Nigeria-Sets-Strict-Guidelines-Licensing-Round_126916, last visited 12 March 2014.

- To grow production capacity by expanding the scope of participation in Nigeria's petroleum industry, especially the upstream sector, through diversification of resources and inflow of investments.
- To increase the oil and gas reserves base through aggressive exploration and development efforts, in particular the deeper hydrocarbon plans.
- Provide an opportunity for portfolio rationalization.
- Promote indigenous participation in the sector, thereby fostering technological transfer.
- Provide an opportunity to gainfully engage high-level technically competent Nigerians in the petroleum sector.
- Promote common usage of assets/facilities to ensure optimal utilization of available capacities.

The guidelines for the 2013 licensing rounds released by the DPR indicate that the application form is to be provided by the DPR and attracts non-refundable application fees of N200,000 (approximately US$1,200) and processing fees of N300,000 (approximately US$1,850) per field. After pre-qualification of the bidders, a pre-qualified applicant may undertake an examination of the data upon payment of US$3,000 per field, or lease the data for US$25,000 per field. Where technical assistance is needed an applicant may pay $2,000 to Schlumberger, which is the designated Technical Adviser to the DPR.

Bids are evaluated on the basis of information on the applicant's shareholding, activities and experiences in petroleum operations of the company, technical and managerial skill, ability to pay the signature bonus of US$300,000, Nigerian content compliance, local corporate social responsibility and community participation in the operations of the company, and experience in holding and operating petroleum fields. The selected farmee is announced following the recommendation made by the Selection Committee to the Minister of Petroleum Resources and then the President. Often the eventual announcement of awardees is politicized and raises issues about the transparency and credibility of the process.

Today three types of licenses are granted to Nigerian citizens or companies incorporated in Nigeria in accordance with the provisions of the Petroleum Act.[46] The OEL is granted for one year and is renewable for another year.[47] The OPL may be granted for a maximum period of

46 Arts 2(1)(a–c) and (2).
47 Paras 2–4 of First Schedule of Petroleum Act.

five years.[48] The third type of lease, the OML, may be granted for a maximum period of 20 years and is renewable. Subject to renewal of the lease, 50 per cent of the area covered by the lease is to be relinquished after ten years.[49] The OML may be granted only to the holder of an oil prospecting license who has satisfied all the conditions imposed on the license or otherwise imposed on them by the Act, and has discovered oil in commercial quantities.[50] Oil is deemed to have been discovered in commercial quantities if the Director of Petroleum Resources 'upon evidence adduced by the licensee is satisfied that the licensee is capable of producing at least 10,000 bbl/d of crude oil from the licensed area'.[51] The Act also provides for assignment, farm-out, termination, surrender/relinquishment and revocation of licenses and leases.[52] A holder of an oil mining lease has 100 per cent control of the area covered by the lease. The lessee is only subject to legislation on tax, environmental protection and the regulatory powers of the DPR.

Unlike the United States and United Kingdom, the Nigerian state, like most resource-rich developing countries, participates in the business of exploration and exploitation of petroleum resources. This is done through various mediums like joint venture, production sharing agreements, service contracts and sole operation through the Nigerian Petroleum Development Company.

## The Licensing Joint Venture

The joint venture (JV) arrangement in Nigeria started in April 1971 when the Nigerian government exercised an option in the 1962 concession agreement with the Nigerian Agip Oil Company and took a 33 per cent equity interest in the company. In a move seen as a punitive measure against France for supporting secessionist Biafra during the Nigerian civil war of 1967 to 1970, the Nigerian government acquired a 35 per cent interest in the operations of Saftrap (now Elf). The Nigerian government also acquired an undivided 35 per cent interest in the operations and assets of Shell-BP Petroleum Development Company of Nigeria Ltd, Mobil Producing Nigeria and Gulf Oil Company (now

---

48 Id., paras 5–7.
49 Id., paras10–12.
50 Id., para 8.
51 Id., para 9.
52 See id., paras14–18 and 24–25.

Chevron Nigeria Ltd). By April 1974, the government had acquired an additional 20 per cent interest in the subsidiaries of the IOCs.[53]

The federal government of Nigeria had in 1972 reserved all rights to explore for and produce mineral oils and natural gas in Nigerian territory to the NNOC (now NNPC) with the exception of those areas in which there were subsisting grants of petroleum rights. The areas reserved included areas subsequently surrendered, relinquished or forfeited to the government. This vested in the NNPC the right, power and authority to enter into contractual joint ventures with IOCs in respect of areas not covered by existing concessions. Pursuant to that, the NNPC entered into joint ventures with IOCs.

Nigeria adopted the participation joint venture (PJV) petroleum development arrangement. Parties to such joint ventures contribute funds for the costs and expenses required for approved operations in the proportion of their respective percentage interests in the joint venture. Each party is required to pay its respective cash call into the joint venture account. Thus federal government, through the Nigeria National Petroleum Corporation (NNPC), controls 60 per cent equity shares in the six joint ventures operational in the country, except that operated by Shell, where it holds 55 per cent shares.[54]

By 1 July 1979, after prolonged negotiations with the IOCs, the government's participation in the interest of the subsidiaries of the IOCs had increased to 60 per cent.[55] The only exceptions to this were the Shell Petroleum Development Company of Nigeria Ltd and those oil companies that are parties with the NNPC to production sharing contracts and service contracts. The participating interest of NNPC in the operations and assets of the former Shell-BP Petroleum Development Company of Nigeria Ltd was increased from 60 per cent to 80 per cent with effect from 1 August 1979 by virtue of the Acquisition of Assets (British

---

[53] Ajomo, M., 'Law and Changing Policy in Nigeria's Oil Industry' in J.A. Omotola (ed.), *Law and Development* (Lagos, University of Lagos Press 1987) p. 84.

[54] NNPC, 'Joint Venture Operations', available at www.nnpcgroup.com/NNPCBusiness/UpstreamVentures.aspx, last visited 26 February 2014; see also Agoro, Bode, 'The Impediments to Expansion: Why is the Upstream Sector of Nigeria's Petroleum Industry Not Growing?', 19 J. Energy & Nat. Resources L. 16, 16–17 (2001).

[55] Etikerentse, Godfrey, *Nigerian Petroleum Law* (Dredew Publishers, Lagos 2004) p. 24.

Petroleum Company Limited) Act.[56] Under section 1 of that Act 'rights, privileges, equities, entitlements and interest of whatever description held by or due to the British Petroleum Company Limited in all concessions in Nigeria including licenses and oil mining leases/granted to the Shell-B.P. Nigeria Limited' were transferred to and vested in the NNPC without further assurance.

The terms arrived at in the course of the negotiations became a participation agreement which: fixes the participation interest and obligations of the parties and sets out the participation level of each party in the operation of the venture; addresses issues of production facilities and working capital applicable to the operations; and makes provisions for rights and obligations of parties in any agreement entered into for the sale and disposal of natural gas obtained from operations of the joint venture as well as other rights and obligations arising from, or incidental to the operations. This agreement also indicates the need for a joint operation agreement to be signed by the parties to clearly establish the parties' obligations and rights in the way the joint venture is to be operated and administered.[57]

The participation agreement for JVs in Nigeria states that the amount paid by the government:

> constitutes a binding full and final settlement of all claims which the Government and/or NNPC on the one hand and the IOC on the other hand may have or alleged against each other in respect of any and all matters arising from or relating to the acquisition by the Government and for NNPC of the said initial participating interests.[58]

In some participation agreements, the payment made to the company (which may be in either local or foreign currency) is exempt from all Nigerian taxes and impositions. The agreements also provide that the 'Oil Mining Leases and all other agreements and arrangements currently existing between NNPC and the Company shall remain and continue in full force and effect in accordance with their terms, save only as modified by the terms of this Agreement and the Operating Agreement'. The governing law of the participation agreement is Nigerian law.[59]

On 30 June 1989, the federal government through the NNPC relinquished 20 per cent of its interest to Shell, Agip and Elf, thus bringing

---

56 Acquisition of Assets (British Petroleum Company Limited) Act, Cap 3 LFN 1990.

57 Etikerentse, G., supra note 55, at p. 26.

58 Omoregbe, Y., supra note 12, p. 277.

59 Id., p. 279.

the NNPC participation joint venture ratio to NNPC 60 per cent, Shell 30 per cent, Agip 5 per cent and Elf 5 per cent.[60] The working relationship between the joint venture partners was based on a gentleman's agreement until July 1991 when a joint operating agreement was signed between the Nigerian government through the NNPC and each of its joint venture partners. In accordance with paragraph 35 of the First Schedule to the Petroleum Act, special terms may be negotiated in relation to conditions for participation by the federal government and special provisions that may apply to any natural gas discovered. The negotiations take place between the Minister of Petroleum Resources or his or her delegate and the applicant for the license or lease.[61]

## Joint Operating Agreements (JOA) in Licenses

The joint operating agreement (JOA) is a contract between two or more parties establishing and setting out the terms of a joint venture between them under which petroleum exploration, development and production operations will be conducted.[62] In Nigeria the activities between joint venture parties in a license are conducted under a JOA.

The JOA establishes and spells out the basis for the sharing of rights and liabilities under the license as between the parties. Each party will have a percentage interest and as such the JOA stipulates that, subject to certain exceptions, all rights and liabilities arising in connection with the license shall be shared between the parties in proportion to their percentage interests. It also lays down the rules and the procedure for the joint development of the area concerned and for property jointly owned by the parties. Joint property under Nigerian JOAs includes expenditure for virtually all services and activities of the IOC (operator), for example, salaries, staff housing schemes, pensions and gratuities. In spite of this, the NNPC hardly ever interferes with the internal management of the oil

---

[60] Etikerentse, G., supra note 55, p. 24.

[61] By section 12 of the Petroleum Act the Minister for Petroleum Resources may, with the exception of the power to make orders and regulations, by writing under his or her hand delegate to another person any power conferred on him or her by the Act.

[62] Taylor, M.P.G., Winsor, T.P. and Tyne, S.M., *The Joint Operating Agreements: Oil and Gas Law* (Longman, London 1992) (MPJFIL) Vol. 4 No. 3 p. 147.

company other than for purposes of monitoring compliance with Nigerian laws.[63]

The conduct of operations under the license is realized by appointing one of the parties to the JV as 'operator', with day-to-day responsibility for the conduct of operations subject to the overall supervision and control of an operating committee upon which the licensees are represented.[64] Therefore, the JOA also deals with change of operatorship. The JOA used in Nigeria has an innovative provision that goes thus:

> [Other parties] acknowledge NNPC's right to operate specified portions of the Joint Venture in due course. Consequently, NNPC wishes to become more actively involved with the responsibilities of the operatorship in respect of Joint Operations. In order to formulate the framework within which NNPC may exercise this right, parties will agree on a programme to progress the build-up of NNPC's operational experience and secure its access to the necessary technology and technical assistance, as required for the exercise of such responsibilities in a cost efficient and professional manner acceptable to the parties. This programme will be reviewed by the Operating committee and adjusted from time to time in the light of progress achieved and experiences gained.[65]

According to Edu the provision is unusual as the right given to NNPC is to operate 'specified portions of the joint venture' and not the entire joint venture operations, unlike the case of the multinational oil company when rights as an operator are unfettered.[66]

The NNPC is a party to every operating agreement in which one or more multinational oil companies in Nigeria participate. All the petroleum licenses jointly owned by NNPC and other parties have in place a JOA.[67] Although the parties are regarded as partners there is no partnership relationship between the parties. It is expressly agreed that the agreement is not to be construed as a partnership. The liabilities of the

---

63 Omoregbe, Y., supra note 12, at p. 279; see also Article 3.1 NNPC/SPDC/NAOC Joint Operating Agreement.

64 Taylor, M.P.G. et al., supra note 62, p. 5; see also NNPC/Shell/Agip/Elf Joint Operating Agreement, Article 3.2.

65 Id., Article 2.7.

66 Edu, Omerionwan Kingsley, 'A Socio-Legal Appraisal of Nigeria's Joint Venture Arrangement in the Petroleum Industry', 22(1) *Sri Lanka Journal of International Law* 13 (2010).

67 Joint venture operations, available at www.nnpcgroup.com/NNPC Business/UpstreamVentures.aspx, last visited 26 February 2014; see also Olisa, M.M,, *Nigeria Petroleum Law and Practice* (Jonia Ventures Ltd, Lagos 1997) p. 97.

parties are certainly not joint or collective but each party is responsible for their obligations.[68]

The operator is liable to the non-operator if it fails to conduct the joint operations in a proper and workmanlike manner and in accordance with good oilfield practice. Thus paragraph 37 of the Nigerian Petroleum (Drilling and Production) Regulations[69] provides that the operator shall maintain all apparatus and appliances in use in their operations and all boreholes and wells capable of producing petroleum in good repair and condition and shall carry out all their operations in a proper and workmanlike manner in accordance with these and other relevant regulations and methods and practices accepted by the Director of Petroleum Resources as good oilfield practice. Unfortunately, the terms 'proper and workmanlike manner' and 'good oilfield practice' are not defined in the regulations.

Neither the Petroleum Act nor the regulations set any standards by which performance of the operator will be judged. These terms are vague, and what constitutes proper and workmanlike manner is problematic in practice. They have been given various interpretations by commentators based on their perspective. For instance, Adeniji opines that the reference to 'proper and workmanlike manner' and 'good oilfield practice' obviously looks to an objective standard to determine what methods and precautions the operator would apply.[70] Ajomo[71] asserts that 'the half-hearted admonition to petroleum operators to conduct their operations in a "proper and workmanlike manner" et cetera has done no more than to merely import into the regulations the common law duty of care by general terms into petroleum operations'.

An IOC or any other operator may take 'good oilfield practice' to mean minimizing the economic cost of production without regard to safety or environmental care. Edu, however, argues that discretion should be employed in the application of these terms by deploying both

---

[68] Edu, O.K., supra note 66, at p. 21.

[69] This was made pursuant to section 9(1)(c) of the Petroleum Act, which provides that the Minister of Petroleum Resources may make Regulations 'regulating the construction, maintenance and operation of installations used in pursuance of this Act'.

[70] Adeniji, K., 'Legal Control of Pollution Hazards from Petroleum Operations: Implications for the Nigerian Oil and Gas Industry', 12 *University of Ghana Law Journal* 112 (1975).

[71] Ajomo, M.A., 'An Examination of Federal Environmental Laws in Nigeria' in Ajomo, M.A. and Adewale, O. (eds), *Environmental Law and Sustainable Development in Nigeria* (Nigerian Institute of Advanced Legal Studies and the British Council, Lagos 1994) p. 20.

objective and subjective standards, which is envisaged by the regulations. Accordingly, the subjective standard would apply when a practice has to be acceptable to the Director of Petroleum Resources as 'good oilfield practice'. Moreover, what amounts to 'proper and workmanlike manner' and 'good oilfield practice' will be in a constant state of change and advancement, in view of new discoveries and advancement in drilling technology and greater environmental concerns.[72] This is because of the provision of regulation 36 which provides that the operator 'shall carry out his operations in a proper and workmanlike manner in accordance with these and other relevant regulations and methods and practices accepted by the Director of Petroleum Resources as good oilfield'. The development and changes in standard are imperative irrespective of the notion, arguable though, that oil companies in Nigeria do not give serious thought to issues of environmental protection.[73]

The operator is not entitled to any remuneration for its duties as operator in any of the JV operations with the NNPC, but costs of services rendered by the parent company of the operator in connection with the joint operations are chargeable as expenses to the joint account.[74] The JOA in use in Nigeria also provides that the 'operator or its affiliate shall not be liable for any loss or damage which result from joint operations' unless such loss or damage results from 'willful misconduct on the part of its Directors or supervisory staff'.[75] Wilful misconduct does not include reservoir damage, damage occasioned by pollution or consequential damage or losses.

An operator may be replaced with the agreement of the parties.[76] In reality they hold their position at the mercy of the majority participating

---

72 Edu, O.K., supra note 66, p. 16.

73 Manby, B., 'The Price of Oil; Corporate Responsibility and Human Rights Violations in Nigeria's Oil Producing Communities', 1 *Environmental Planning Law Review* (EPLR) 9 (2005).

74 Olisa, M.M., supra note 67, p. 89.

75 Article 2.2.2. NNPC/Shell/Agip and Elf Joint Operating Agreement 1991.

76 The operator may be replaced in any one of the following circumstances:

(a) Where the operator assigns or purports to assign its general powers and responsibilities of supervision and management as operator of joint operations except to an affiliate as specified in the agreement;

(b) Where the operator ceases or threatens to cease to carry on its business or becomes bankrupt or insolvent or commits or suffers any act of bankruptcy or insolvency or makes any assignment for the benefit of creditors;

(c) Where the operator defaults in its duties or obligations or any of them and fails to commence to rectify the default within the number of days

interest holders and may be removed with or without cause or wrong-doing on their part by the majority.[77] This is acceptable as it is in accord with a democratic management style. The justification behind this development is that the operator is not really working for themselves alone but that they are ultimately the representative of the participants, who may decide to relieve them of their responsibility if the interests of the parties are not well served. A foreign investor demands the operator position for the purpose of protecting and enhancing their investment in developing countries. However, the situation where the operator holds their position at the mercy of the majority interest may not be in the interest of the foreign investor as the investor may feel insecure.[78]

**The operating committee**

The operating committee or management board is established by the parties to the operating agreement to provide orderly overall supervision, control and direction of all matters pertaining to the JV operations. The membership of the committee is a reflection of the percentage of the parties' participating interest. In all the participation agreements in Nigeria, the NNPC representatives are in the majority, by virtue of its majority participation interests. For instance, under the NNPC/Gulf (now Chevron) Participation Agreement signed in 1984, the management board consists of six government representatives and four from Gulf. The supervisory role of the operating committee is vital for the effective functioning of the JOA. Though NNPC representatives are in the majority in all operating committees in Nigeria, all major decisions of joint ventures are taken unanimously by representatives of the parties regardless of the number of parties and their proportionate interests in the venture.[79]

---

specified in the agreement after written notice from other co-venturers specifying the default; and

(d) Where the affiliate to whom the operator has assigned powers and responsibilities of supervision and management as operator ceases to be an affiliate of the operator; or

(e) Where the party acting as the operator (or any affiliate of the operator which is a party) assigns or otherwise disposes of, other than to an affiliate, all its participating interest in the joint venture.

77 Maclean, J.A., 'The Canadian Association of Petroleum Landman Operating Procedure: An Overview of the Revisions', *Alberta Law Review*, Petroleum Law Edition Vol. No. 1, 1992, p. 133 at 143, cited in Edu, O.K., supra note 66, p. 29.

78 Edu, O.K., supra note 66, p. 20.

79 Id.; see also Omoregbe, Y., supra note 12, p. 279.

**Heads of agreement**

These are generally short but exceedingly crucial agreements between the NNPC and the JV partner, delineating the general principles intended to govern offtake, scheduling and lifting agreements for crude oil. This agreement provides, inter alia, that there shall be undivided interests in the rights granted by the applicable leases with respect to petroleum in the license area and that each interest owner will share therein to the extent of its equity participation. Each interest owner may nominate, lift and separately dispose of its share of crude oil. A make-up right shall accrue to the interest owner to nominate the balance of its equity share in the future. These are saving provisions for the NNPC that very often fails to lift its full quota of oil, in which case the oil is left underground.[80]

**Costs under joint venture operations**

Parties contribute to the operations in proportion to their participating interest. The contribution of funds process is begun by the operator calling on each party to pay its participating interest share of the costs and expenses incurred into the joint account. This process is termed 'cash call'.[81] The Nigerian government under joint operating agreements participates in the ownership and authorized operations and shares in crude oil and other products.

Nigeria holds as high as 55–60 per cent equity interest in joint ventures. As a result Nigeria finds it difficult to meet its cash call obligations. Operators are sometimes forced to source funds with high interest charges, leading to inflated costs of production.[82] The government's failure to meet its cash call obligations towards its JV partners has an adverse impact on the operations of the JV. This affects the work plan of operators and slows down the pace of exploitation of petroleum under the JOA.[83]

---

80 Omoregbe, Y., supra note 12, 279.

81 Cash call means the amount in all currencies which the operator estimates a party must pay into the Joint Account in any given month to meet such party's participating interest share of the costs and expenditures to be paid for the Joint Account in such month after adjusting for balances or deficits in such bank account as well as any credit receipt anticipated during each month.

82 Edu, O.K., supra note 66, p. 26.

83 Id., p. 23.

**Sole Risk or Independent Operators**

In recent years, in addition to the above forms of licensing for oil exploration and production which are predominately with the IOCs, petroleum licenses are being granted to companies on a sole risk basis as independent operators. Most indigenous lessees in Nigeria operate as sole risk or independent operators. Some undertake these ventures in conjunction with foreign companies as technical partners. This is the case with most indigenous operators who gained marginal fields through processes made possible by the Petroleum (Amendment) Decree No. 23, 1996.

## 6. LICENSING, LOCAL CONTENT AND DEVELOPMENT OF INDIGENOUS TECHNOLOGY

Since independence Nigeria has had a policy to ensure IOCs who dominate the petroleum industry do not operate in Nigeria without developing indigenous capacity and transfer of technology. As early as the first decade of her independence, OMLs granted under the Mineral Oils Act 1914[84] required lessees 'to start a training scheme for the technical training of suitable Nigerians' at the inception of their operations. This training scheme was to train Nigerians for employment as 'tradesmen and craftsmen'. Lessees were also required 'within a reasonable time' to commence the training of Nigerians in supervisory posts and the recruitment and training of professional Nigerian staff for employment in senior managerial grades. These early local content requirements neither set targets for the training or recruitment of Nigerians nor provide any specific mechanism for assuring compliance with these provisions.[85]

The Petroleum Act 1969 replaced the Mineral Oils Act, and provisions in the First Schedule to the Petroleum Act required that within ten years the holder of an OML would ensure that:

> the number of citizens of Nigeria employed by him in connection with the lease in managerial, professional and supervisory grades (or any other grades

---

[84] The Mineral Oils Act (formerly Mineral Oils Ordinance of 1914) was the law governing petroleum operations at the time.

[85] Oguine, Ike, 'Nigerian Content in the Nigerian Petroleum Industry: Legal and Policy Issues', 29 J. Energy & Nat. Resources L. 405, 406–7 (2011).

designated by him in a manner approved by the Minister) shall reach at least 75% of the total number of persons employed by him in those grades.[86]

There was a further requirement that the number of Nigerians in any one of those grades should not be less than 60 per cent of the total.[87] Paragraph 26 of the Petroleum (Drilling and Production) Regulations 1969, which were made under the Petroleum Act, provided that the lessee of an OML would, on the grant of the lease, submit for approval by the Minister of Petroleum Resources a detailed programme for the recruitment and training of Nigerians. Under paragraph 29, such a lessee was to provide a report on its compliance with the approved programme in June and December of every year. Perhaps not satisfied with the implementation of the above provisions, the government took a step further by ensuring that indigenous Nigerian companies become operators.

### Awards of Oil Prospecting Licenses and Marginal Fields to Indigenous Companies

This indigenization policy was introduced between 1989 and 1990 and some oil concessions were granted specifically to Nigerians. Basic criteria for the allocation were that the grantee company must be truly owned by Nigerians who are not merely agents or fronts for foreign companies and must be the operator. However, the grantee may have foreign technical partners with no more than a 40 per cent participating interest but the managing director must be a Nigerian.

#### Objectives of the indigenization policy

The strategic objectives of the policy were to promote 'Nigerianization' of the upstream sector as it became evident that even with the increased number of Nigerians in management and operations positions in the multinational oil companies this did not guarantee the Nigerianization and technology transfer objective of government unless Nigerians owned and operated these ventures. Secondly, it sought to create a strategic national production capacity, which was not only necessary but imperative for security of energy. Therefore, it is vital for the nation to achieve a minimum level of productivity from indigenous oil companies. The third objective was the need to establish a group of small-sized producers to complement the IOCs. The growth of this group becomes even more

---

[86] Para 38(a)(i) of First Schedule to the Petroleum Act; see also clause 17(1) OML 92.

[87] Para 38(a)(ii) of First Schedule to the Petroleum Act.

important as Nigeria's primary petroleum basin approaches maturity; their absence would otherwise create a vacuum in the mature stages of petroleum development and planning.[88] These policy efforts have given birth to dozens of indigenous exploration and production companies of varying sizes, sophistication and ambition.[89]

The earliest implementation of these policies was the award of OPLs to Nigerian business people in the late 1980s by the military regime of the time.[90] Grantees were permitted to assign not more than 40 per cent of their interest to foreign technical partners. That policy has been criticized because the process of the awards was not transparent[91] and the policy did not enable true Nigerian participation in the upstream industry since, in most cases, the work on the leases awarded to Nigerian entrepreneurs was managed by their foreign technical partners. However, companies such as Conoil's upstream arm and Zebra Energy, which emerged as a result of that policy, have, through participation in subsequent bid rounds or through joint ventures and acquisitions, sought to expand their roles in the petroleum industry.

Perhaps responding to the criticism about the lack of transparency in the early awards of OPLs to indigenous companies, in the more recent bid rounds foreign bidders have been required to enter into joint ventures with Nigerian-owned companies (referred to as 'local content vehicles' – LCVs). Bidders are required to state in their bids how much of the work they undertake in their work programmes would be performed by their LCVs. This newer policy was criticized following experiences during the 2005 licensing round, on the ground that the LCVs were 'a rash of paper companies causing bidders serious difficulty with due diligence'.[92] Only 10 per cent of the 100 LCVs pre-qualified by the DPR for the 2005 licensing round had any previous experience of oil exploration and production. Many of the LCVs were said to be linked to prominent political figures. In spite of these shortcomings, the emphasis on Nigerian content in the licensing rounds has provided opportunities for some Nigerian companies, already well established in the downstream sector like Oando and Sahara Oil, to acquire upstream interests and thus increase Nigerian participation in that sector of the economy.[93]

---

88 Agoro, B., supra note 54, p. 21.
89 Oguine, I., supra note 85, p. 406–7.
90 Etikerentse, G., supra note 55, p. 95–6.
91 Id.
92 Oguine, I., supra note 85, p. 411.
93 Id.

## Award of Marginal Fields to Indigenous Companies

Another policy to encourage Nigerian participation in the upstream sector is the award of marginal fields to Nigerian companies. Pursuant to the relevant provision of the Petroleum (Amendment) Decree No. 23, 1996, it is the intention of government to carry out in collaboration with operating companies the farm-out of un-produced, unapprized, abandoned or producing fields on existing OML to independent leaseholders and indigenous companies on a periodic basis.[94] The marginal fields awarded are carved out from OMLs held by the IOCs. Chevron Nigeria Limited pioneered marginal field farm-outs when it assigned the Ogbelle field in OML 54 to Niger Delta Petroleum Resources in 1994.[95]

Fields are considered 'marginal' under the applicable laws when, among other factors, they have been left unattended for a significant period of time after discovery. The characteristic features of a marginal field according to the 2013 guidelines are as follows:

- Fields not considered by license holders for development because of assumed volatile economics under prevailing fiscal and market conditions;
- Fields with at least one exploration well drilled and reported as oil and or gas discovery for more than 10 years with no follow up appraisal or development effort;
- Fields with crude oil characteristics different from current streams (such as crude with very high viscosity and low API gravity), which cannot be profitably produced through conventional methods or current technology;
- Fields with high gas and low oil reserve;
- Hitherto producing fields that have been abandoned by the leaseholders for upwards of three years for economic or operational reasons; or
- Fields that the present leaseholders may consider for farm-out as part of portfolio rationalization programmes.[96]

Some IOCs in the early years of the marginal fields programme argued that the delay in developing certain fields in their leases was not due to the fact that those fields were marginal but because of OPEC production quota restrictions enforced in Nigeria by the NNPC, and budgetary shortfalls because the NNPC could not always provide the required

---

[94] Para 3.0, *Guidelines for Farm Out and Operations of Marginal Fields – 2013*, supra note 44.

[95] Etikerentse, G., supra note 55, p. 97.

[96] Para 5.0, *Guidelines for Farm Out and Operations of Marginal Fields – 2013*, supra note 44.

funding to meet its share of joint venture work programmes.[97] Even though there have still been occasional debates between the IOCs and the regulatory authorities over which fields can be properly categorized as marginal, the IOCs have largely accepted the marginal fields programme and have on the whole co-operated with the DPR and the NNPC in implementing the programme. Twenty-four marginal fields delimited from fields held by the NNPC joint ventures with Shell, Chevron, Texaco, Elf Petroleum and Eni were offered to pre-qualified Nigerian companies for bid in 2001. Thirty-one winners emerged and some fields were jointly awarded to some bidding companies.[98]

The same technical and commercial factors that delayed the development of these fields by the IOC leaseholders have dogged the indigenous companies, and many have struggled to bring their fields to production. However, some companies, such as Brittania U and Platform, have been relatively successful. These more serious, and perhaps more fortunate, marginal field awardees have begun production in their fields that will probably serve as a springboard to further opportunities. For instance, Platform was part of the consortium that acquired the interests held by the IOCs in the NNPC/Shell/Total/Eni joint venture's OMLs 4, 38 and 41.[99] Some other Nigerian companies that have emerged through government policies to promote Nigerian participation in the upstream sector have also reportedly been invited by Shell to bid for other OMLs that the joint venture is offering for sale.[100]

In terms of production, these indigenous companies remain insignificant as the IOCs, through their joint ventures and PSCs with the NNPC, continue to produce more than 90 per cent of Nigeria's oil and gas. With time, better-managed and ambitious indigenous companies in the upstream industry are likely to obtain a progressively greater share of the business. Despite the criticism of government policies in this area due to the number of short-lived companies it produced, and although Nigerian upstream companies continue to face problems of funding and access to commercially attractive acreage, these policies have clearly contributed to the emergence of some potentially significant indigenous upstream companies.[101]

---

97 Agoro, B., supra note 54, pp. 18–19.

98 Etikerentse, G., supra note 55, p. 100.

99 Ayankola, Martin, 'Niger Delta Crisis: Shell Divests from Three Nigerian Oil Blocks', *The Punch,* 10 January 2010.

100 Oguine, I., supra note 85, p. 413.

101 Oguine, I., supra note 85, pp. 412–13.

The marginal field allocation to indigenous Nigerian companies has been variously criticized. First, that Decree No. 23 has draconian characteristics. Its provisions are such that the government can compulsorily acquire the IOCs' assets without compensation from the government.[102] Secondly, the bidding guidelines exclude the IOCs from operating or acquiring participatory interests in any marginal field. In other words, only indigenous companies can apply for, or operate, any marginal field. Thirdly, many of the marginal fields are co-owned by parties to various joint operating agreements, pursuant to which the fields were discovered.[103] Some have argued that the assignment of interests contemplated under the farm-out agreement must be made with each co-venturer as a party to the extent of its undivided interest. As such the assignment cannot be validly made by the operator of the joint venture, as envisaged by the Decree.[104] However, the successful award of the field and the commencement of operations by some awardees indicates the process has no such legal bottlenecks. Agoro contends that the decrees and guidelines promulgated by the government have not had the desired effect of propelling the development of these fields, rather they have impeded growth in the upstream sector.[105]

## Limits of the Indigenization Policy

Although the objective of the policy is laudable in every respect, there are certain inherent limitations that may inhibit the success of the policy.

### Lack of legal and fiscal framework

A major concern is the fact that the indigenization programme is not formally written into the Petroleum Act, or any other petroleum legislation in Nigeria. It exists merely pursuant to a policy statement issued by the Nigerian government in 1990 and is regulated by the Department of Petroleum Resources (DPR).[106] The allocation of such concessions is on a discretionary basis, lacks transparency in implementation and is consequently open to abuse. No indigenous company operating under the indigenization policy has been able to obtain any written statement of the

---

[102] Agoro, B., supra note 54, p. 19.

[103] Id., 19–20.

[104] Olisa, M.M., supra note 67, p. 154.

[105] Agoro, B., supra note 54, p. 20.

[106] Etikerense, Godfrey, 'Oil and Gas Exploration in Nigeria: Challenges for the 21st Century', paper presented at the Nigerian Institute of Advanced Legal Studies, Lagos, on 17–18 March 1999, p. 18.

1990 policy issued by the Nigerian government, if any exists.[107] In terms of title the grantee is simply issued a letter of allocation of an OPL over a concession, and unlike the multinationals, no license or lease agreement was executed in favour of any of the Nigerian grantees. On the other hand, IOCs enjoy different degrees of supplemental legislation that guarantees security and certainty of tenure and also addresses certain peculiarities such as crude oil price, terrain of operation and capital investments in addition to providing a specific legal umbrella that covers each category of operations.

For the IOCs, an industry minimum rate of return (ROR) of 15 per cent on their investment is assured in the supplemental legislation and avenues for redress in the event of any contractual breach are more clearly specified. In contrast, the indigenous companies, which require more assistance and incentives in order to survive in an industry already dominated by the IOCs, have no guaranteed ROR and their profits are subject to taxation under the Petroleum Profits Tax Act (PPTA) at the rate of 65 per cent.[108] Another implication of the absence of a specific fiscal regime is that at low oil prices the indigenous operators pay higher taxes than the IOCs as large producers (who consequently have a lower cost per barrel) due to the memorandum of understanding (MOU) protection they enjoy. Similarly, deep water PSC operators pay lower taxes and royalties than their indigenous continental shelf counterparts.

**High technical cost and low profitability**

Most of the acreage operated by indigenous companies is acreage relinquished by the IOCs. Consequently, it often consists of high-risk blocks containing small to marginal fields, which are in most cases considered uneconomic to develop. Though substantial commercial finds have been made in some of them, the argument is that it is impractical and unfair for indigenous operators to be expected to operate such fields under fiscal terms that are less favourable than those enjoyed by the IOCs that relinquished them.[109] Furthermore, because the fields operated by the indigenous operators are usually small, production volumes are low so the unit technical cost is high. Profitability is therefore much more difficult to achieve.

---

[107] Agoro, B., supra note 54, p. 21.

[108] Etikerense, G., supra note 106, p. 24.

[109] Agoro, B., supra note 54, p. 22.

**Sole risk venture**

The indigenous companies operate primarily on the basis of the sole risk venture, unlike the IOCs that have the benefit (or otherwise) of joint ventures with the NNPC or the PSC. Consequently, the indigenous operators have to rely on foreign technical partners to finance the venture after obtaining millions of dollars in participation fees. In these circumstances, joint operating agreements between foreign technical partners and indigenous operators always tilt heavily in favour of the foreign technical partners. Where the indigenous operators are unable to secure a foreign technical partner to finance the development of the concession it lies fallow as many of the indigenous operators, who perhaps obtained the concessions due to political patronage, would rather not risk their own resources.[110] For instance, Bayelsa Oil and Gas, which is owned by the oil-rich Bayelsa State of Nigeria, was awarded the Atala Field with Shell as the farmor. Bayelsa shopped for a foreign technical partner with the necessary resources to develop the field for several years. In the meantime, the field was left fallow.

## 7. ENVIRONMENTAL REGULATION

There are several provisions under Nigerian law for environmental protection in the petroleum industry. Whether they are diligently implemented and are effective are different issues. In alignment with practice in most parts of the world, under the Environmental Impact Assessment Act 1992, an environmental impact assessment (EIA) is required for any petroleum activities. This is applicable even to farm-outs in existing fields. It is required prior to the commencement of operations in the farm-out area of a marginal field, with the farmee required to conduct an EIA that will ensure minimal impact of their operations on the environment.[111]

---

110 Id., p. 22–3

111 Clause 27, *Guidelines for Farm-Out and Operation of Marginal Fields* (Office of the Presidential Adviser on Petroleum and Energy, Abuja 2001); See also *Guidelines for Farm Out and Operations of Marginal Fields – 2013*, supra note 44.

The Petroleum (Drilling and Production) Regulations 1969 (as amended) granted rights and powers over land covered by a lease to holders of petroleum prospecting licenses and petroleum mining leases.[112]

Provisions on pollution and conservation are vaguely and imprecisely drafted. The main provision on pollution is regulation 25 of the Petroleum (Drilling and Production) Regulations 1969, requesting all licensees and lessees to 'adopt all practicable precautions' to prevent pollution, and where any such pollution occurs, to take 'prompt steps to control and, if possible, end it'. Nowhere is it stated in precise terms what type of legal duty is imposed on the companies.[113] Conservation is covered by provisions referring to good conservation practice,[114] while the lack of provisions on pollution seems to be mere negligence, the fact that nothing concrete is said on conservation stems from the fact that Nigeria has not practised any conservation policy in the management of oil resources,[115] and production has followed the demands of the economy.

The Oil in Navigable Waters Act 1968 prohibits polluting the navigable waters of Nigeria with crude oil, fuel oil, lubricating oil, heavy diesel oil and any mixture of oil. This is applicable both to petroleum and maritime industries as vessels are required to be fitted with equipment to prevent oil pollution. However, the DPR issues the Environmental Guidelines and Standards for the Petroleum Industry (EGASPIN). The EGASPIN contains a permit regime for operational emissions and discharge of effluent produced in the petroleum industry. No discharge is to take place without the requisite permit and the discharge sites must be registered with the DPR. Some of the required permits are the Industrial Waste Discharge/Disposal Permit relating to all forms of waste discharged, the Production Waste Injection Well Permit for drilling to dispose of waste, the Storage Treatment and Disposal Facilities Permit for facilities that store, treat and dispose of on-site hazardous waste and the Gas Flaring Permits/Certificates granted by the Minister of Petroleum Resources under the

---

112 Reg. 15, Petroleum (Drilling and Production) Regulations 1969 (as amended); see paras 36 and 37 of First Schedule to the Petroleum Act.

113 See Olisa, M., 'Legal Framework for Pollution Control' in *The Petroleum Industry and the Nigerian Environment* (NNPC/Thomopoulos Environmental Consultants, Lagos 1983).

114 Reg. 43, Petroleum (Drilling and Production) Regulations 1969 (as amended).

115 Amu, L.A.O., 'Energy Policy for Nigeria'. see Vol. No. 3 Third Quarter 1983 (*NAPETCOR*). This is the quarterly magazine of the NNPC. Mr Amu was Managing Director of NNPC at the time.

Associated Gas Re-Injection Act (AGRA), permitting an exploration company to flare gas in specified oilfields. EGASPIN also sets out monitoring programmes and schedules to ensure environmental quality control for the oil and gas industry.

Under AGRA, gas flaring is illegal and the oil producing companies were given a deadline of 1 January 1984 to stop all gas flare or be made to pay a penalty for breach of the law. The ban of flaring was subject to Ministerial discretion: section 3(1) of AGRA which provides that 'Subject to subsection (2) of this section, no company engaged in the production of oil or gas shall after 1 January, 1984 flare gas produced in association with oil without the permission in writing of the Minister.' The Minister's permission is granted on a field-by-field basis and on fulfilment of certain conditions. Also, under section 4 of AGRA the penalty for breach of section 3 is forfeiture of the concessions granted. Unfortunately, flaring of associated gas has continued unabated, with its attendant health and environmental hazards. The government's promise to reduce flaring or enact laws banning flaring altogether are more wishful thinking than a reality.

Unfortunately, there are attempts to legalize gas flaring in the new Petroleum Industry Bill (PIB) 2012 by virtue of section 201 which provides that 'the lessee shall pay such gas flaring penalties as the Minister may determine from time to time' and to that end 'install all such measurement equipment as ordered by the Inspectorate to properly measure the amount of gas being flared'. This simply mandates companies to pay what could be seen as gas flaring tax and indefinitely flare as much gas as they want. However, section 275 of the PIB provides that gas shall not be flared after a date to be given later by the Minister through subsidiary legislation, although such a date being imposed in the near future is doubtful. The implication is that with the repeal of the Associated Gas Re-Injection Act by the PIB 2012, gas flaring will become legal in line with section 201, pending the date to be prescribed by the Minister. This is a grave uncertainty. Moreover, section 283 provides that all certificates issued for gas flaring under AGRA will continue to have effect until they lapse. The implication is that communities host to on-going flaring by oil companies will continue to suffer from the obvious health hazards associated with gas flares.

States within the Federal Republic of Nigeria have their own environmental laws and regulations for the protection of their environments. Though petroleum activities are completely within the regulatory powers of the federal government, in terms of environmental issues an operator would be expected to, in addition to national environmental regulations, observe the laws and regulations of the state in which their operations are located. Especially when the Land Use Act 1978 vests ownership of land in the governor of the state.

## 8. SAFETY FRAMEWORK

The Minister of Petroleum Resources is responsible for making regulations on matters relating to OPLs and OMLs, including matters relating to safety at work. All operations carried out by operators of concessions and leases and the farmee in the farm-out area containing the field(s) shall be in accordance with the existing safety regulations applied in the industry. Towards this end, any field development plan proposed for the production of a marginal field shall include provisions that will provide for the safe conduct of all operations.[116] The main regulations on safety are the Mineral Oils (Safety) Regulations 1963 which prescribe standard safety measures and impose duties relating to safety of operations on holders of OPLs and OMLs. Also, the Oil Pipelines Act prohibits a pipeline licensee from constructing any works within 50 yards of any public roads, dam, reservoir or building or over any watercourse required for domestic or irrigational use. Furthermore, the Petroleum Refining Regulations 1974 require measures be taken to prevent and control environmental pollution and to ensure that all on-site personnel have suitable protective clothing, equipment and appliances as approved by the DPR.

## 9. PETROLEUM LAW REFORM IN NIGERIA

The Nigerian petroleum industry is undergoing reform. The Petroleum Industry Bill 2012 (PIB) is, as at June 2014, before the National Assembly for an Act that:

> provides for a legal, fiscal and regulatory framework for the Nigerian petroleum industry and establishes institutions, regulatory and commercial entities for the proper administration and coordination of the operations of the upstream and downstream sectors of the petroleum industry as well as providing for the imposition, assessment and collection of the Nigerian Hydrocarbon tax.[117]

---

[116] See *Guidelines for Farm Out and Operations of Marginal Fields – 2013*, supra note 44.

[117] Explanatory memorandum of the PIB.

This bill is, first and foremost, in line with the government policy objective to increase government revenue from the petroleum industry, the acquisition of the necessary technological and managerial skills for petroleum development and the attainment of de facto control of the petroleum industry that still is largely in the hands of the IOCs. However, the PIB is also a response to the agitation of the country's oil producing region, the Niger Delta, for resource control.

## Licensing under the 2012 PIB

The Minister of Petroleum Resources (MPR) is to be empowered, on the advice of the Upstream Petroleum Inspectorate, to 'grant, amend, renew, extend or revoke upstream petroleum licenses and leases' in accordance with the Act.[118] Functions, assets and liabilities of the DPR in relation to the upstream petroleum sector are to be vested in an Upstream Petroleum Inspectorate established under the PIB.[119] One striking function of the proposed Inspectorate is to 'ensure adherence to national and applicable international environmental and other technical standards' in upstream petroleum operations.[120]

Under the new Act, exploration licenses would be for three years' non-exclusive right and without a right to win, get or carry away any petroleum.[121] Petroleum prospecting licenses would have an exclusive right to explore for petroleum within the area covered by the license. There is a difference in duration between onshore and deep waters. The duration shall be three years and a renewal period of two years for onshore and shallow water areas, while it will be a five year initial period and three year renewal for deep waters and frontier acreage.[122] Once there is a commercial discovery of petroleum under a petroleum prospecting license and all conditions imposed on the licenses have been met and approval for the related development plan is granted by the Inspectorate, a petroleum mining lease (PML) may be granted.[123] The development period under the PML shall be an initial five years for onshore and shallow waters, seven years for deep waters and frontier leases and ten

[118] S 6(1)(g), PIB 2012. This power seems to have also been erroneously vested in the Inspectorate by s 16(a).

[119] Id., s 13.

[120] Id., s 15(1)(d).

[121] Id., s 175.

[122] Id., s 176 and 177.

[123] Id., s 181.

years for the subsequent PML derived from the license.[124] There is a reduction in the life of the lease to avoid unnecessary delay in the development of oilfields.

One significant reform is the imposition of a domestic gas supply obligation on lessees.[125] This may have informed the change in the name of the mining lease from oil mining lease (OML) to petroleum mining lease (PML). Perhaps the intent is to ensure that lessees are conscious of the fact that the lease is not to win oil alone and flare the associated gas but to also make plans for proper utilization of the associated gas.

Under the PIB it will be mandatory that the award of petroleum prospecting licenses and petroleum mining leases shall be through an 'open, transparent and competitive bidding process'.[126] There shall be no discretionary award of licenses and leases save for the powers of the President to grant licenses and leases under special circumstances.[127]

## Fiscal Regime under the PIB

Subject to the enactment of the PIB, the main fiscal regime on petroleum production is the Petroleum Profits Tax (PPT) which is applied to the chargeable profits of companies engaged in upstream petroleum operations as provided in the Petroleum Profits Tax Act. This involves a general PPT of 85 per cent and a PPT rate of 65.75 per cent for the first five years of petroleum operations. PPT applicable to PSCs is a 50 per cent flat rate for the contract area. In addition regulations and the Deep Offshore and Inland Basin Production Sharing Contracts Decree No. 9, 1999 as amended by Decree No. 26, 1999 (DOI-PSC Act) deals with royalties. For instance the DOI-PSC Act has a fixed royalty rate of 20 per cent for onshore PSCs (with production above 10,000 bpd) and a sliding scale of royalties (based on water depth) for offshore PSCs.[128] Rents and fees are dealt with in the regulations in which signature and production bonuses are required for PSCs and other concessions.

A significant fiscal reform is the introduction of the Nigerian Hydrocarbon Tax (NHT) to replace the current Petroleum Profit Tax (PPT). This is in addition to Company Income Tax (CIT) on upstream petroleum

---

124 Id., s 183(3)(d).

125 Id., s 183(1)–(3).

126 Id., s 190.

127 Id., s.190(3) and 191.

128 See ss 3, 4, 5 and 16 of Deep Offshore and Inland Basin Production Sharing Contracts Decree No. 9 of 1999 as amended by Decree No. 26 of 1999 ('DOI-PSCs Decree').

operations. The amount of CIT is not deductible when calculating NHT. While the proposed NHT rate is lower than the current PPT, upstream operators would incur additional income tax because NHT is not deductible in computation of CIT. Also, the NHT and CIT are applicable to all operators in the industry irrespective of size. This may negatively affect the growth of indigenous operators as they do not have the advantage of economies of scale.

Another fiscal issue is the payment of a renewal bonus based on renegotiated lease terms that reflect economic realities and is not deductible when calculating NHT. Payment of royalties is now based on value, water depth and production volume so that the government earns more from royalties when product prices go beyond the estimated benchmark. Furthermore, the penalty for gas flaring shall not be calculated based on a fixed sum against volume of gas flared but the penalty shall be no less than the value of gas flared and this also will not qualify as a tax deductible expense. Also a General Production Allowance (GPA) was introduced to replace Investment Tax Credit (ITC) and Investment Tax Allowance (ITA). Like the ITA, it decreases assessable profit which will have an adverse effect on the returns of pre-July 1998 PSC operators.

In addition to the NHT introduced in the bill and other changes to the fiscal terms for upstream operators, there are additional financial obligations expected of lessees. Part II (G) of the PIB provides for the establishment of the Petroleum Host Communities Fund (PHC Fund), its purpose and beneficial entitlements are for the communities. According to section 117, the PHC Fund shall be 'utilized for the development of the economic and social infrastructure of the communities within the petroleum producing area'. Every upstream petroleum operator is to remit on a monthly basis 10 per cent of its net profits[129] derived from upstream petroleum operations in onshore areas and in the offshore and shallow water areas directly into the PHC Fund, while those from upstream petroleum operations in deepwater areas should be remitted directly into the fund for the benefit of petroleum producing littoral states.[130] The contribution to the PHC Fund is a tax deductible expense that will have no effect on an operators' profit. It is therefore expenditure to the benefit of host communities. This is in addition to a similar remittance of 3 per cent of profit to the Niger Delta Development

129 'Net profit' for the purpose of the PHC Fund means 'the adjusted profit less royalty, allowable deductions and allowances, less Nigerian Hydrocarbon Tax less Company Income Tax', see s 118(2) of PIB.

130 S 118(1) of PIB.

Commission (NDDC) under the NDDC Act 2000. Since the NDDC is not included in the statutes to be repealed by PIB this seems like an additional obligation.

However, despite the laudable idea of the PHC Fund, section 118(5) and (6) creates doubt that communities will actually receive benefits from the PHC Fund. Section 118(5) states that 'where an act of vandalism, sabotage or other civil unrest occurs that causes damage to any petroleum facilities within a host community, the cost of repair of such facility shall be paid from PHC Fund entitlement unless it is established that no member of the community is responsible'. However, there is no provision for community representation in the process of determining community members' complicity in vandalism or acts of sabotage, which would result in abuse, and the malicious blaming and criminalization of communities resulting in collective punishment. In Nigeria, IOCs are quick to blame the majority of oil spills on 'sabotage' and, if not checked, the PHC Fund may become an industry and government fund for repair of facilities and environmental remedies rather than for the benefit of long-suffering host communities.

Section 118(6) makes the administration, management and disbursement of the PHC Fund political, non-transparent and open to abuse and manipulation as it is entirely within the discretion of the Minister for Petroleum. The uncertainty created by this provision may bring about abuse in the application of the Fund, resulting in chaos and restiveness in the communities. What is needed is certainty and an absolute guarantee that can only be provided by inserting the actual distribution and management mechanism of the 10 per cent in the PIB itself and not through subsidiary legislation by the Minister.

## Environmental Protection under the PIB

Part VII of the bill is on health, safety and the environment. All licensees and lessees are to have an environmental management plan for approval by the Inspectorate. The development plan is to meet adequate health, safety and environmental standards. In addition, it must contain an approved environmental management plan and provide for the elimination of routine gas flaring and must contain information on the environmental and social impact of the proposed oil and gas exploration and exploitation.[131] The Inspectorate shall have an environmental remediation fund to which every licensee or leaseholder is to make contributions

131 Id., ss 200 and 179.

that shall be prescribed by the Inspectorate.[132] Section 289(3) provides that the Inspectorate and the Downstream Petroleum Regulatory Agency shall, in consultation with the Federal Ministry of Environment, make regulations and issue directives specifically relating to environmental aspects of the petroleum industry. In regulating environmental management of petroleum upstream operations, the Inspectorate is to consult with the Federal and State Ministries of Environment within which the operations are based and any other relevant bodies.[133] These are bodies like the National Oil Spill Detection and Response Agency (NOSDRA) whose Act[134] prescribes regulations for dealing with waste emanating from oil production and exploration and its potential effect on the environment.

Operators in the industry have always complained of the duplication of regulations and inspections they are subjected to by the present DPR and the Federal Ministry of Environment.[135] Perhaps the PIB provisions would help to eliminate such duplication and conflict between sister government agencies.

As usual, all operations must be in line with 'good oilfield practice' which the bill defines as operations being in line with all relevant regulations, use of equipment that results in effective and efficient exploration, appraisal and development with minimal impact on the environment while adhering to all standards set by relevant international professional bodies like the American Petroleum Institute, the British Standards Institution and the International Organisation for Standards among others.[136] One of the grounds for revocation of a license or lease, provided in section 195(1)(g), is where a licensee or lessee fails to implement its environmental management plan in line with good oilfield practice.[137] However, 'good oilfield practice' should also take into consideration the application of best available technology and best environmental practice if the environment is to be adequately protected.

Furthermore, section 290 makes it mandatory for all oil and gas companies to comply with all environmental health and safety laws,

---

[132] Id., s 203.

[133] Id., s 202.

[134] National Oil Spill Detection and Response Agency (Establishment) Act 2006.

[135] These sentiments were expressed by management staff of both IOCs and indigenous operators at the 15th Biennial International HSE Conference on the Oil and Gas Industry, Abuja, Nigeria, 5–7 November 2012.

[136] Ss 295 and 362, PIB.

[137] Id., s 195(1)(g).

regulations and directives while, section 291 makes it mandatory for such companies to conduct their operations 'in accordance with internationally acceptable principles of sustainable development which includes the necessity to ensure that the constitutional rights of present and future generations to a healthy environment are protected'. Section 292 places an obligation on oil and gas companies to use the 'precautionary approach to environmental challenges' and to develop and use 'environmentally friendly technologies for exploration and development in Nigeria'. These provisions (sections 290, 291 and 292) are standard terms and should not be whittled down in any form while passing the bill into law. These sections incorporate internationally accepted environmental law principles of sustainable development, inter-generational and intra-generational equity and the precautionary principle. The challenge, however, is the application of these principles in day-to-day petroleum operations.

However, the bill still needs additional clauses to be effective. There is no incorporation of the polluter-pays principle, and it fails to jettison the monstrous provision about sabotage. Section 293(2), (3) and (4) is completely anti-environment and anti-people. It imposes an unbearable burden on innocent poor rural communities. The implication of the provision is that host communities will have to bear the burden of securing oil pipelines and equipment in their domain. The bill seems to hold them responsible once the oil operators allege sabotage until the contrary is proved. What the bill fails to put into consideration is that even where there is sabotage it does not necessarily follow that it was an act of a member of the community.

## Land Acquisition and Compensation under the PIB

The PIB introduces nothing significant to guarantee the payment of fair, adequate and reasonable compensation to those whose land must be compulsorily acquired for petroleum activities. Sections 198 and 199 that prohibit the destruction of trees and objects of commercial value and religious veneration have not changed the existing unfair situation. The PIB provision for compensation excludes the active participation of landowners in oil producing areas in the determination of what is adequate as compensation for them. Section 199(1) provides that 'the amount of compensation payable under section 198 shall be determined by the inspectorate in consultation with designated persons and representatives of the persons whose protected objects have been damaged and the licenses or lessees, in accordance with regulation made by the Minister on the advice of the Inspectorate'.

Unlike the position in the Minerals and Mining Act 2007 where the landowner has a say on what their compensation should be, section 296 of the PIB provides that the owner of land covered by a license will be compensated on terms to be determined by the Agency through its guidelines that shall be issued annually to reflect the rate of inflation and other relevant factors. The 'other relevant factors' are unqualified, and may pose a problem to the annual reviews and create uncertainties.

Section 298 on penalties is a far cry from what is expected of petroleum legislation, as penalties and sanctions are left to the Inspectorate and Agency to determine without specifying the criteria to be used for such determination. It is important for the companies to know the penalties they will face in case of environmental damage and it is also important for all citizens to know the penalties and sanctions for environmental damage under a new petroleum law. This will aid communities and environmental activists to have a basis on which to check operators' actions and government responses.

Furthermore, environmental regulations in the oil and gas sector of Nigeria appear to be insensitive to human rights, health, safety and the environment of host communities when compared with provisions in the 2007 Mineral and Mining Act (Mining Act). While the Petroleum Act 1969 was enacted at a time when environmental awareness was low, it is hard to explain why a 2012 Petroleum Industry Bill will contain provisions short of the 2007 Mining Act which regulates solid minerals. Under the Mining Act, in addition to land rent, communities and individual landowners are entitled to compensation for disturbance of surface rights.[138] The amount of the compensation payable is 'determined by the Mining Cadastre Office after consultation with the State Minerals

---

[138] See s 107 of the Mining Act which states:

> A holder of the Mineral title may, in addition to any other amounts payable under the provision of this Act and subject to valuation report by a Government licensed valuer, pay to the occupier of the land held under a State lease or the subject of right of occupancy –
>
> (a) reasonable compensation for any disturbance of the surface rights of the owner or occupier and any damage done to the surface of the land on which the exploration or mining, is being or has been carried; and
> (b) in addition pay to the owner of any crop, economic tree, building or work damaged, removed or destroyed by the holder of the mining title or by any of its agents servants, compensation for the damage, removal or destruction of the crop, economic tree, building or work.

Resources and Environmental Management Committee and a Government licensed Valuer'.[139] This creates more certainty on how the amount of compensation would be arrived at than the provisions of sections 198 and 296 of the PIB.

Section 19 of the Mining Act provides for a Mineral Resources and Environmental Management Committee for each state of the Federation to, inter alia, consider issues affecting compensation and make necessary recommendations to the Minister, discuss, consider and advise the Minister on the matters affecting pollution and degradation of any land on which any mineral is being extracted and advise the local government areas and communities on the implementation of programmes for environmental protection and sustainable management of mineral resources. There seems to be no equivalent of this committee in the petroleum industry under the PIB. Considering the host community concerns and issues in the oil producing Niger Delta of Nigeria, such a committee is relevant and needed in the petroleum industry and should be provided for in the PIB with clear inclusion of representation for host communities. Also, similar provisions to sections 3(1)(c), 100, 102 and 107 of the Mining Act on the process of acquisition of land and the grant of mining title and compensation ought to be incorporated into the PIB.

## 10. CONCLUSION

The Nigerian petroleum industry has grown over the years since 1956 when the commercial discovery was made at Oloibiri in Bayelsa State. Licensing has moved gradually from concessions over almost limitless expanses of land to carefully mapped-out oil blocks. Competitive bidding is gradually and increasingly taking over the discretionary award of oil mining licenses. This has opened up the industry to different categories of oil and gas companies from various countries. Indigenous companies have also emerged as a result of the deliberate government policy of Nigerianization.

Nigeria has practised virtually every form of petroleum licensing from concession to JV to PSC and sole risk, settling on the grant of licenses in a transparent bid process. There are on-going changes to the legal framework for the petroleum industry in Nigeria in a bid to attract investment and increase government revenue, and make the industry competitive, transparent and accountable. There has been a lot of

---

139 S 108, Mining Act.

opposition to some of the changes, especially the fiscal terms, but it is safe to say that the terms are still competitive, considering the investment climate in Nigeria.

# 10. Regulation of the Russian Federation petroleum licensing regime

**Kristoffer Svendsen and Andrey Kompaniets**

## 1. INTRODUCTION

Petroleum production has been, and still is, a large part of the Russian economy and society. In 2013 Russia became the top oil producer [in the world], producing 523.2 million tonnes of oil. This chapter takes a look back at petroleum development in Russia, giving a summary of the history and an overview of the petroleum activities from Tsarist Russia, through the Soviet Union era, to today's Russia. It continues with an outline of the legal framework for offshore petroleum activities in Russia and the historical development, reform, and current petroleum laws. In section 5 we examine the subsoil user's requirement to obtain a license, a product sharing agreement, or a service agreement to conduct oil and gas activities in Russia and the conditions for granting such a license. The chapter continues to look at assessment of petroleum field development and the strong government involvement in the petroleum sector. The environment and safety aspects of the industry are important parts of oil and gas operations, and are discussed in the final part of this chapter.

## 2. HISTORY AND OVERVIEW OF RUSSIAN PETROLEUM ACTIVITIES

Petroleum activities in Russia extend back over several decades. The petroleum industry has always been very closely connected to the country's economic and political life. Scientific and technological advances played an important role in the development of the industry, as it did in other countries.

Over a period of many centuries, signs of oil have been registered on the Apsheron peninsula in the Caspian Sea. 'Everlasting belches of fire', which were worshipped by the Zoroastrians, were located in the Baku region. In the year 1702, in the days of Emperor Peter I, the press published reports about oil on the river Sok in Povolzhye of the Volga River Basin. Official news also emerged about oil deposits in other parts of the country; such as in the Northern Caucasus, especially in the vicinity of Grozny, Priobye, and the riverside of Irtysh.[1] In 1745, oil extraction started from the Ukhta riverbed, at which point the world's first oil refinery started delivering oil products to Moscow and Saint Petersburg.[2]

In the modern understanding the birth of the world's, and incidentally the Russian, oil industry can be attributed to the year 1846 when, more than a decade before the pioneering well in the USA, the world's first oil well was drilled close to Baku, which at that time was part of the Russian Empire. Previously, oil was extracted through primitive hand-dug pits.[3] In 1864 a well was completed in the Krasnodar Territory, and at the riverside of Ukhta four years later. The growth of oil production was accompanied by the construction of oil processing factories, which resulted in more than 20 refineries in Baku by 1873.[4] The change from state monopoly to oil industry at that time facilitated the emergence of private industrial enterprises, which led to the sector actively growing. Fifteen years later the amount of extracted oil in Russia increased ten fold to 23 million barrels.

The famous Russian chemist Mendeleev turned his attention for the first time to the importance of oil as the source of chemical raw materials

---

[1] Andrey Kuz'michev (2003) 'Istoriya hefti v Rossii' ('История нефти в России'), 7 (35) July 2003 *Energetika i promyshlennost' Rossii (Энергетика и промышленность России)*, retrieved 30 October 2014 at www.eprussia.ru/epr/35/2371.htm.

[2] Andrey Kuz'michev (2003) 'Istoriya hefti v Rossii' ('История нефти в России'), 7 (35) July 2003 *Energetika i promyshlennost' Rossii (Энергетика и промышленность России)*, retrieved 30 October 2014 at www.eprussia.ru/epr/35/2371.htm.

[3] One of the first references to procuring oil in Russia emerged in the 19th century. Residents by the Ukhta riverside, in the northern part of the Timan-Pechora region, collected oil from the surface of the river and used it in household activities.

[4] A.S. Fedorov (1982) 'Glava VI Razvitiye khimicheskoi tekhnologii' ('Глава VI Развитие химической технологии') in *Tekhnika v ee istoricheskom razvitii* (*Техника в ее историческом развитии)* (editors S.V. Schukhardin, N.K. Laman, A.S. Fedorov), paragraph: 6. Tekhnologiya pererabotki nefti I kamennogo uglya (Технология переработки нефти и каменного угля).

and not just fuel material with the statement '*mozhno topit' i assignatziyami*' in 1885,[5] meaning that oil can be burnt as fuel as well as used as a specific type of paper money used in Russia during the period 1770–1840. Mendeleev dedicated a body of work to oil processing.

Foreign investors played an important role in the growth of the Russian oil industry during the Russian Empire period. Branobel (the Nobel brothers) was one of the largest oil companies in the world during the late 19th century. Around the same time, the Rothschild family owned one of the largest Russian oil companies extracting Russian oil. Before the 1917 revolution, about 50 per cent of the invested capital in the Caucasus oil industry came from foreign investors. The Bakinskiy region accounted for 90 per cent of the all-Russian oil extraction. At the end of the 19th century, more than 30 per cent of the world oil production took place in Russia. Fast growth in oil output necessitated new market outlets. In this period Russian oil entered the global market with force.[6]

The 1917 revolution caused a drop in Russian oil extraction due to the nationalization of oil deposits following the revolution. Many companies were unsatisfied with the Soviet government's politics. The Nobel brothers sold a significant part of their Russian assets, while the companies Vacuum and Standard Oil continued investing in Russia. By 1923 the export of oil returned to its pre-revolutionary levels.[7]

The Caspian and North Caucasus played the leading role in Soviet oil industry until the Second World War. During the war, Germany strived to cut off the Soviet Union's extraction in this region due to Germany's and its allies' lack of substantial oil deposits. Therefore, seizure of Caucasian oil deposits was one of Germany's overall objectives for its advancement to the East. Caspian oil extraction rose again after the war and hit a record level in 1951 at 850 000 barrels per day (bbl/d). Baku was

---

[5] D.I. Mendeleev (1949) *Sochineniya: V 25 t – T.10: Neft* (Менделеев Д. И. Сочинения: В 25 т. – Т. 10: Нефть. Л.-М., 1949). p. 463.

[6] Andrey Kuz'michev (2003) 'Istoriya hefti v Rossii' ('История нефти в России'), 7 (35) July 2003 *Energetika i promyshlennost' Rossii (Энергетика и промышленность России)*, retrieved 30 October 2014 at www.eprussia.ru/epr/35/2371.htm.

[7] Andrey Kuz'michev (2003) 'Istoriya hefti v Rossii' ('История нефти в России'), 7 (35) July 2003 *Energetika i promyshlennost' Rossii (Энергетика и промышленность России)*, retrieved 30 October 2014 at www.eprussia.ru/epr/35/2371.htm.

established as a large industrial centre producing about two-thirds of Soviet oil equipment in this region.[8]

Paralleling this development, prospecting work was carried out in the Volga-Ural region. From the 1950s, extraction from new oilfields averaged about 45 per cent of the gross output of the Soviet Union. Significant amounts of money were invested to facilitate the growth of oil extraction in the USSR.[9] With these investments, it became possible for the Soviet Union to significantly increase oil exports. The Soviet Union actively worked for an increase in its share of the world market to receive the maximum inflow of foreign currency. In the early 1960s, the Soviet Union superseded Venezuela as the second largest oil producer in the world. At the same time, many western oil companies had to lower their price on oil extracted in the Middle East due to the competition caused by the market entry of Soviet oil. This led to a decline in the Middle Eastern countries' oil extraction taxes and royalties, substantially contributing to the creation of OPEC.[10]

In the beginning of the 1960s, the first giant oilfields were discovered and developed in Western Siberia. Chief among which was the opening in 1965 of the supergiant Samotlor field with recoverable resources of about 14 billion barrels (Bbbl) (2 billion tonnes). Western Siberia became one of the largest global oil sources, which was most welcome to the Soviet Union as production in the Volga-Ural region, having reached 4.5 million barrels per day in 1975, fell by two-thirds from the 1975 level. The Soviet Union quickly set up production in Western Siberia notwithstanding the extreme climatic conditions and its enormous permafrost territories located above the polar circle. Western Siberia supported the growth of oil production in the Soviet Union from 7.6 million bbl/d

---

8 Andrey Kuz'michev (2003) 'Istoriya hefti v Rossii' ('История нефти в России'), 7 (35) July 2003 *Energetika i promyshlennost' Rossii (Энергетика и промышленность России)*, retrieved 30 October 2014 at www.eprussia.ru/epr/35/2371.htm.

9 Andrey Kuz'michev (2003) 'Istoriya hefti v Rossii' ('История нефти в России'), 7 (35) July 2003 *Energetika i promyshlennost' Rossii (Энергетика и промышленность России)*, retrieved 30 October 2014 at www.eprussia.ru/epr/35/2371.htm.

10 Andrey Kuz'michev (2003) 'Istoriya hefti v Rossii' ('История нефти в России'), 7 (35) July 2003 *Energetika i promyshlennost' Rossii (Энергетика и промышленность России)*, retrieved 30 October 2014 at www.eprussia.ru/epr/35/2371.htm.

(more than a million tonnes) in 1971 to 9.9 million bbl/d (about 1.4 million tonnes) in 1975.[11]

The Soviet oil industry's advances in extraction from gigantic and relatively cheap field development in Western Siberia paved the way to a period of decline. The drive to produce the maximum amount of oil in the shortest time possible led to low reservoir pressure, a fall in well production capacity, and increased water cutting. The reason being the drilling of multiple wells, in which too much water was injected. Nevertheless, the Soviet leadership addressed the problems and constrained the decrease in production up to the beginning of 1990, at the expense of increased financing. Significant investments were made in drilling and the Soviet Union reached a production all-time high in 1988, extracting about 11 million bbl/d. The country became the largest oil producer in the world with production quantities significantly higher than the USA and the Saudi Arabia.[12]

The fall of the Soviet Union caused an economic collapse, which brought about a breakdown in oil extraction. The subsequent economic crisis caused a slump in domestic oil demand, while the ability to export oil was limited. New prospecting work and the amount of drilling and repair work to existing wells continued to decline. As a result, continuous and unavoidable falls in production took place.

During the 1990s Russian industries returned to private ownership and entrepreneurial activities, which gave rise to growth in joint-stock companies, partnerships, and other forms of enterprises. Vertically integrated oil and gas companies were created during the privatization of the oil industry,[13] which played a leading role in Russian economic life. In the second half of the 1990s the Russian oil production decline was halted as a result of increased demand, rising oil prices on the global market, and renewed development of the industry in Russia.

In 2013, Russia set a new oil production record by extracting 523.2 million tonnes of oil, a number not seen by the industry since the

---

[11] For detailed information about the Soviet oil and gas industry, see Office of Technology Assessment, United States Congress, 'The Soviet Oil and Gas Industry' in *Technology & Soviet Energy Availability* (U.S. Government Printing Office, Washington D.C. 1981), Chapter 2.

[12] Irina Kezik (11 July 2013) 'Russia's Daily Oil Production Hits 25-Year High' in *Russia beyond the Headlines*, retrieved 28 October 2014 at http://rbth.com/business/2013/07/11/russias_daily_oil_production_hits_25-year_high_27975.html.

[13] Gazprom, Lukoil, YUKOS, Surgutneftegaz, Sibneft and others.

1990s.[14] The same year, Russia again became the top oil producer in the world, surpassing Saudi Arabia by a million bbl/d. High oil prices encouraged companies to invest part of their profits in the exploration and development of new oilfields.[15]

Significant merger and acquisition activities have been characteristic of Russian oil producing companies in recent years (Gazprom Neft acquired Sibneft, Rosneft purchased YUKOS, TNK-BP was created, Rosneft acquired TNK-BP). Foreign investors continue to enter Russia and participate in joint ventures with Russian companies to develop oilfields in different parts of the country, such as in the Caspian region and on Sakhalin. In summary, the oil industry continues to occupy a leading place in the Russian economy and is an engine for other sectors, from the service industry to the shipbuilding industry.

## 3. LEGAL FRAMEWORK FOR OFFSHORE PETROLEUM ACTIVITIES IN RUSSIA

The largest part of Russia's known oil and gas reserves is concentrated on dry land, of which about 75 per cent are already developed. The Russian continental shelf consists of a huge resource potential and contains the more lucrative areas for hydrocarbon production. The Russian continental shelf is the largest in the world and consists of more than 6 million km$^2$, which is greater than 20 per cent of the surface of the world's oceans.[16] However, the amount of discovered deposits is few. Several large fields have been discovered on the Russian shelf (Shtokmanovskoye, Rusanovskoye, Leningradskoye, Dolginskoye, Prirazlomnoe, and others) with oil and gas deposits of about 10 billion tonnes oil

---

[14] Statement by the Minister of Energy Aleksandr Novak to RiaNovosti reported by Taras Fomchenkov (10 Jan 2014) 'Россия установила рекорд по добыче нефти в 2013 году', retrieved 28 October 2014 at www.rg.ru/2014/01/10/neft-site-anons.html.

[15] Statement by the Minister of Energy Aleksandr Novak to RiaNovosti reported by Taras Fomchenkov (10 Jan 2014) 'Россия установила рекорд по добыче нефти в 2013 году', retrieved 28 October 2014 at www.rg.ru/2014/01/10/neft-site-anons.html.

[16] V.I. Bogoyavlenskii (2012) 'Perspektivi i problem osvoeniya mestorozhdenii nefti i gaza shel'fa Arktiki 11/2012 *Burenie i neft'*, analiz i prognoz. (В.И. Богоявленский (2012) 'Перспективы и проблемы освоения месторождений нефти и газа шельфа Арктики' 11/2012 *Бурение и нефть*, Анализ и прогноз).

equivalents.[17] The enormous Shtokmanovskoye field in the Arctic area consists of 3.8 trillion cubic metres of gas and 54.3 million tonnes of condensate.[18] The commencement date of production for this field has been delayed indefinitely due to economic considerations, and demanding climatic and environmental conditions.

Currently, offshore deposits account for less than 3 per cent of Russian oil production due to the difficulty of extracting natural resources on the shelf compared to natural resources on dry land. Russian offshore oil and gas deposits are concentrated on the Arctic shelf, which suffers from a harsh climate and a difficult navigation climate due to the presence of ice. The climatic conditions also account for a shortage in coastal infrastructure. As a result of these specific features, field development on the Russian shelf requires significant investments, particularly in expensive technology to facilitate extraction in a complex environment.

Currently, active offshore petroleum development occurs in the Sakhalin Islands and the Caspian Sea. In December 2013, oil extraction also started at the Prirazlmnoe oilfield in the Pechora Sea. About 85 per cent of the explored hydrocarbon resources are located on the Arctic shelf (Barents Sea and the Kara Sea), while 12–14 per cent of the resources are located in the Russian Far East.[19] Additionally, oil deposits are located on the Baltic, Black, and Azov shelfs.

A plethora of international and Russian law regulates the legal status and relations of the Russian continental shelf. The main legal instruments governing Russian offshore (aqua-territorial) activities are: international conventions and agreements, such as the United Nations Convention on the Law of the Sea 1982 (UNCLOS); the Constitution of the Russian Federation; the Subsoil Law;[20] the Federal Continental Shelf

---

[17] V.I. Bogoyavlenskii (2012) 'Perspektivi i problem osvoeniya mestorozhdenii nefti i gaza shel'fa Arktiki 11/2012 *Burenie i neft'*, analiz i prognoz. (В.И. Богоявленский (2012) 'Перспективы и проблемы освоения месторождений нефти и газа шельфа Арктики' 11/2012 *Бурение и нефть*, Анализ и прогноз).

[18] Gazprom (2014) 'Shtokmanovskii proyekt' ('Штокмановский проект') retrieved 10 March 2014 at www.gazprom.ru/about/production/projects/deposits/shp/.

[19] A.V. Filipov (2013) 'Neft' i gaz – morskoye prodolzheniye zemnoi istorii' 12/2013 *Neftegazovaya Vertikal'* (А.В. Филиппов, (2013) 'Нефть и газ – морское продолжение земной истории' 12/2013 *Нефтегазовая Вертикаль*) p. 53.

[20] Zakon RF 'O nedrakh' (Закон РФ 'О недрах' от 21.02.1992 N 2395-1).

Law;[21] the Federal Exclusive Economic Zone Law;[22] the Federal Internal Sea Waters, Territorial Sea and Territorial Waters Law;[23] the Water Code;[24] and other normative legal acts and regulations. The Russian marine zone is divided into: internal waters, territorial sea, the exclusive economic zone, the continental shelf, and the international area of the ocean floor.[25]

Subsoil on the Russian continental shelf is under state control and appropriate government authorities assign subsoil for use. The Russian Federation and the subjects (constituent entities: oblasts, republics, krais, autonomous okrugs, federal cities, and autonomous oblasts) of the Russian Federation jointly carry out possession, use, and disposal of subsurface resources within the limits of the Russian Federation located in the State Fund. However, subsoil blocks on the continental shelf are within the exclusive jurisdiction of the Russian Federation. The Russian Federation has the right to establish legally binding rules about the use of natural resources in state ownership. Russia's right to the continental shelf is embedded in UNCLOS, and Article 77 sets forth that Russia exercises sovereign rights over the continental shelf for the purposes of exploring and exploiting its natural resources. As expressed in Article 77, the sovereign rights are exclusive in the sense that if Russia does not explore the continental shelf or exploit its natural resources, no other entity may undertake these activities without the express consent of Russia.

The Federal Continental Shelf Law also confirms the UNCLOS position that Russia exercises sovereign rights over the continental shelf for the purposes of exploring the continental shelf and exploiting its

---

21 Federal'nii zakon 'O kontinental'nom shel'fe Rossiiskoi Federatsii' (Федеральный закон от 30.11.1995 N 187-ФЗ 'О континентальном шельфе Российской Федерации').

22 Federal'nii zakon 'Ob isklyuchitel'noi ekonomicheskoi zone Rossiiskoi Federatsii' (Федеральный закон от 17.12.1998 N 191-ФЗ 'Об исключительной экономической зоне Российской Федерации').

23 Federal'nii zakon 'O vnutrennikh morskikh vodakh, territorial'nom morye i prilezhaushshei zone Rossiiskoi Federatsii' (Федеральный закон от 31.07.1998 N 155-ФЗ 'О внутренних морских водах, территориальном море и прилежащей зоне Российской Федерации').

24 Vodnii kodeks RF (Водный кодекс РФ от 03.06.2006 N 74-ФЗ).

25 A.I. Perchik (2002) *Gornoye pravo: Uchebnik* (Перчик А.И. Горное право: Учебник. Изд. 2-е – М.: Филология три, 2002).

natural resources.[26] Some examples of exclusive rights are the right to: authorize and regulate drilling on the continental shelf for all purposes; and construct, authorize, and regulate the erection, operation, and use of artificial islands, installations, and structures (the Russian Federation exercises jurisdiction over artificial islands, installations, structures, customs, tax, health and immigration laws, and safety laws and regulations).[27]

The Federal Subsoil Resources Management Agency, a federal government agency under the Ministry of Natural Resources and Environment of the Russian Federation, issues all licenses. The right to use a subsoil block on the continental shelf arises at the moment of receiving the license, issued on the grounds of a tender or an auction. However, in practice, it is possible for a state company to receive a license without holding an auction.

A license for use of blocks on the continental shelf, compared to a subsoil license (see more in section 5 of this chapter) should additionally contain information about the environmental security of the used blocks including: the environmental monitoring organization, agreement on mitigation and compensation for inflicted harm to aquatic bioresources, preventive measures and damage control and mitigation, insurance, conservation, and liquidation of installations and constructions upon completion of the work.[28] Furthermore, the agency should write into the license and agree with the block users regarding investment into community development of the coastal territories and order placement with Russian companies (local content provisions).[29]

Under the civil legislation of the Russian Federation the right to use blocks on the continental shelf cannot be transferred from the original user to a third party. The Subsoil Law sets forth specific procedures for

[26] Federal'nii zakon 'O kontinental'nom shel'fe Rossiiskoi Federatsii' (Федеральный закон от 30.11.1995 N 187-ФЗ 'О континентальном шельфе Российской Федерации'), art. 5.

[27] Federal'nii zakon 'O kontinental'nom shel'fe Rossiiskoi Federatsii' (Федеральный закон от 30.11.1995 N 187-ФЗ 'О континентальном шельфе Российской Федерации'), art. 5.

[28] Federal'nii zakon 'O kontinental'nom shel'fe Rossiiskoi Federatsii' (Федеральный закон от 30.11.1995 N 187-ФЗ 'О континентальном шельфе Российской Федерации'), art. 8.

[29] I.B. Kalinin (2001) *Pravovoe regulirovaniye resursopol'zovaniya* (Калинин И. Б. Правовое регулирование ресурсопользования. – Томск: Изд-во Научно-технической литературы, 2001), p. 88.

the transfer of rights to use blocks.[30] Subsoil users understand that current technical and economic constraints prohibit the quick development of difficult offshore deposits. In this regard, Gazprom and Rosneft, as companies authorized by the government to operate on the shelf, strive their utmost to obtain licenses for offshore fields for subsequent development in the long term.

The Federal Continental Shelf Law of the Russian Federation is the main regulatory act. However, other acts also regulate issues arising on the continental shelf. The Subsoil Law,[31] the Federal Environmental Protection Law,[32] the Federal Environmental Examination Law,[33] the Federal Production Sharing Agreement Law,[34] the Civil Code,[35] and the Tax Code[36] are some examples of acts regulating oil extraction issues on the shelf.

According to the Subsoil Law, a subsoil user engaged on the continental shelf can only be a company in which the state directly or indirectly controls more than 50 per cent of the total number of voting shares. In other words, the Russian Federation's shareholders' equity holding in legal entities engaged in oil production on the shelf should be more than 50 per cent. Additionally, a company operating on the Russian shelf is required to have five years of work experience from shelf operations. Currently, only Gazprom and Rosneft fulfill these requirements. The law does not, however, mention whether a mother company's experience may extend to the daughter company and vice versa. Therefore, it is difficult to assess the ability of other companies with direct or indirect state participation in developments on the Russian shelf (for example, Gazprom Neft). How a company can gain similar work experience when such a five-year requirement is in place is incomprehensible. Take, for example, the state company Zarubezhneft (100 per cent state owned). The company holds more than 20

---

30 Zakon RF 'O nedrakh' (Закон РФ 'О недрах' от 21.02.1992 N 2395-1), art. 17.1.

31 Zakon RF 'O nedrakh' (Закон РФ 'О недрах' от 21.02.1992 N 2395-1).

32 Federal'nii zakon 'Ob okhrane okruzhayushshei sredi' (Федеральный закон 'Об охране окружающей среды' от 10.01.2002 N 7-ФЗ).

33 Federal'nii zakon 'Ob ekologicheskoi ekspertize' (Федеральный закон 'Об экологической экспертизе' от 23 ноября 1995 г. N 174-ФЗ).

34 Federal'nii zakon 'O soglasheniyakh o razdele produktsii' (Федеральный закон от 30.12.1995 N 225-ФЗ (ред. от 19.07.2011) 'О соглашениях о разделе продукции').

35 Grazhdanskii kodeks RF (Гражданский кодекс РФ (ГК РФ) от 30.11.1994 N 51-ФЗ).

36 Nalogovoi kodeks RF (Налоговый кодекс РФ (НК РФ) часть 1 от 31.07.1998 N 146-ФЗ).

years of work experience from foreign shelves, but has not obtained five years of work experience from the Russian continental shelf. As a result, the company does not have the right to access and develop the Russian shelf. Consequently, the only option for participation in Russian continental shelf development is by entering into joint ventures with state companies, which remain the license holders.

The largest offshore oil and gas development projects in Russia, Sakhalin-1 and Sakhalin-2, occurred under production sharing agreements (PSA). These respective agreements were signed on 30 June 1995 and 22 June 1994 accordingly, before the Federal Production Sharing Agreement Law entered into force (the law was approved on 30 December 1995). Therefore, the provisions of that law do not apply to the investors of these two projects.

## 4. PETROLEUM LAW: HISTORICAL DEVELOPMENT, REFORM AND CURRENT LAWS

Exploration and development of natural resources constitutes Russia's economic foundation, which is unlikely to change for a very long time, as the subsoil legislation and its development substantially inform the country's politics. The historical development of the oil extraction legislation in Russia is closely connected to the historical development of the oil industry (see section 2 of this chapter). An identifying regulatory feature of the Russian oil sector, compared to many other countries, is the historical non-existence and current lack of a special act regulating oil extraction activities. The government authorities throughout history considered it sufficient to have a general act about subsoil use.

The formation of legislation regulating subsoil use relations in Russia started in the mid-17th century. However, attempts to systemize and take a comprehensive approach to the issue occurred during the reign of Peter I (2 November 1721 – 8 February 1725). The development of mineral law was one of the conditions when developing the mineral industry. In this regard, Russian mineral law can be said to be what stimulated and motivated the permitted entities to engage in mineral extraction.

Among the host of regulatory acts in Russia's history, it is possible to extract four main pieces of legislation in the area of mineral law from distinct time periods: the Mineral Statute of the Russian Empire;[37] the

---

[37] Ustav Gornii Rossiiskoi Impreii (Устав Горный Российской Империи 1832).

Mineral Act[38] and the Subsoil Code[39] of the Russian Soviet Federative Socialist Republic (RSFSR); and the Subsoil Law of the Russian Federation.[40]

The first fundamental publication of Russian mineral legislation is considered to be the *Compilation of Authorities and Regulations of Mineral Management*[41] published in 1832. It was compiled on the basis of decrees, orders, instructions, and other legal acts at the beginning of the 19th century. It is the first official publication of the Mineral Statute of the Russian Empire, and did not consist of legal norms regulating issues regarding oilfield development.[42]

The 'Instruction about the oil industry' issued by the Ministry of Finance on 14 November 1839[43] is the first regulatory act in Russia regulating oil extraction. This date can be considered as the birth of oil legislation in Russia. The document regulated the management of state oilfields in the Bakinskiy and Schemakhanskiy districts of the Caspian region.

The second edition of the Mineral Statute was published in 1842, in which two sections were devoted to oil production and concerned: oil extraction and the management of state oilfields in the Bakinskiy and Schemakhanskiy districts of the Caspian region (as in the Instruction), as well as oil extraction on the lands of the Chernomorsk Cossack forces. The edition clarified through amendments and additions the order of developing different kinds of mineral resources and strengthened state influence in subsoil use. Priority was determined by the subsoil resources and not by land. Additionally, ownership in produced oil was defined accordingly: ownership to oil produced on the lands of Chernomorsk Cossack forces vested in the forces, while the state took ownership in oil from the lands of the Bakinskiy and Schemakhanskiy districts.

In the latter editions of the Mineral Statute (1857 and 1893), the statute systemized and improved the submission order of subsoil blocks for exploration and production of various types of mineral resources. Later

---

38 Gornii zakon RSFSR (Горный закон РСФСР 1928).

39 Kodeks RSFSR o nedrakh (Кодекс РСФСР о недрах 1976).

40 Zakon RF 'O nedrakh' (Закон РФ 'О недрах' от 21.02.1992 N 2395-1).

41 Svod uchrezhdenii i ustavov gornogo upravleniya («Свод учреждений и уставов горного управления»).

42 Russkoe gornoe zakonodatel'stvo s raz''yasneniyami. (Русское горное законодательство с разъяснениями. Часть 1. Устав горный. – СПб. 1905.// Справочно-правовая система «Гарант»).

43 Instruktsiya o neftyanom promisle Ministerstva finansov (Инструкция о нефтяном промысле Министерства финансов от 14 ноября 1839 г.).

government agencies received a range of acts regulating oil production and exploration, as well as the submission of oil-bearing parcels of land, based on the first edition of the Mineral Statute. Therefore, 'rules about the oil industry'[44] were included in the last edition of the Mineral Statute of 1893 under the title 'about private oil industry'.[45]

Landowners could conduct oil exploration and production. That said, a seizure-of-necessary-lands mechanism came into play when a landowner did not want to independently conduct oil development. Oil producing subsoil was considered state property; therefore, the seizure mechanism resulted in the state seizing the use and the enjoyment of the subsoil. However, provisions facilitated the transfer of subsoil blocks to private individuals through bargaining. Bargaining was conducted through a lump-sum payment accompanied with a yearly payment for use of the block. Later, simpler procedures to determine rights in oil resources were established.

After the 1917 revolution the oil industry experienced decay. The new politico-economic reality abolished private ownership, which necessitated new oil regulation. The Soviet Government Decree about land from 26 October 1917 nationalized subsoil.[46] The Decree about subsoil of land from 30 April 1920[47] contained principles, which among others, separated subsoil from the surface, changed all acts and contracts with private individuals regarding subsoil, and established exclusive rights in the Soviet state (RSFSR) to exploit subsoil and sell extracted mineral products.

The Mineral Act of the Russian Soviet Federative Socialist Republic (RSFSR Code) was enacted in continuance of the Mineral Enactment of the USSR of 1920 and includes: state ownership of subsoil and subsoil resources; subsoil user's duty to inform the state about geological information; assignment of subsoil for use through tendering or without tendering (paid use); subsoil user's right; assignment and transfer of rights; and state control and supervision.

Russian subsoil legislation is national legislation within the federal framework of the Russian state penetrating both levels of the two-level system: federal legislation and legislation of the subjects (similar to a

---

44 Pravila o neftyanom promisle («Правила о нефтяном промысле»).

45 'O chastnom neftyanom promisle' («О частном нефтяном промысле»).

46 DekretCovetskogo gosudarstva 'O zemle' (Декрет Советского государства от 26 октября 1917 г. 'О земле').

47 Dekret CNK RSFSR 'O nedrakh zemli' (Декрет СНК РСФСР от 30.04.1920 г. 'О недрах земли').

state) of the Russian Federation. Article 72 of the Russian Constitution[48] outlines the two-level system of subsoil legislation, stating that questions of subsoil possession, use, and disposal are within the joint jurisdiction of the Russian Federation and the subjects of the Russian Federation. The fundamental principles of subsoil legislation for the USSR and the Union Republics were passed on 9 July 1975,[49] and the year following the Subsoil Code of the RSFSR.[50] Subsoil users could be state, cooperative, or community enterprises, as well as citizens of the USSR. The Code was silent on foreign legal entities' and foreign private individuals' participation, but it was not prohibited. As earlier, the legislation clearly underlined that the state held ownership to subsoil and strengthened the principle that subsoil was a national resource as well as an interest for current and future generations.

The change in the Russian political system in the early 1990s required amendments and reform of the subsoil legislation. The RSFSR Code was replaced in 1992 by the current Subsoil Law of the Russian Federation,[51] which has recently undergone significant amendments due to the return of private ownership. Thus, supplements were introduced to address rights held by subsoil users and work under PSAs.

The Federal Production Sharing Agreement Law was one of the first Russian regulatory acts passed to attract Russian and foreign investors to prospect, explore, and produce hydrocarbons on PSA conditions. PSA constitutes a contract between the Russian Federation on the one hand and one (or several) investor(s) on the other. According to the terms, the Russian Federation is a subject of commercial activity in return for consideration. The state obtains the exclusive rights to prospect, explore, and produce hydrocarbons on the subsoil block as set forth in the contract and control the related work for a specific time period. The investors are hired in as contractors, while the state retains ownership. The investors are obligated to carry out the required work at their own expense and risk, with the government part of the production output in accordance with the agreement. Further discussion of the contractual

48 Konstitutsiya Rossiiskoi Federatsii (Конституция Российской Федерации от 25 декабря 1993 года).

49 Zakon SSSR 'Ob utverzhdenii Osnov zakonodatel'stva Soyus SSR i soyus nikh respublik o nedrakh' (Закон СССР от 09.07.1975 г. 'Об утверждении Основ законодательства Союза ССР и союзных республик о недрах').

50 Zakon RSFSR 'Ob utverzhdenii Kodeksa RSFSR o nedrakh' (Закон РСФСР от 09.07.1976 г. 'Об утверждении Кодекса РСФСР о недрах' (вместе с 'Кодексом РСФСР о недрах')).

51 Zakon RF 'O nedrakh' (Закон РФ 'О недрах' от 21.02.1992 N 2395-1).

terms of subsoil use is undertaken in section 5 of this chapter. At the present time, realizing oil production projects on the Russian shelf through PSAs are not a preferred or prioritized model.

Notwithstanding that Russian mineral resource use legislation is predominantly based on licensing and is subject to legal-administrative rules, the legislation provides for contractual relationships in this area of law. Contractual relationships as an alternative to the licensing system are laid down in Article 12 of the Subsoil Law. Therefore, subsoil use can formally be divided into contractual (civil-legal) and administrative (licensing) use. Unfortunately, Russian legislation does not contain regulation of the contractual subsoil use regime. The Subsoil Law mentions two types of subsoil use contracts: PSA and service contracts. However, only PSA has a legislated base for realization and is formally considered the only existing contract for subsoil use in Russian law.

## 5. GRANT OF LICENSES FOR PETROLEUM ACTIVITIES

The license is the basis for subsoil use in accordance with the Subsoil Law. A license is a document confirming the right of the owner to license a specifically defined block of subsoil in accordance with the expressed determination during the stated period of time subject to the previously agreed-upon requirements and conditions.[52]

The issuing of a license occurs through the state licensing system and the Federal Subsoil Resources Management Agency. Subsoil is granted in accordance with the subsoil license as blocks, which represent geometric areas of subsoil. Subsoil blocks are granted for use through mining or geological permits. Licenses issued for the right to explore subsoil blocks of mineral resources are issued as mining permits. The subsoil users tender payments, which include the right to use the subsoil; deductions for the replacement of the mineral raw material base; and charges for the right to use the water and seabed areas.

In accordance with the types of subsoil use, licenses are granted for: geological study of subsoil; extraction of mineral resources; construction and exploitation of underground facilities not connected to mineral resource extraction; and the formation of specially protected sites.

[52] Postanovlenie VS RF 'O poryadke vvedeniya v deistvie Polozheniya o poryadke litsenzirovaniya pol'zovaniya nedrami' (Постановление ВС РФ от 15.07.1992 N 3314-1 «О порядке введения в действие Положения о порядке лицензирования пользования недрами»).

Russian legislation does not clearly define all terminology used in the regulation of subsoil use. Some examples of this terminology are: 'geological study' and 'exploration and production'. The Subsoil Law uses the terms 'exploration' and 'production' (Art. 6 'types of subsoil use') and 'exploration' (Art. 41 'subsoil use payments') about the activities conducted during the period of geological study without defining them. To confuse this example further, many normative legal acts also use the terms 'prospecting and exploration' or 'geological prospecting works'.[53]

Licenses for geological study of the subsoil grant a right to conduct search and valuation of mineral resource deposits, construction and objects used for construction, and construction of underground facilities not connected with the extraction of mineral resources. A license for an in-depth study (exploration) of mineral resource deposits is not issued separately as the right to explore is provided for in the extraction license. Importantly, a license to geological study grants the right to study the type of mineral resources specified in the license. The license does not grant the licensee any priority in receiving an exploitation of mineral resources license over the same block. However, a joint license, or a license to explore and extract a subsoil block where a license to geological study is active, may be issued after a state inspection of the mineral reserves' geological information. Licenses for geological studies are issued for a period of up to 5 years, while licenses for extraction of mineral resources, or unrelated to extraction, are issued for a period of up to 20 years. Joint licenses last no longer than 25 years.

The Supreme Court Decree of 1992 sets forth the requirements to a license. A license must consist of: the recipient subsoil user's information; information about the specific purposes of the work connected to the geological study or other subsoil use; description of the delimitations of the issued subsoil block; description of the delimitations of the land parcel used for performing work connected to geological study and other subsoil use; the license's period of validity and the date for commencement of the work; tax payment conditions for subsoil, land, and water areas; the agreed-upon level of extraction of mineral resources and its sharing arrangement; agreement about the rights to geological information received in the process of subsoil use; legal requirements for executing such a project (norms, rules, and standards regarding environmental and subsoil protection and work safety regulations); conditions to

---

53 A.I. Perchik (2002) *Gornoye pravo: Uchebnik* (Перчик А.И. Горное право: Учебник. Изд. 2-е – М.: Филология три, 2002).

extend the validity of the license; and the amount and type of production and industrial waste.[54]

Commercial activities undertaken under a license may be performed by entities independent of incorporation. Thus, legal entities and citizens of other states may be licensees unless otherwise stated in legislation.[55] They are granted through competitive tenders and auctions. Subsoil blocks are also granted without any competitive tenders. Parties interested in receiving a license for subsoil use should engage with the Federal Subsoil Resources Management Agency for the required information about deadlines and conditions for license areas of interest. The entity or person applying for a license must within the required time period submit the following: information about the commercial applicant including its main place of business and financial position, information about the management or owners of the entity applying for the license, information about the applicant's and its contractors' technical and technological abilities, information about the applicant's previous activities for the last 5 years, and the applicant's offer for the subsoil use terms.[56]

Applications for participation in an auction go through an examination assessing the applicant's financial and technical competence. Applications for participation in a competitive tender do not go through a preliminary examination as the examination occurs throughout the tender process. The granted license, awarded to the winner of an auction or a tender, is then sent to the federal or the territorial geological funds for registration, which occurs within the period of one month from the time of its grant. These funds, like the Russian Federal Geological Fund of the Russian Federal Agency for Subsoil Use are registers of, among others, licenses and other geological information related to, for example, oil reserves.

---

[54] Postanovlenie VS RF 'O poryadke vvedeniya v deistvie Polozheniya o poryadke litsenzirovaniya pol'zovaniya nedrami' (Постановление ВС РФ от 15.07.1992 N 3314-1 «О порядке введения в действие Положения о порядке лицензирования пользования недрами»), Part 8.

[55] Postanovlenie VS RF 'O poryadke vvedeniya v deistvie Polozheniya o poryadke litsenzirovaniya pol'zovaniya nedrami' (Постановление ВС РФ от 15.07.1992 N 3314-1 «О порядке введения в действие Положения о порядке лицензирования пользования недрами»), Paragraph 9.1.

[56] Postanovlenie VS RF 'O poryadke vvedeniya v deistvie Polozheniya o poryadke litsenzirovaniya pol'zovaniya nedrami' (Постановление ВС РФ от 15.07.1992 N 3314-1 «О порядке введения в действие Положения о порядке лицензирования пользования недрами»), Paragraph 11.

Within the internal and territorial seas, which are a part of the Russian state's territory, licensing is conducted according to the standard procedure with the participation of the subjects of the Russian Federation. The Federal Subsoil Resources Management Agency by the decision of the Government of the Russian Federation conducts licensing in respect of the existing norms of international law of the sea within the continental shelf and the maritime exclusive economic zone (outside the outer limits of the territorial sea).[57] A license for the right to subsoil use on the continental shelf and the maritime exclusive economic zone stipulates a special regime to conduct work, which is agreed upon by the various government agencies such as border services, the shipping regulator, the fish and trade regulator, and environmental protection bodies.

Nonetheless, additional requirements may be added to the license for the right to subsoil use on the continental shelf and a maritime exclusive economic zone as long as they do not contradict Russian legislation, which may relate to infrastructure facilities (artificial islands, and installations and constructions) related to the scope of the license or granted subsoil blocks under the license when required for work safety reasons and the protection of workers' health when executing work under the license.

The right to subsoil use is terminated upon the expiration of the license's period of validity, the rejection of the license for the right to subsoil use by the licensee, or upon the occurrence of expressed conditions excluding further execution of the granted rights under the license.[58] The Federal Subsoil Resources Management Agency can make decisions to prematurely terminate, temporarily suspend, and limit the rights of the licensee under the license. The most common license breach leading to a termination of the validity of the license is caused by a systematic failure to pay the subsoil use payments. Another common breach of the license conditions is missing the date to commence the mining operations as expressed in the license. Thus, the agency has the

---

[57] Postanovlenie VS RF 'O poryadke vvedeniya v deistvie Polozheniya o poryadke litsenzirovaniya pol'zovaniya nedrami' (Постановление ВС РФ от 15.07.1992 N 3314-1 «О порядке введения в действие Положения о порядке лицензирования пользования недрами»), Paragraph 12.

[58] Postanovlenie VS RF 'O poryadke vvedeniya v deistvie Polozheniya o poryadke litsenzirovaniya pol'zovaniya nedrami' (Постановление ВС РФ от 15.07.1992 N 3314-1 «О порядке введения в действие Положения о порядке лицензирования пользования недрами»), Paragraph 15.1.

right to terminate a license when the subsoil user breaches its obligations under the license.[59]

Russian legislation concerning the use of natural resources is mainly licensing legislation and is regulated according to norms of an administrative nature.[60] This is characteristic of the Subsoil Law, the Continental Shelf Law, the Exclusive Economic Zone Law and others. At the same time, the legislator facilitated subsoil use based on contractual relations, as evident by the Product Sharing Agreement Law. Article 2 of this law states that an agreement about product sharing is a contract, of which the Russian Federation is represented as the subject of business activity in return for a consideration and for a defined period of time with exclusive rights to search, exploration, and extraction of mineral resources on the subsoil block as defined in the contract, as well as work related to this. The investor is obligated to execute the specified work at his or her own expense and risk.

Federal law lists subsoil blocks available under the PSA regime. The lists take into account the interests of the subjects. Generally, the subsoil blocks entered in the PSA regime lists are high cost blocks, which were not purchased on auction. Subsoil blocks under the PSA regime may not exceed more than 30 per cent of the explored and recorded mineral reserves in the state balance sheets. Subsoil users under the PSA regime may be investors such as: Russian citizens; foreign citizens; legal entities; joint venture structures, partnership structures, and other non-legal entities and associations of legal entities conditioned on them being jointly and severally liable for the PSA conditions. When an investor does not hold legal entity status, the license is granted to one participant of the association as a representative of the association specifying the other members of the association.

The rights and obligations of the subsoil user under the Subsoil Law arise at the moment the license for subsoil use is registered in the state register, while these rights and obligations arise for the parties to a PSA at the time when the agreement enters into force.

Contractual relations regarding subsoil use are an alternative to the licensing system under Russian law, see Subsoil Law Article 12. The

---

59 Zakon RF 'O nedrakh' (Закон РФ 'О недрах' от 21.02.1992 N 2395-1), art. 21.

60 A.Ya. Bolyukh, A.O. Shshakhmov, V.V. Silkin (2006) *Pravovoi rezhim innostrannikh investitsii v Rossii: Uchebnoe posobiye* (Болюх А.Я., Шахов А.О., Силкин В.В. Правовой режим иностранных инвестиций в России: Учебное пособие. – М.: Московский государственный институт международных отношений (Университет) МИД России, 2006).

formal separation of types of subsoil use are: (1) contractual (civil-normative), and (2) administrative (licensing). Regulation of a contractual subsoil use regime does not, however, exist in Russian legislation. The Subsoil Law states two types of subsoil use contracts: (1) PSA, and (2) service contracts (with risk and without risk). Only the PSA regime has its own legislative basis, and is currently the only subsoil use contractual regime used for subsoil resources development.

## 6. ASSESSMENT OF PETROLEUM FIELD DEVELOPMENT

A subsoil user is obligated to implement the legislative requirements and the confirmed standards, norms, and rules for conducting technological work related to subsoil use.[61] State monitoring controls whether the stated norms are complied with to a significant degree. The state, as the property owner of subsoil, always sets general requirements and conditions, expressed in the license or license agreement, for the subsoil user.

The tasks of the state when monitoring geological studies, rational subsoil use, and protection of the subsoil are to ensure: implementation by all subsoil users of the legislative requirements in accordance with the specific standards, norms, and rules for geological studies, rational subsoil use, and subsoil protection; implementation of the monitoring rules;[62] and compliance with the technical requirements of the projects, plans, and schemes regarding the development of the mining works with zero tolerance for any losses above the set standards.

The main normative legal acts directly regulating the technical aspects of subsoil use are: the Subsoil Protection Rules from 6 June 2003[63] (of the former 'Federal Mining and Industrial Supervision of Russia') and the Subsoil Protection Rules for Mineral Processing from 6 June 2003[64] (of the former 'Federal Mining and Industrial Supervision of Russia').

---

[61] Zakon RF 'O nedrakh' (Закон РФ 'О недрах' от 21.02.1992 N 2395-1), art. 22.1. part. 2.

[62] Zakon RF 'O nedrakh' (Закон РФ 'О недрах' от 21.02.1992 N 2395-1), art. 37.

[63] Postanovlenie Gosgortekhnadzora 'Ob utverzhdenii "Pravil okhrani nedr" (Постановление Госгортехнадзора РФ от 6 июня 2003 г. N 71 'Об утверждении 'Правил охраны недр').

[64] Postanovlenie Gosgortekhnadzora 'Ob utverzhdenii "Pravil okhrani nedr pri pererabotke mineral" ogo sirya' (Постановление Госгортехнадзора РФ от 6 июня 2003 г. N 70 'Об утверждении Правил охраны недр при переработке минерального сырья').

The current agency is the Federal Environmental, Industrial and Nuclear Supervision Service (the Service) of Russia.

The Federal Environmental, Industrial and Nuclear Supervision Service of Russia conducts, among others, state mining, monitoring, and compliance control with norms and rules governing initiation and implementation of projects for the extraction of natural resources and the coordination of such projects. The Service can carry out expert examination with respect to subsoil use and protection, as well as impose measures when norms, rules, and standards are breached during the initiation and implementation of projects, protection of subsoil, and unsafe conduct of work during geological studies. When a state inspection detects a breach of the rules, norms, or standards the project documentation must be amended. Construction and exploitation objects for extraction, primary processing of natural resources, geological, or survey work cannot proceed without corrected project documentation. The Service can acquire objects in breach. In accordance with Article 13 of the Federal Law on Industrial Safety of Hazardous Production Facilities,[65] documentation concerning construction expansion, reconstruction, technical upgrading, and conservation and liquidation of the hazardous production facility is subject to expert appraisal of industrial safety hazardous production facilities. Hazardous construction faculties are where mining, mineral processing, and underground operations are conducted.

The Service also coordinates technical projects for the use of subsoil blocks and supplements to them. Such projects may consist of: construction and exploitation of underground installations, technical and economic rationale, technological schemes, development programs (experimental programs, production testing, reconstruction, liquidation, and conservation), field infrastructure development, projects and technical schemes for mineral processing, survey production and geological work projects, and other project documentation for subsoil block use and relevant technical design assignments. The key data for design choices are rational and complex subsoil block use, and protection of subsoil and the environment. This is located in the technical part of the project 'protection of subsoil and the environment'. Supporting the safety of life and health of people, and the protection of building and structures, air, land, forest, water, animals, and other parts of the environment is a

---

[65] Federal'nii zakon 'O promishlennoi bezopastnosti opasnikh proizvodstvennikh ob''ektov' (Федеральный закон от 21 июля 1997 г. N 116-ФЗ 'О промышленной безопасности опасных производственных объектов').

requirement during subsoil use. When the subsoil user detects a necessity for implementation of a more effective measure to protect the environment, the project documentation must be amended.

Subsoil extraction and processing projects are executed in accordance with the conditions of the subsoil use license or the production sharing agreement. When developing the project documentation, the organization or individual businessman compare different technical and economic solutions concerning facility locations, technological approaches, and optimization. The quantity and content of these materials and propositions depend on the difficulty of the geological and technical mining conditions. The projects should be executed according to the following principles: rational multiple use of the mineral reserves; safe mining; and protection of subsoil and the environment.[66]

The project documentation must properly set forth the sequence or priority to mine blocks, horizons, and diggings (and extraction of share interests of such blocks, horizons, and diggings), which are intended to lead to an unsound loss of mineral resources, premature exhaustion and devaluation of the fields' reservoirs. This is as a result of field development of blocks, horizons, and diggings of commercial character and quality in situations where a sharp distinction occurred in the formation's quality of the technical mining conditions and other parameters. The project documentation must specify the sequence or priority from year to year as set forth in the field's mining schedule and plans for the development of the mining work.[67] The project documentation includes the justification and technical (engineering) solutions to the questions set forth in the rules concerning the protection of subsoil.[68]

The state is also given general requirements with respect to the project documentation. Accordingly, the oilfield infrastructure development project designs should provide for:

---

66 Postanovlenie Gosgortekhnadzora 'Ob utverzhdenii "Pravil okhrani nedr" (Постановление Госгортехнадзора РФ от 6 июня 2003 г. N 71 'Об утверждении 'Правил охраны недр'), art. 16–17.

67 Postanovlenie Gosgortekhnadzora 'Ob utverzhdenii "Pravil okhrani nedr" (Постановление Госгортехнадзора РФ от 6 июня 2003 г. N 71 'Об утверждении 'Правил охраны недр'), art. 19.

68 Postanovlenie Gosgortekhnadzora 'Ob utverzhdenii "Pravil okhrani nedr" (Постановление Госгортехнадзора РФ от 6 июня 2003 г. N 71 'Об утверждении 'Правил охраны недр'), art. 20 sets forth detailed questions for reply by the applicant.

- sustainable and efficient construction of facilities required for production;
- a field development system that supports a more complete, complex, and more economically attractive extraction of subsoil from the mineral reserves;
- rational use of overburden rocks when developing a field;
- storage, preservation (for example, reinjection into the reservoir), and accountancy of extracted but not used mineral resources;
- geological study of the subsoil block while developing a field; and
- safety measures for people, production facilities, as well as environmental protection.[69]

Plans and schemes for the development of mining works are generally written in accordance with the project documentation of the oil company. Therefore, mining, geological, and geotechnical conditions concerning the implementation of new technology and workers' safety improvement according to the legislative requirements are taken into account. However, certain inevitable losses (pollution) occur during oil extraction. These are incorporated into the projects, plans, and schemes of field development, subsoil production and processing. The legislation permits inevitable losses, also called normative losses, as they are preplanned. The subsoil user can rightfully conduct extraction of natural resources with allowances made for such losses. Nonetheless the user is obligated to prevent non-preplanned losses above the set standard for the preplanned losses.

The Mining Plan Instructions[70] from 24 November 1999 are instructions on the coordination of yearly scheduled work plans for the development of hydrocarbon deposits by the subsoil user and the Federal Environmental, Industrial and Nuclear Supervision Service of Russia (formerly Federal Committee for Mining and Industrial Supervision). Furthermore, the Instructions require the subsoil user to: protect subsoil; ensure industrial safety; and prevent unfavourable effects of mining on the environment. A yearly work program is a document comprised by the subsoil user and specifies: the direction of the mining works; the amount

---

[69] A.I. Perchik (2002) *Gornoye pravo: Uchebnik* (Перчик А.И. Горное право: Учебник. Изд. 2-е – М.: Филология три, 2002), sub-chapter 8.1.

[70] Postanovlenie Gosgortekhnadzora 'Ob utverzhdenii Instruktsii po soglasovaniu godovikh planov razvitiya gornikh rabot' (Постановление Госгортехнадзора РФ от 24.11.1999 № 85 «Об утверждении Инструкции по согласованию годовых планов развития горных работ» (Зарегистрировано в Минюсте РФ 10.12.1999 № 2000)).

of extracted mineral resources; other work conditioned in the license; measures to protect the subsoil; and so on. The Service considers and approves the yearly programs. Any mining work without the Service's approval of yearly plans or any departure from the approval is not allowed.[71] Every subsoil user, regardless of form of incorporation or nationality, planning to do any surveying must obtain a license from the Service.[72] Scientific and expert organizations may conduct subsoil protection and survey expertise with the right qualifications. Subsoil protection expertise may be conducted at the same time as other types of expertise.[73]

## 7. GOVERNMENT INVOLVEMENT IN THE EXTRACTION OF PETROLEUM

Possession, use, and disposal of natural resources are in most countries carried out with the consent of the state. This is established practice recognized by international norms regarding national possession, use, and disposal of natural resources, which is treated as an integral element of each country's sovereignty.[74] As a result, state ownership of subsoil is the dominant position in most countries, including throughout Russia's history.

The state, having ownership of subsoil and realizing sovereign power on its territory, formulates a single state policy in the field of subsoil use regulating the many relationships arising from subsoil use. The main tasks of state subsoil regulation are to secure sustainable, safe, and rational subsoil use. State regulation of subsoil relationships includes the activities of authorized governmental bodies for the administration of

---

[71] Postanovlenie Gosgortekhnadzora 'Ob utverzhdenii Instruktsii po soglasovaniu godovikh planov razvitiya gornikh rabot' (Постановление Госгортехнадзора РФ от 24.11.1999 № 85 «Об утверждении Инструкции по согласованию годовых планов развития горных работ» (Зарегистрировано в Минюсте РФ 10.12.1999 № 2000)), art. 30.

[72] Postanovlenie Gosgortekhnadzora 'Ob utverzhdenii "Instruktsii po proizvodstvu marksheiderskikh rabot" (Постановление Госгортехнадзора РФ от 06.06.2003 N 73 'Об утверждении "Инструкции по производству маркшейдерских работ" (вместе с 'РД 07-603-03…')).

[73] Постановление Госгортехнадзора РФ от 6 июня 2003 г. N 71 'Об утверждении "Правил охраны недр", art. 28.

[74] B.D. Klyukin (2000) *Gornie otnosheniya v stranakh Zapadnoi Evropi i Ameriki* (Клюкин Б.Д. Горные отношения в странах Западной Европы и Америки. – М.: Городец, 2000). p. 44.

subsoil block use, licensing of subsoil use, management of subsoil blocks, subsoil users, and work concerning subsoil block use.

State administration of subsoil use relationships is conducted by state executive bodies of the Russian Federation – the President and the Government, bodies of executive power of the subjects, the federal body for the administration of the state fund of subsurface resources executed by the Ministry of Energy, and state mining supervision bodies. The Federal Environmental, Industrial and Nuclear Supervision Service of Russia (Rostechnadzor) is the main body conducting state supervision concerning safe work practices in the subsoil use sector. This Service is the central body of federal executive power implementing state normative regulation of questions relating to ensuring industrial safety on Russian territory, as well as special licensing, supervisory, and control functions.

The social and economic significance of subsoil use, and specifically in the oil sector, is explicitly stated in the Russian Constitution and legislation. The disposal of subsoil should be carried out in the interest of the people living in the area of the field location as well as all Russian people. The Russian Constitution sets forth norms regulating the subsoil use, underlining its high priority, and confirms through constitutional provisions the stability and protection of the norms enacted in the subsoil legislation.[75]

Article 9 of the 1993 Russian Constitution states that 'other natural resources shall be utilized and protected in the Russian Federation as the basis of the life and activity of the peoples living on the territories concerned', and that 'natural resources may be subject to private, state, municipal and other forms of ownership.'[76] In addition, Article 1.2 of the Subsoil Law establishes exclusive state ownership of subsoil.

The state's involvement and role in the Russian oil sector is significant. Various state bodies have authority over petroleum activities, including:

- the grant of access to the resource through the grant of a license;
- development and improvement of subsoil legislation;
- determination and implementation of federal subsoil use policy;
- determination of subsoil use strategy, speed of reproduction, expansion and qualitative improvement of the mineral resource base through development and implementation of federal programs;
- the establishment of a common order for subsoil use, protection,

[75] A.I. Perchik (2002) *Gornoye pravo: Uchebnik* (Перчик А.И. Горное право: Учебник. Изд. 2-е – М.: Филология три, 2002), sub-chapter 12.1.

[76] Konstitutsiya Rossiiskoi Federatsii (Конституция Российской Федерации от 25 декабря 1993 года), art. 9.2.

and development conforming to standards such as reserves classification and projected commercial mineral resources;

- creation and maintenance of the unified system of federal and territorial fund of geological subsoil information, and disposal of information received at the expense of state funds;
- state expert review of explored mineral resources reserves, and other characteristics of subsoil defining their value and danger;
- determination of regional mineral resource lists with the subjects, as well as allotment of subsoil blocks of federal, regional, and local significance, and the creation of subsoil block lists for use based on product sharing agreements;
- creation of the State Register of Mineral Reserves; state recording-keeping of subsoil blocks used for extraction of natural resources and construction of underground facilities not connected to extraction; maintenance of the national map system for natural resource fields and their occurrence; state registration of work related to geological study of subsoil;
- disposal of subsoil on the continental shelf;
- introduction of limitations on subsoil use to ensure national safety and protect the environment;
- maintenance of the state subsoil fund jointly with the subjects, except the blocks located in the Russian Federation's exclusive jurisdiction;
- determine the conditions and order to collect subsoil use payments with the subjects, including the determination of their ultimate levels by natural mineral groups, forms, and payment sizes for use of subsoil blocks of federal significance;
- coordination of research and development activities connected to subsoil use;
- settle disputes of subsoil use between the subjects; state control over geological studies, rational use, and subsoil protection, as well as establishing procedures for carrying them out; and
- enter into product sharing agreements for use of a subsoil block.[77]

[77] Konstitutsiya Rossiiskoi Federatsii (Конституция Российской Федерации от 25 декабря 1993 года), art. 72 and Zakon RF 'O nedrakh' (Закон РФ 'О недрах' от 21.02.1992 N 2395-1), articles 3–5.

The Ministry of Energy (the regulating body for subsoil activities) may pass certain powers to regulate subsoil relations to the subjects.[78,79] For this purpose, duly authorized bodies and various branches of power preserve the function of state regulation of the web of specific relationships emerging through the use of distinct subsoil characteristics both connected and not connected to extraction of natural resources.

The primary goals of state regulation of subsoil use relations are: guarantee the reproduction of the mineral resources base, its rational use and subsoil protection in the interests of the current and future generations of people of Russia.[80] The federal executive bodies and executive bodies of the subjects are given the authority to solve the primary goals. State regulation of subsoil use relationships occur through a system of agencies determining tasks, functions, and powers with bases in legislation limiting the jurisdiction of the executive bodies of the federation, the subjects, and the bodies of local self-governing authorities.

## 8. ENVIRONMENTAL REGULATION FRAMEWORK

International agreements of the Russian Federation, the Russian Constitution, constitutional and federal legislation, presidential and government decrees, normative legal acts of federal ministries and departments, and legislation of the subjects govern issues of environmental protection, protection of rational natural resource use, and their preservation and recovery.

The oil extracting industry is a major polluter of the environment and adversely affects all environmental components: land, subsoil, forest, air, water, animals, landscape, and the ecosystem as a whole. Breach of the legislation protecting the environment imposes civil (property), disciplinary, administrative, and criminal liability. Legal entities and physical individuals, having inflicted harm to the environment, are obligated to compensate the environment 'to the full extent' of the damage. The legislation stipulates compensation for the following forms of pollution:

---

[78] There are currently 85 subjects (two of these are disputed with Ukraine). Subjects are: oblasti, republics, krais, autonomous okrugs, federal cities, and autonomous oblasts.

[79] I.B. Kalinin (2001) *Pravovoe regulirovaniye resursopol'zovaniya* (Калинин И. Б. Правовое регулирование ресурсопользования. – Томск: Изд-во Научно-технической литературы, 2001).

[80] Zakon RF 'O nedrakh' (Закон РФ 'О недрах' от 21.02.1992 N 2395-1), art. 35.

spoiling; abolition; degradation; destruction of the natural ecological system, natural complexes, and natural landscapes; exhaustion and irrational use of natural resources; or other breaches of the legislation.

According to the Russian Constitution, every person has the right to a favourable environment and is therefore obligated to protect nature and the environment, and take good care of the environmental riches, which are fundamental to the sustainable development, life, and activity of all peoples in the Russian Federation. Article 71 of the Constitution establishes the exclusive jurisdiction of the Russian Federation, in which the foundation of federal policy and federal programs for environmental development is located. Environmental protection and environmental protection legislation are part of the joint jurisdiction between the Russian Federation and the subjects (oblasts, republics, krais, autonomous okrugs, federal cities, and autonomous oblasts) of the Russian Federation.[81]

The Federal Environmental Protection Law[82] sets forth the fundamental legal principles of state policy for environmental protection, ensuring a balancing resolution of the social and economic objectives while preserving a favourable environment, biological diversity, and natural resources for the purposes of meeting the needs of the current and future generations and strengthening law and order in environmental protection as well as ensuring environmental safety.[83] This law regulates the interactive relations of society and nature occurring in economic and other activities affecting the natural environment within the territory of the Russian Federation and on the continental shelf and in the exclusive economic zone of the Russian Federation.[84] Furthermore, the law contains fundamental definitions of environmental law and fundamental principles of environmental protection. The statute grants authority to state executive bodies of the federation and the subjects as well as local self-governing authorities, and rights and obligations to legal entities,

---

[81] Konstitutsiya Rossiiskoi Federatsii (Конституция Российской Федерации от 25 декабря 1993 года), art. 72.

[82] Federal'nii zakon 'Ob okhrane okruzhayushshei sredi' (Федеральный закон 'Об охране окружающей среды' от 10.01.2002 N 7-ФЗ).

[83] Federal'nii zakon 'Ob okhrane okruzhayushshei sredi' (Федеральный закон 'Об охране окружающей среды' от 10.01.2002 N 7-ФЗ), preamble.

[84] Federal'nii zakon 'Ob okhrane okruzhayushshei sredi' (Федеральный закон 'Об охране окружающей среды' от 10.01.2002 N 7-ФЗ), preamble.

non-legal entities, and physical individuals in the area of environmental protection.[85]

The Federal Environmental Protection Law defines methods of economic regulation, limits and their order of establishment, quality limits, limits for admissible exposure to the environment, limits for acceptable emissions and discharges of matter and microorganisms, limits and establishment of production and household waste generation, limits on acceptable physical impacts on the environment, limits on the admissible elimination of environmental components, and limits of acceptable anthropogenic stress on the environment. Furthermore, the law establishes requirements in the course of economic and other activity to protect rare and endangered growths, animals and other organisms, urban and rural vegetation, rare and endangered soils, and the tasks of the state environmental monitoring and environmental control.[86]

In the course of allocation, design, construction, reconstruction, installation and exploitation of petroleum effective measures are required to clean up and process production waste, manage associated petroleum gas and salt water, rehabilitate damaged and polluted lands, reduce negative impacts of the activity on the environment, and compensate for environmental harm.

Construction and exploitation of oil and gas production, transportation, conservation, and realization of oil, gas, and other processing products are allowed if followed by the implementation of rehabilitation projects for the polluted lands, positive findings of the state environmental expertise, fulfilment of other legislative requirements, and a financial guarantee for rehabilitation projects. Construction and exploitation of these objects when located in the surface water areas, on the continental shelf, or in the exclusive economic zone are allowed with a positive finding of the state environmental experts and other legislative requirements after the rehabilitation of the polluted land. When opening up offshore oil and gas fields close attention is paid to protection of the environment and oil spill prevention on land and at sea. The financing of necessary environmental actions should, however, be specified in the yearly work programs and the hydrocarbon operation schemes.[87]

---

[85] Other chapters of the law are dedicated to scientific research in the area of environmental protection, the fundamentals of environmental cultures, liability for environmental violations, and international environmental cooperation.

[86] A.Yu. Puryaeva (2012) *Ekologicheskoye pravo. Uchebnik* (Пуряева А.Ю. Экологическое право. Учебник. – М.: 'Юстицинформ', 2012).

[87] A.I. Perchik (2002) *Gornoye pravo: Uchebnik* (Перчик А.И. Горное право: Учебник. Изд. 2-е – М.: Филология три, 2002), sub-chapter 9.2.

Similar requirements and obligations for the subsoil user exist for the protection of the environment in the following federal legislation: the Subsoil Law, the Federal Specially Protected Natural Areas Law,[88] the Water Code, the Federal Ecological Examinations Law, the Forest Code,[89] the Federal Production and Consumption Wastes Law,[90] the Federal Air Protection Law,[91] the Federal Environmental Protection Law, and other laws.

## 9. SAFETY FRAMEWORK

The organization and execution of safe work practices are essential in the oil sector as the technological processes alone generate danger. Some examples of dangerous elements in the oil industry are: seismic investigation, well drilling, oil combustion, landslides, explosions, and uncontrolled discharges (blowouts).

The Russian legislation currently defines the fundamental principles of organization as well as a list of basic requirements assuring the safe execution of mining work. The terminology 'safe' is defined in the Federal Safety Law[92] as relating to environmental, manufacturing, raw materials, energy, and other fields of safety.

The construction and operation of natural resource businesses, underground structures, or the execution of geological studies are acceptable only when the health and safety of people living in the area and the workers are protected. The bodies of executive power and the subsoil users are obligated to implement the standards, norms, and rules on safe work practices in the area of subsoil use. The state mining supervision

---

[88] Federal'nii zakon 'Ob osobo okhranyaemikh prirodnikh territoriyakh' (Федеральный закон от 14 марта 1995 г. N 33-ФЗ 'Об особо охраняемых природных территориях').

[89] Lesnoi kodeks RF (Лесной кодекс РФ от 04.12.2006 N 200-ФЗ).

[90] Federal'nii zakon 'Ob otkhodakh proizvodstva i potrebleniya' (Федеральный закон от 24 июня 1998 г. N 89-ФЗ 'Об отходах производства и потребления').

[91] Federal'nii zakon 'Ob okhrane otmosfernogo vozdukha' (Федеральный закон от 4 мая 1999 г. N 96-ФЗ 'Об охране атмосферного воздуха').

[92] Federal'nii zakon 'O bezopastnosti' (Федеральный закон от 28.12.2010 № 390-ФЗ «О безопасности»).

authority, as part of its jurisdiction, implements state normative regulations concerning work safety in the oil industry and maintains a monitoring position.[93]

The leadership of a company involved in subsoil use carries personal liability to provide for a safe working environment regardless whether the company conducts work in accordance with an issued license or under a contract.[94] All dangerous work related to subsoil development must be conducted according to the license and its relevant activity category. The licensee is required to contract specialist mining rescue services as well as oil spill response providers. Upon the occurrence of a dangerous incident, the licensee is required to suspend work and transport people to a safe location. They must also immediately inform the appropriate state executive and local self-governing bodies about the incident.[95]

The federal and appropriate territorial agencies for geological information and the state statistical authorities collect data regarding explored, extracted, and non-recoverable reserves to ensure safe work practices.[96] It is obligatory for the subsoil user to comply with the approved and established procedure of environmental protection against adverse events. A breach of these requirements incurs a temporary stay until the breach is corrected, or a complete halt in activities until approval is granted by the relevant authority.[97]

The laws regulating the energy system's safety from accidents and catastrophic events are: the Federal Safety Law, the Federal Law on

---

[93] Zakon RF 'O nedrakh' (Закон РФ 'О недрах' от 21.02.1992 N 2395-1), art. 24.

[94] Zakon RF 'O nedrakh' (Закон РФ 'О недрах' от 21.02.1992 N 2395-1), art. 24.

[95] A.I. Perchik (2002) *Gornoye pravo: Uchebnik* (Перчик А.И. Горное право: Учебник. Изд. 2-е – М.: Филология три, 2002), sub-chapter 8.4.

[96] Postanovlenie Gosgortekhnadzora 'Ob utverzhdenii Edinikh pravil bezopastnosti pri razrabotke mestorozhdenii poleznikh iskopaemikh otkritim sposobom' (Постановление Госгортехнадзора РФ от 9 сентября 2002 г. № 57 'Об утверждении Единых правил безопасности при разработке месторождений полезных ископаемых открытым способом'), art. 47.

[97] P.A. Mushnikov (2001) 'Pravo gosudarstvennoi sobstvennostoi na nedra' *Problemi gornogo i ekologicheskogo prava v neftegazovom komplekse* (Мушников P.A. Право государственной собственности на недра. // Проблемы горного и экологического права в нефтегазовом комплексе. Научные труды кафедры горного права. М., 2001. Вып. 1).

Industrial Safety of Hazardous Production Facilities,[98] the Federal Hydro-Technical Installation Safety Law,[99] the Federal Environmental Examination Law,[100] and the Law on the Protection of the Population and Areas against Natural and Man-Made Emergencies.[101]

The Federal Law on Industrial Safety of Hazardous Production Facilities (the Safety Law) defines the legal, economic, and social basis for the safe operation and management of hazardous production facilities. It seeks to prevent accidents on hazardous production facilities and support the licensee in identifying and correcting the causes of incidents. The provisions of this federal law extend to all organizations, regardless of their form, conducting activities within the scope of the law in Russian territory.

Hazardous production facilities are companies, their production facility, land, ground, and other production facilities as specified in Annex 1 of the Safety Law.[102] Annex 1 expressly states that the category of hazardous production facilities includes mining, mineral processing, underground work, and oil platforms. As such, oil producing companies, as an asset group, are also hazardous production facilities. The law further states that amendments to the project documentation for construction, expansion, reconstruction, technical upgrades, conservation, and liquidation of hazardous production facilities are subject to industrial

---

98 Federal'nii zakon 'O promishlennoi bezopastnosti opasnikh proizvodstvennikh ob''ektov' (Федеральный закон от 21 июля 1997 г. N 116-ФЗ 'О промышленной безопасности опасных производственных объектов').

99 Federal'nii zakon 'O bezopastnosti gidrotekhnicheskikh cooruzhenii' (Федеральный закон от 21.07.1997 № 117-ФЗ «О безопасности гидротехнических сооружений»).

100 Federal'nii zakon 'Ob ekologicheskoi ekspertize' (Федеральный закон 'Об экологической экспертизе' от 23 ноября 1995 г. N 174-ФЗ).

101 Federal'nii zakon 'O zashshite naseleniya i territorii ot chrezvichainikh situatsii prirodnogo i tekhnogennogo kharaktera' (Федеральный закон от 21.12.1994 № 68-ФЗ «О защите населения и территорий от чрезвычайных ситуаций природного и техногенного характера»).

102 Federal'nii zakon 'O promishlennoi bezopastnosti opasnikh proizvodstvennikh ob''ektov' (Федеральный закон от 21 июля 1997 г. N 116-ФЗ 'О промышленной безопасности опасных производственных объектов'), art. 2.

safety expert review in agreement with the specially authorized federal executive body or its territorial body.[103]

A variety of other acts also regulate issues of safety in the oil extraction sector. The Oil and Gas Industry Safety Rules[104] establish requirements for industry safety for the organization and the workers conducting petroleum activities on hazardous production facilities, requiring dangerous work during exploration and production to be performed in accordance with the requirements set by the organization for the safe performance of such work and approved by the technical manager of the organization.

The oil-and-gas-exploiting organization is obligated to develop plans for identifying and liquidating consequences from accidents. The subsoil user should also develop plans for the prevention and liquidation of oil and oil-product spills to avoid the advancement of accidental oil and oil-product spills. The following regulations govern the development of plans for the prevention and liquidation of petroleum spills: the Prevention and Liquidation Measures of Accidental Oil Spill Regulation,[105] the Action Management for Prevention and Liquidation of Oil Spill Regulation,[106] and several decrees from relevant ministries regulating the content of the field development plans and how to define the negative impact of an oil spill.[107]

---

[103] Federal'nii zakon 'O promishlennoi bezopastnosti opasnikh proizvodstvennikh ob''ektov' (Федеральный закон от 21 июля 1997 г. N 116-ФЗ 'О промышленной безопасности опасных производственных объектов'), art. 8.

[104] Pravila bezopastnosti v neftyanoi i gazovoi promishlennosti (Правила безопасности в нефтяной и газовой промышленности» (утверждены Приказом Федеральной службы по экологическому, технологическому и атомному надзору от 12 марта 2013 г. №101)).

[105] Postanovlenie Pravitel'stva RF 'O neotlozhnikh merakh po preduprezhdeniyu i likvidatsii avariinikh razlivov nefti i nefteproduktov' (Постановление Правительства РФ от 21.08.2000 № 613 «О неотложных мерах по предупреждению и ликвидации аварийных разливов нефти и нефтепродуктов»).

[106] Postanovlenie Pravitel'stva RF 'O poryadke organizatsii meropriyatii po preduprezhdeniyu i likvidatsii razlivov nefti i nefteproduktov na territorii RF' (Постановлением Правительства РФ от 15.04.2002 «О порядке организации мероприятий по предупреждению и ликвидации разливов нефти и нефтепродуктов на территории РФ»).

[107] Prikaz MPR RF 'Ob Ukazanii po opredeleniyu nizhnego urovnya razliva nefti i nefteproduktov dlya otneseniya avaruunogo razliva k chrezvichainoi situatsii' (Приказ МПР РФ от 03.03.2003 N 156 'Об утверждении Указаний по определению нижнего уровня разлива нефти и нефтепродуктов для

## 10. CONCLUSION

This chapter has briefly outlined the substantive law, the procedure, and historical development and overview of petroleum activities and petroleum law in the Russian Federation. It focused on the legal framework for offshore petroleum activities and some areas directly related to these activities, such as license grants, assessment of petroleum field development, environmental regulations, safety regimes, and the state's involvement in the industry.

---

отнесения аварийного разлива к чрезвычайной ситуации' (Зарегистрировано в Минюсте РФ 08.05.2003 N 4516)); prikaz MChS Rossii 'Pravila razrabotkii i soglasovaniya planov po preduprezhdeniyu i likvidatsii razlivov nefti i nefteproduktov na territorii RF' (приказ МЧС России от 21.12.2004 «Правила разработки и согласования планов по предупреждению и ликвидации разливов нефти и нефтепродуктов на территории РФ»).

# 11. Licensing and regulation of Japan's offshore resources

**Souichirou Kozuka**

## 1. INTRODUCTION

Procurement of petroleum has always been a critical issue in the modern history of Japan. Soon after opening up the country to the world in the mid-19th century, the demand for paraffin oil increased due to the penetration of imported oil lamps. In the 20th century, as the primary energy source shifted from coal to petroleum, the Japanese government attempted to promote production of petroleum through public funding of operators in the upstream sector. After the oilfields overseas were all lost as a result of the defeat in the Second World War, the government and industry of Japan again made various attempts to secure interests in the upstream of the petroleum industry.

Still, the regulatory regime remained relatively underdeveloped. The Mining Act,[1] applicable to mining of various kinds of minerals, also includes the exploration for and production of petroleum. It was only in 2011 that amendments to the Mining Act established a special regime for petroleum and other important minerals. The introduction of the new regime was motivated, among others, by the intent to promote exploration of petroleum (as well as other resources, such as undersea minerals) within the Exclusive Economic Zone (EEZ) and on the Continental Shelf.[2]

In this chapter, the Japanese regulation of the upstream petroleum sector having just entered a new phase is analyzed, with some insights into the historical background. First, a brief overview of the history of

[1] Law no.289 of 1950.

[2] Kazuhiro Nakatani, 'Japan's Undersea Resources and Its New Mining Act', in: Donald N. Zillman, Aileen McHarg, Lila Barrera-Hernández and Adrian Bradbrook (eds.), *The Law of Energy Underground*, p. 203 (Oxford University Press, Oxford, 2014).

petroleum production and the relevant regulation is given. Then the current regulation is examined in the following order: the licensing regime after the 2011 amendments, the law relevant to the operators, namely the nationality restrictions, the government's funding of the upstream operators and the provisions applicable to joint venture agreements, the rules on the liability of the operator for environmental damages caused by its operation, and finally the regulation of safety in mining.

## 2. HISTORY OF PETROLEUM PRODUCTION AND ITS REGULATION

The high demand for petroleum within the country emerged immediately after Japan opened up to the Western world in the 1850s. After several years of manual production, Nihon Sekiyu Co. Ltd. succeeded in the extraction of crude oil from the offshore of Amaze (Niigata prefecture) facing the Japan Sea in 1891. The attempt by Standard Oil to extract petroleum in Japan followed.

However, the petroleum production within the territory of Japan declined from the 1910s.[3] Since then, imported oil has dominated the Japanese market. After the embargo on petroleum triggered the country to plunge into war in 1941, Japan developed several oilfields in its occupied territories of Southeast Asia. Since those activities came to an end as a result of defeat in the Second World War, the petroleum industry in Japan has concentrated mostly on the downstream business. Yet, attempts to secure interests in the upstream were never abandoned. Among the few successes was the production at the Khafji oilfield off the Neutral Zone between Saudi Arabia and Kuwait from 1960 until the expiry of the concessions in 2000 and 2003.

### Regulation of Mining

Throughout these periods, the exploitation of oil had been regulated as 'mining of a mineral'. The initial regulation of mining in modern Japan was the Mining Manual of 1872,[4] followed by Japan's Mining Law[5] the

[3] Masanari Koike, Gento Mogi and Walheed H. Albedaiwi, 'Overseas Oil-Development Policy of Resource-Poor Countries: A Case Study from Japan', *Energy Policy* vol.36, p. 1764, at p. 1765 (2008).

[4] Proclamation by Dajokan (Great Council of State) no.100 of 1872.

[5] Proclamation by Dajokan no.259 of 1873.

next year. The latter law was modelled on the mining law of Spain and Australia. It adopted the principle that the mineral resources (including the petroleum) belong to the state and its exploitation is deemed to be contracted by the state. However, in 1890 the government enacted the Law on Mining[6] to replace the previous law of 1873. The new Law on Mining was modelled on the French and Prussian mining laws and introduced a license system. While ownership of unextracted minerals was vested in the state, a private party exploiting the mineral was to be licensed, not to lease the mining district. The Law on Mining of 1890 also introduced the 'first-come, first-served' rule for granting the license, which was termed as 'the right of mining' (*Berkwerkseigentum*). These principles were confirmed by the Mining Act of 1905.[7]

A new Mining Act was enacted in 1950 to accommodate the requests from the Allied Forces that exercised administrative power over Japan at that time. However, the changes from the Mining Act of 1905 were only technical. The licensing system and the principle of first-come, first-served remained as they used to be under the preceding law.

## Focus on the Production of Petroleum in the Ocean

Japan ratified the United Nations Convention on the Law of the Sea (UNCLOS) in 1996. Upon ratification, minimal adaptations to existing laws were made by extending the scope of application of Japan's existent statutes to the EEZ and Continental Shelf relating to enumerated activities, including petroleum.[8] As a result, the Mining Act became applicable to the production of oil in the EEZ or on the Continental Shelf without any modification.

It was only during the first decade of the 21st century that Japan took seriously the development of natural resources, including petroleum, in

---

6 Law no.87 of 1890.

7 Law no.45 of 1905.

8 Art.3(1) of the Law on the Exclusive Economic Zone and the Continental Shelf, Law no. 74 of 1996 (Provisional translation in 39 *Japanese Annual of International Law* 354 (1996)). The enumerated types of activities to which Japanese law shall be applicable are: (1) exploring and exploiting, conserving and managing the natural resources; the establishment, construction, operation and use of artificial islands, installations and structures; the protection and preservation of the marine environment; and marine scientific research in the EEZ or on the continental shelf; (2) the activities for the economic exploitation and exploration of the EEZ; and (3) drilling on the continental shelf. Apparently the law largely replicated the wording of Articles 56(1) and 77(1) of the UNCLOS.

the ocean surrounding its territory.[9] It enacted the Basic Act on Ocean Policy[10] in 2007. The Basic Act mandated the government to formulate the Basic Plan on Oceans, which is to be reviewed every five years. Further, the Basic Act named the development and exploitation of ocean resources as the first item of the policy measures to be taken by the government.[11] In response, the first Basic Plan was published in 2008,[12] which mentioned, as one of the measures to promote the development of the EEZ and the Continental Shelf, 'intensive investment in the exploration and development of petroleum … in the EEZ and on the Continental Shelf'.

The exploitation of minerals in the ocean, in particular deep-sea resources, differs greatly from surface mining. The facilities needed are so costly that a serious operator must be equipped with a huge amount of capital. Such costs rendered the existing first-come, first-served principle of the Mining Act inappropriate for offshore petroleum exploitation. In addition, exploration activities (with the exception of exploratory drilling) were not regulated under the Mining Act until 2011. Several cases of foreign vessels allegedly engaged in scientific research in the EEZ of Japan, which were suspected of conducting exploration of undersea resources, raised an argument in Japan for the regulation of disguised exploration under the appearance of scientific research. Against this background, amendments to the Mining Act were made in 2011.[13]

---

[9] Souichirou Kozuka and Hideyuki Nakamura, 'The Law Applicable on the Continental Shelf and in the Exclusive Economic Zone: The Japanese Perspective', *Ocean Yearbook* vol.25, pp. 357–78 (2011).

[10] Law no.33 of 2007.

[11] Art.17, the Basic Act on Ocean Policy.

[12] 'Basic Plan on Ocean Policy (provisional translation)', March 2008, available at www.kantei.go.jp/jp/singi/kaiyou/kihonkeikaku/080318kihonkeikaku_E.pdf (accessed on 8 December 2014). A subsequent Basic Policy was published five years later, on 26 April 2013 (for which no English translation is available), available at http://japan.kantei.go.jp/96_abe/actions/201304/26kaiyou_e.html (accessed on 8 December 2014).

[13] The English translation of the text of the Mining Act after the 2011 amendments is available at www.isa.org.jm/files/documents/EN/NatLeg/JapMiningAct.pdf (accessed on 8 December 2014). The terms used in this chapter are not the same as those used in this translation, though.

## 3. LICENSE REGIME AFTER 2011 AMENDMENTS

### Types of License

The Mining Act provides for two types of license: a license for prospecting and a license for extraction.[14] The system may have worked for traditional mining, such as mining of coal or limestone. However, when applied to the exploration of petroleum, it results in covering only a part of the activities. While exploratory drilling is covered by a prospecting license, neither type of license covers seismic exploration. The license for prospecting is valid for four years with regard to petroleum, renewable twice, each time for two years. The license for extraction has no limited period.[15]

The license is deemed to be a property right and governed by the law on real property unless otherwise regulated in the Mining Act.[16] The license is to be registered in the Registry when granted[17] and becomes effective upon registration.[18] The license is exclusive in the sense that only one license may be granted in one area for the same kind of mineral.[19]

### Criteria for Granting a License

The general rule for the grant of a license in the Mining Act is 'first-come, first-served', namely a license is granted to the applicant that makes the application for an area earlier than anyone else,[20] on condition that the

---

14 Art.11, Mining Act.

15 Art.18, Mining Act. With regard to minerals other than petroleum and combustible natural gas, the initial period for a license for prospecting is two years.

16 Art.12, Mining Act. As a result of a license being deemed as a property right, the licensee is entitled to exclude the interference with its exercise of the license (Great Court of Judicature, 28 Feb. 1935, *Minshû* vol.14, p. 238). When the licensee has no right to the use of the land necessary for exploitation, however, the claim for excluding the interference by a third party is admissible only when there is a serious prospect of commencing the exploitation soon (District Court of Kumamoto, Tamana branch, 25 November 1969, *Hanrei Jihô* no.599, p. 67).

17 Art.59, Mining Act.

18 Art.60, Mining Act.

19 Art.16(1), Mining Act.

20 Art.27(1), Mining Act.

applicant satisfies certain required criteria.[21] However, under the amended Mining Act of 2011, petroleum is a 'specified mineral',[22] and the first-come, first-served rule is excluded. Instead, the government (Minister of Economy, Trade and Industry (METI)) invites competing applications through a licensing round, and grants a license to the applicant with the best capability to exploit petroleum in the area.[23] To be eligible for the competition, the applicant must satisfy the following criteria:

(i) that the applicant has the financial and technical feasibility of mining;
(ii) that the applicant has the social credibility;
(iii) that the applicant has no history of being punished under the Mining Safety Act or having a license revoked under the Mining Act;
(iv) that there is no overlap with an area reserved for a specified mineral and that the production under the license will not interfere with the exercise of a license in a neighbouring area;
(v) that the production under the license is neither without economic value, nor is harmful to the public health, nor destroys a facility for the public use, nor disrupts conservation of a cultural heritage, park or hot spring, nor impairs the interest of agriculture, forestry or other industry, nor is contrary to the public welfare;[24] and
(vi) that the exploitation of the mineral is not otherwise likely to be extremely improper in light of social and economic situations and to hinder the promotion of public interest.[25]

The criterion (vi) is expected to be useful in securing a stable supply of the resource. The text is carefully drafted to avoid imposing a duty on the licensee to deliver the produced mineral for the domestic use in Japan, with the relevant provisions of the World Trade Organization rules in mind.[26]

---

21 The criteria are provided for in Art.29(1), Mining Act.

22 Art.6-2, Mining Act. Also included in specified minerals are combustible natural gas and other kinds of minerals named in the Cabinet Order.

23 Art.38(1), Mining Act.

24 Criteria (v) requires the balancing of public interest achieved by the exploitation of the mineral and the public interest to be impaired due to the exercise of the mining license (Supreme Court, 12 April 1962, *Minshû* vol.16, no.4, p. 781).

25 Art.40(1), Mining Act.

26 Nakatani, supra note 2, at 205, 207.

The competitors that are found eligible are then assessed according to the business plan submitted upon application.[27] The government publishes the criteria for assessing the business plan prior to the invitation to apply for a license.[28] The government grants the license for the area to the applicant who is found, through the examination of the business plan, to be the most capable of exploiting the petroleum in the area and rejects other applications.[29] If, however, the applicant chosen as the licensee fails to pay the fee within thirty days, the license is not granted and the applicant with the second best capability is chosen as the licensee.[30]

## Monitoring of the Licensee by the State

The licensee must commence its operation within six months of the registration of the license.[31] Prior to doing so, the licensee shall prepare an operations plan and have it approved by the government. The operations plan needs to conform to the business plan submitted when applying for the license.[32] When the government finds that the licensee's operations plan will not ensure the full exploitation of petroleum in the area, it may recommend modifications to the operations plan and, if not accepted, order the licensee to modify its plan.[33]

As deemed property, a license is transferable.[34] However, the transfer is subject to the approval of the government.[35] The government will examine the transferee according to the criteria equivalent to those for granting a license.[36] A license can also be mortgaged, which becomes

---

27 Art.40(2), Mining Act.

28 Art.38(3), Mining Act.

29 Art.40(3) and (5), Mining Act.

30 Art.40(6) and (7), Mining Act.

31 Art.62(1), Mining Act.

32 Art.63-2, Mining Act. In the case of a license for mining of minerals other than specified minerals, the licensee for prospecting needs only to file its operations plan with the government. The approval by the government is necessary at the stage of production, based on a license for extraction (Art. 63, Mining Act). However, for mining of specified minerals, including petroleum, a licensee of a license for prospecting is also required to have its operations plan approved.

33 Art.100, Mining Act.

34 Art.13, Mining Act.

35 Art.13-2, Mining Act.

36 Art.51-2, Mining Act.

effective upon registration.[37] An agreement alone is not sufficient in creating an interest for the mortgagee.[38]

**Permits for Exploration**

The 2011 Mining Law amendments introduced regulation over exploration of petroleum and other specified minerals, requiring a permit by the government (METI) for exploration that does not accompany drilling, such as seismic exploration.[39] In order for a permit to be issued, the planned exploration must conform to certain criteria. The applicant must have a system to undertake the exploration properly and ensure that the exploration does not damage underground pipelines or otherwise cause danger to the area.[40] Many of the criteria for eligibility of an applicant for a license are also relevant.[41] Once a permit is issued and the exploration is carried out, the government may request reports of the outcome of the exploration, if necessary.[42]

The new regulation is designed to qualify as the exercise of the right of the coastal state under Articles 56 and 77 of UNCLOS. It is also noted that, while UNCLOS distinguishes the scientific research from the exploration and exploitation of natural resources and requires the coastal state to grant consent to scientific research projects by other states, the consent may be withheld if the scientific research project is of direct significance for the exploration and exploitation of natural resources.[43] When the scientific research is suspected to be a disguised exploration of natural resources, it will fall under this exception and the Japanese government will treat it as exploration without a permit.[44] The government then issues an order requesting the exploration to be discontinued, the facilities and equipment used for the exploration to be removed and the original state to be restored.[45]

---

[37] Art.60, Mining Act.

[38] Court of Appeals of Tokyo, 16 July 1960, *Kaminshû* vol.11, no.7, p. 1504 (alleged mortgage over a license to extract limestone denied).

[39] Art.100-2, Mining Act.

[40] Art.44-8, Ministerial Order to Implement the Mining Act (Ministry of International Trade and Industry Order no.2 of 1951).

[41] Art.100-3, Mining Act.

[42] Art.100-11, Mining Act.

[43] Art.246, UNCLOS.

[44] Nakatani, supra note 2, p. 208.

[45] Art.100-6, Mining Act.

### Ownership of the Extracted Petroleum

The current Mining Act provides that the state has the power to grant licenses for producing and acquiring minerals.[46] Unlike its predecessor of 1905, the Mining Act is drafted to avoid reference to the ownership of the resources prior to being extracted. Therefore, the current Mining Act assumes that there is no ownership of the underground natural resources.[47] Despite the difference in drafting, however, the current Mining Act has retained the same principle as under the previous Mining Act that the underground minerals do not belong to the owner of the land but are to be exploited by the licensee.

When the licensee extracts the mineral (including petroleum) under the Mining Act, the licensee acquires ownership of that mineral. If someone other than the licensee extracts the mineral, the licensee still acquires the ownership of the extracted mineral.[48] 'Extraction' in this provision means that the extracted mineral is sufficiently independent as to be controlled through the ownership.[49] If no license had been granted in the area, the extracted mineral shall be property without an owner.[50]

## 4. THE PETROLEUM EXPLOITATION OPERATOR AND JOINT VENTURES

### Nationality of the Licensee

A licensee is granted only to a Japanese national or a corporation incorporated under Japanese law.[51] It is a restriction on the nationality of a licensee, but not a restriction on investments in the mining sector.[52] A foreign investor can incorporate a subsidiary under Japanese law, which may then apply for a license.

---

46 Art.2, Mining Act.

47 Sakae Wagatsuma and Noboru Toyoshima, *Kôgyô Hô* (The Law on Mining) p. 66 (Yûhikaku, 1958) [in Japanese].

48 Art.8(1), Mining Act.

49 Appeals Court of Fukuoka, 9 August 1982, *Hanrei Jihô* no.1068, p. 77 (the silica stone removed by the constructor held not to be seen as 'extracted'); Appeals Court of Fukuoka, 26 August 2008, *Hanrei Taimuzu* no.1175, p. 294 (the fireclay removed by the constructor together with the sand held not to be seen as 'extracted').

50 Art.8(2), Mining Act.

51 Art.17, Mining Act.

52 Nakatani, supra note 2, p. 204.

## Operators Funded by the Government

In order to secure a stable supply of petroleum without restricting foreign investment, the Japanese government has occasionally established an entity with the public funding and mandated it with the mission as the primary operator.

The oldest of such attempts was the establishment of Teikoku Oil KK (*Teikoku Sekiyu Kabushiki Kaisha*) by enacting the Teikoku Oil KK Law[53] in 1941. The government contributed half of the initial capital, while the rest was contributed by the private sector. The mission of Teikoku Oil was to exploit the petroleum resources within the Japanese territory as well as overseas.[54] After the Second World War, the Teikoku Oil KK Law was abolished and the company was privatized in 1950.

As the demand for securing the supply of petroleum became all the greater after the war, the Law on the Exploitation of Petroleum and Combustible Natural Gas Resources[55] was enacted to fully introduce the conservation techniques. The law also established the Exploitation of Petroleum and Combustible Natural Gas Resources Council, which in its 1953 report endorsed the promotion and exploitation of domestic hydrocarbon resources.[56]

With the aim of carrying out the endorsement in the report, the Japan Petroleum Exploration KK Law[57] was enacted in 1955 and the Japan Petroleum Exploration KK (*Sekiyu Shigen Kaihatsu Kabushiki Kaisha*) was established. The government made a contribution in kind by contributing the shares of Teikoku Oil KK that it held. It contributed more than half of the initial capital, with industry members, including Teikoku Oil KK, contributing the remainder.[58] Later, the Japan Petroleum Exploration KK expanded its scope of business to overseas, including the exploration, followed by the production, in North Sumatra by the production sharing agreement through its subsidiary, North Sumatra Petroleum Exploration Co., Ltd. The production was successful and the subsidiary was renamed

---

[53] Law no.73 of 1941.

[54] *Teikoku Sekiyu 50-nenshi: Keiei Hen* (Fifty Years of Teikoku Sekiyu: Book on the administration), pp. 23–6 (Teikoku Sekiyu, Tokyo, 1992) [in Japanese].

[55] Law no.162 of 1952, abolished in 2011.

[56] *Teikoku Sekiyu 50-nensh: Keiei Hen*, supra note 54, p. 76

[57] Law no.152 of 1955, abolished in 1967.

[58] *Sekiyu Shigen Kaihatsu Kabushiki Kaisha 10-nenshi* (Ten Years of Japan Petroleum Exploration KK), pp. 8–10 (Sekiyu Shigen Kaihatsu Kabushiki Kaisha, Tokyo, 1967) [in Japanese].

as the INPEX Corp. (*Kokusai Sekiyu Kaihatsu Kabushiki Kaisha*), which merged with Teikoku Oil KK in 2008, forming a new INPEX Corp. (Japanese name: *Kokusai Sekiyu Kaihatsu Teiseki Kabushiki Kaisha*).[59] As a significant portion of the capital of INPEX had been funded by the public finances, the government (METI on its behalf) holds a golden share with a veto on important management decisions in the new INPEX Corp.[60]

Back in 1967, faced with the ever-growing demand for petroleum, the government considered it useful to introduce a financing mechanism for the investment in the exploration of petroleum. The idea developed into incorporation of the Japan Petroleum Development Corporation (*Sekiyu Kaihatsu Koudan*) by enacting the Japan Petroleum Development Corporation Law[61] in 1967. The Japan Petroleum Exploration KK was merged into this newly established corporation. The new corporation was not a joint stock company but a corporation for public purposes. It was given the mandate to finance the exploration of petroleum through equity holding or loan.[62]

However, the Japan Petroleum Development Corporation, later renamed as the Japan National Oil Corporation (*Sekiyu Koudan*), had not achieved much success. Towards the end of the 1990s, it was revealed that the unsuccessful outcome of its activities was not properly disclosed in the balance sheet of the corporation. The finding developed into a political controversy and led to the abolishment of the corporation. Instead, the Japan, Oil, Gas and Metals National Corporation (JOGMEC; *Dokuritsu Gyôsei Hôjin Sekiyu, Tennen Gasu, Kinzoku Koubutsu Shigen Kikou*) was incorporated by the Japan, Oil, Gas and Metals National Corporation Law[63] and succeeded the mandate of the Japan Petroleum Corporation.[64] JOGMEC, similar to its predecessor, is equipped with the power to invest in, or guarantee a loan for, the exploration and production

---

[59] See the section on corporate history on the INPEX Corp.'s website, available at www.inpex.co.jp/english/company/history.html (accessed on 8 December 2014).

[60] Kenichi Osugi, 'Transplanting Poison Pills in Foreign Soil', in: Hideki Kanda, Kon-Sik Kim and Curtis J. Milhaupt (eds.), *Transforming Corporate Governance in East Asia*, p. 36 at p. 40 (Routledge, London and New York, 2008).

[61] Law no.99 of 1967, abolished in 2002.

[62] Koike et al., supra note 3, at p. 1767. See for the details, *Teikoku Sekiyu 50-nenshi, Kaigai hen*, (Fifty Years of Teikoku Sekiyu: Book on the overseas activities), pp. 31–3 (Teikoku Sekiyu, Tokyo, 1992) [in Japanese].

[63] Law no.94 of 2002 (JOGMEC Law).

[64] Koike et al., supra note 3, at pp. 1772–3.

of petroleum.[65] It can also acquire an interest in the exploration of petroleum abroad by itself when it is difficult for private businesses to do so.[66]

## Joint Venture Agreements

In the case of exploration and production of petroleum, a joint venture (JV) is usually formed. An unincorporated JV is based on an agreement that is characterized as an *agreement of association* (*kumiai keiyaku*) under the Japanese Civil Code,[67] which creates an 'association'. An agreement of association is an equivalent of *Gesellschaftsvertrag* in the German Civil Code or *société civile* in the French Civil Code. Most of the provisions in the Japanese Civil Code are non-mandatory and, therefore, joint venture agreements (JVAs) can validly derogate from them.

The operation of an association is either carried out by all the members or delegated to an operator, who can be one or more of the members or a non-member. In the former case, the majority of the members make decisions regarding the operation.[68] In the latter case, which is usually the case with JVAs, the majority of the delegated operators make the decision.[69] The Japanese law distinguishes the executive power within the association and the authority to trade with a third party, but delegating the executive power to the operator usually entails the granting of the authority vis-à-vis the third party.[70]

The members of an association jointly own the assets and liabilities.[71] Even if a member sells or transfers its share in the association's asset, such sale or transfer cannot be valid against the association and the association's creditor.[72] This rule implies that a creditor of an association member cannot attach the member's share in the association's asset.[73] For the debts arising from the operation of the association, the primary

---

65 Art.11, items 1 and 3, JOGMEC Law.

66 Art.11, item 4, JOGMEC Law.

67 Law no.89 of 1896.

68 Art.670(1), Civil Code.

69 Art.670(2), Civil Code.

70 Great Court of Judicature, 8 March 1911, *Minroku* vol.17, p. 104; Great Court of Judicature, 27 September 1919, *Minroku* vol.25, p. 1669.

71 Art.668, Civil Code.

72 Art.676(1), Civil Code.

73 Great Court of Judicature, 1 September 1931, *Hôritsu Shinbun*, no.3313, p. 9.

liability should be on the association's asset.[74] However, the members also have unlimited liability for them,[75] unless otherwise agreed.

If the association has no specified terms, a member may terminate its participation at any time.[76] Even if the agreement has a defined period, a member may terminate its participation when there is a compelling reason to do so.[77] The latter rule is mandatory, and a derogation clause is null and void.[78] The association is dissolved and wound up when its goals are achieved or the goals become decisively unachievable.[79] A member may request the dissolution of an association when there is a compelling reason.[80]

### Co-License and Association by the Force of Law (Deemed Joint Venture)

In principle, an association is formed by an agreement. However, the Mining Act deems an association to be formed when two or more persons hold a license jointly.[81] As a result, a co-licensee is entitled to basic rights over the association, including the right to monitor and the right to veto changes in the membership, without any explicit agreement. Further, the assets used for, and the liabilities arising from, the operation are deemed to be shared by all the co-licensees.

Because the license belongs to the asset of the association, one of the co-licensees cannot sell or transfer its 'share' in the license. It is not allowed to mortgage the share in the license, either.[82] For the same reason, the creditor of one of the co-licensees cannot attach the latter's share in the license.[83] A co-licensee may, as a member of the association, request the dissolution of the association when there is a compelling reason. Then a liquidator is appointed to wind up the association, including the sale of the association's license.[84]

These rules do not exclude the possibility of the licensee forming an agreement of association (JVA) with a party without a license. It is

---

74 Great Court of Judicature, 25 February 1936, *Minshû* vol.15, p. 281.
75 Art.675, Civil Code.
76 Art.678(1), Civil Code.
77 Art.675(2), Civil Code.
78 Supreme Court, 23 February 1999, *Minshû* vol.53, no.2, p. 193.
79 Art.682, Civil Code.
80 Art.683, Civil Code.
81 Art.43(5), Mining Act.
82 Supreme Court, 22 June 1962, *Minshû* vol.16, no.7, p. 1389.
83 Appeals Court of Tokyo, 19 January 1960, *Kaminshû* vol.11, no.2, p. 370.
84 Great Court of Judicature, 14 July 1923, *Minshû* vol.2, p. 491.

argued, however, that the operator of such a JVA must be the licensee, as the Mining Act assumes that a licensee operates the exploration and production by itself.[85]

## 5. PROTECTION OF ENVIRONMENT AND LIABILITY FOR DAMAGES

The prevention of environmental damage is an important duty of the licensee exploring or extracting petroleum. The licensee is required to include in its operations plan the measures to be taken to prevent pollution.[86] The Mining Safety Act (discussed below in section 6) also imposes on the licensee the duty to prevent pollution from its operations according to the Ministerial Order.[87] Further, the technical standard of the equipment used in the production of petroleum requires that the mechanism to prevent blowouts be installed.[88]

In case damage is caused by the exploration or production of petroleum, the general rules of tort or special rules under the Mining Act are applicable, depending on the cause of the damage. The general rules of tort are codified in the Civil Code. When the party commits an act or omission with intent to cause damage or by negligence, that party is liable for the resulting damage as long as there is a causal link (proximity) between the act or omission and the damage.[89]

The Mining Act provides for a special rule of strict liability for damages caused by the drilling of land, discharge of pit water or wastewater, rock piles, mine slag-piles, or the discharge of smoke. For these types of damage, the first of which may in particular be relevant to the exploration or production of petroleum, the licensee at the time of occurrence of the damage shall be strictly liable.[90] Further, in case the license has been transferred after the occurrence of the damage, the licensee at the time of occurrence of the damage and the subsequent

---

85 Wagatsuma and Toyoshima, supra note 47, p. 36.

86 See Form 20-5 (Attachment to the Ministerial Order Implementing the Mining Act).

87 Art.8, Mining Safety Act.

88 Art.17, 19 and 20, Ministerial Order on the Technical Standards of Equipment Used in Mining (METI Order no.97 of 2004).

89 Art.709, Civil Code.

90 Art.109(1), Mining Act. If there is currently no licensee, the last licensee shall be liable.

licensees are jointly and severally liable.[91] Whether the licensee has ever operated or not does not matter, since the special liability under these provisions arises from the fact of holding a license, not causing the damage.[92]

The remedy for the damage is pecuniary compensation and it must be fair and equitable. Only when the restoration does not entail significant increase in the cost as compared with the pecuniary compensation, restoration may be requested. The liable licensee may offer restoration instead of pecuniary compensation, notwithstanding.[93] The licensee is not exempted from the liability even if a natural disaster or other force majeure contributes to the damage. The court may merely take it into consideration when determining the liability or scope of compensation.[94]

## 6. SAFETY OF THE WORKPLACE IN OIL EXPLORATION

Besides the prevention of pollution and environmental damages, the safety of the workplace is also a big issue in mining in general, and the exploitation of petroleum in particular. To address this problem, the Mining Safety Act[95] introduces the duty of the licensee as well as the responsibility of workers. To the extent the Mining Safety Act regulates that the safety of the workplace in the mines (defined as including the production site of petroleum[96]), the Workers' Safety and Health Act[97] is excluded.[98]

The licensee (as the operator) must take necessary measures to prevent harm to workers in the workplace according to the Ministerial Order.[99] It must be emphasized that the METI, not the Minister of Health, Labour

---

91 Art.109(3), Mining Act.

92 District Court of Miyazaki, Nobeoka Branch, 28 March 1984, *Hanrei Jihô* no.1111, p. 3; Appeals Court of Fukuoka, Miyazaki Branch, 30 September 1988, *Hanrei Jihô* no.1292, p. 29; District Court of Miyazaki, Nobeoka Branch, 26 March 1990, *Hanrei Jihô* no.1363, p. 26 (all the decisions concerning the arsenic poisoning due to the production of arsenous acid).

93 Art.111, Mining Act.

94 Art.113, Mining Act.

95 Law no.70 of 1949.

96 Art.2(2), Mining Safety Act.

97 Law no.57 of 1972.

98 Art.115(1), Workers' Safety and Health Act.

99 Art.5, Mining Safety Act.

and Welfare (MHLW), enacts the order.[100] The licensee must also provide safety education to the workers.[101] The machines, equipment and explosives that are of serious danger must be in conformity with the technical standards provided for in the Ministerial Order.[102] These regulations are applicable whether or not the licensee directly employs the workers. In other words, the Mining Safety Act and the Ministerial Order also protect workers employed by the contractor.

Conversely, the workers have a responsibility to comply with the requirements necessary to prevent harm to workers and ensure security of the facility.[103] Further, workers' representatives sit in the Central Mine Safety Committee together with the representatives of the operators and the academics[104] and play an important role in the promulgation of safety regulations. The government must consult this Committee when enacting or amending the Ministerial Order under the Mining Safety Act.[105]

Apart from the duty of the operator stipulated in the statutes, Courts have developed a doctrine regarding employer's duty of care for the safety of the workers. The doctrine derives from the principles of good faith and fair dealing under the Civil Code,[106] and now is codified in the Labour Contract Act.[107] Therefore, the employer can be held liable for failure to ensure safety of the employee even when there is no specific regulation to comply with or when the administrative regulation is fully complied with.[108]

If the operator (licensee) has fully complied with the regulation and is still held liable for not caring sufficiently about the safety of workers, it means that the regulation was too outdated to secure the safety of the workplace. In such a case, the responsibility of the government may be implied. The Supreme Court affirmed the states' liability for failing to amend the Ministerial Order implementing the Mining Safety Act in a timely manner even after the threat of pneumoconiosis to the workers in

100 The Ministerial Order to Implement the Mining Safety Act (METI Order no.96 of 2004).

101 Art.10, Mining Safety Act.

102 Art.11, Mining Safety Act.

103 Art.9, Mining Safety Act.

104 Art.54, Mining Safety Act.

105 Art.52, item 1, Mining Safety Act.

106 Art.1(2), Civil Code.

107 Art.5, Labour Contract Act (Law no.128 of 2007).

108 As cases concerning the mining operator, Supreme Court, 22 February 1994, *Minshû* vol.48, no.2, p. 441; Supreme Court 22 March 1994, *Rôdô Hanrei* no.652, p. 6.

the coalmines became apparent.[109] Because the order is enacted or amended by METI, not MHLW, there is a danger that the consideration to the operator prevails over ensuring the safety and health of the workers. The same principle may be extended to apply to the regulation of the petroleum production.

## 7. CONCLUSION

Since the early days of its modern history, Japan has always been aware of the importance of petroleum. However, the legal regime has not been tailored to the exploration and production of petroleum. For both the licensing regime and the safety regulation, the laws are applicable to mining in general, and the provisions appear to be more suitable to the mining of coal and other metals than to the exploitation of petroleum.

The 2011 reform to the Mining Act was a significant step in tailoring the regulatory regime. Petroleum and some other resources are now classified as 'specified minerals' and the competitive application system was introduced, excluding the traditional first-come, first-served system. Still, other provisions, such as the rule on the association by the force of law (deemed JVA) among the co-licensees, the strict liability of the licensee for damages caused by the operation, and the safety regulation under the Mining Safety Act, have remained unchanged. Such reluctance in changing the law may, at least partly, be due to the unfortunate reality that the production of petroleum on Japanese territory has not enjoyed much success. As a result, there has not been great inconvenience in practice.

However, the 2011 reform was made on the prospect that recent technological developments may enable the exploitation of hydrocarbon in the deep-sea, and that the EEZ of Japan is a promising petroleum-producing province. If such prospects turn out to be more than mere optimism, the practical issues this chapter has raised will need reconsideration. Then the law on the upstream petroleum sector in Japan will have to go through further modernization.

---

[109] Supreme Court 27 April 2004, *Minshû* vol.58, no.4, p. 1032.

# PART IV

# Legal issues in petroleum regulation

# 12. Comparison of access to petroleum in developed and developing licensing and concession systems

**Tina Hunter**

## 1. INTRODUCTION

The licensing and concession system (LCS) has been developed by many states to interact with and govern the conduct of participating oil companies, and to maximize financial gains.[1] The LCS is defined as a system of petroleum regulation where a license is granted over a 'concession', or area. That license grants proprietary rights to the license holder, which are often also imbued as contractual rights between the participating parties through agreements such as joint operating agreements (JOAs). Unlike the production sharing contract (PSC), the concession system assumes that the operating oil companies obtain a license from the state on certain terms and conditions, most of which are fixed by legislation and some of which are negotiated case by case between the state and the relevant oil companies. An important characteristic of the concession system is that, since legislative power is a state prerogative, the state remains at considerable liberty to modify at any time those terms and conditions that are not negotiated but fixed by legislation.[2]

The LCS grants specific contractual and proprietary rights to the participants who have been awarded petroleum licenses. Although the LCS is used in many countries to regulate the exploitation of petroleum, the level of government control over petroleum activities differs. Furthermore, there are two main recognizable systems of regulation in licensing

---

[1] Guiditta Cordero Moss, 'Contract or Licence? Regulation of Petroleum Investment in Russia and the Role of Foreign Legal Advice' (1998) 3–11 *CEPMLP Internet Journal*, accessed 14 July 2014 at www.dundee.ac.uk/cepmlp/gateway/index.php?news=28136.

[2] Ibid.

and concession systems – the long established 'North American' model and the 'North Sea' model, developed by the UK and Norway when exploiting the petroleum resources in the North Sea.

The North American system refers to the regulatory framework that had been implemented in the management of petroleum resources in the United States and Canada.[3] Generally, the North American model of petroleum exploitation is typified by a minimalist approach to state intervention.[4] It originated in the United States, and is categorized by minimal government involvement in the development of petroleum resources, with a preference for free market forces to direct the exploitation of these resources.[5] In addition, this model tends to award petroleum licenses through a bid system (both cash and work program), again allowing capitalist forces to influence the exploitation of petroleum.[6] The state regulates the award of the petroleum license as well as establishing and enforcing laws and regulations that protect workers and the environment.[7] The private oil companies are given de facto control over the license area, with companies retaining autonomy over issues relating to the development of petroleum resources.[8] This includes field development, rates of depletion, and other issues relating to production.[9] Usually, these licenses are awarded using an auction bidding system. This may include either cash or work program bidding (WPB).[10] This North American model of petroleum exploitation has been adopted by Australia for the exploitation of its offshore petroleum resources.

The North Sea model was established by the licensing and concession activities of the United Kingdom and Norway in the exploitation of

---

[3] See Jerome Davis, *Does One Size Fit All: Reflecting on Governance and North Sea Licencing Systems* (2004) Background Paper: BC Offshore: Potential and Problems, a MASC Workshop for Lawyers, Dunsmuir Lodge, Sidney BC, 18–21 March 2004, 2.

[4] See Brent F Nelsen, *The State Offshore: Petroleum, Politics and State Intervention on the British and Norwegian Continental Shelves* (Praeger, Westport, 1991), 8.

[5] Bernard Taverne, *Petroleum Industries and Governments: A Study of the Involvement of Industry and Governments in the Production and Use of Petroleum* (2nd ed. Kluwer, Netherlands, 2008), 165–8.

[6] Ibid, 168.

[7] Nelsen, above n 4, 8.

[8] Øystein Noreng, *The Oil Industry and Government: Strategy in the North Sea* (Croom Helm, London, 1980), 31.

[9] Davis, above n 3, 2.

[10] Kenneth Dam, *Oil Resources: Who Gets What How?* (University of Chicago Press, Chicago, 1976), 6–7.

petroleum resources in the North Sea.[11] The primary difference in this regulatory model is the high level of state intervention,[12] as the state controls the award of petroleum licenses through the administrative allocation of petroleum licenses. Generally, the allocation of licenses is through the use of established criteria, although allocation is at the discretion of the state.[13] The state also exerts high levels of control over the development of a petroleum field for the life of the field.[14]

This chapter draws upon the characteristics of the award of licenses in the LCS that have been outlined in the preceding eleven chapters to highlight some of the characteristics of licensing and concession systems in both developed and developing regimes.

## 2. PREQUALIFICATION FOR APPLICATION FOR LICENSE

The idea of prequalification is to provide for a single pre-application assessment of an applicant prior to applying for one or more licenses. This is because it is costly to apply for a license, and prequalification ensures that the applicant qualifies to apply before expending financial resources in an application where they may not qualify. Not all jurisdictions require prequalification, although interestingly prequalification is not confined to a developed or underdeveloped jurisdiction. Rather, it is a requirement in both developed and developing petroleum jurisdictions.

In order to apply for a license in Brazil, there is a requirement to fulfil the basic requirements for qualification to undertake petroleum activities. In order to qualify, the applicant must meet technical, financial and legal criteria. Once an applicant is assessed to have met these criteria, they are qualified to participate in the licensing round. The prequalification is based on points, where the more points a company obtains, the more likely it is able to participate in any area offered (offshore and onshore).

---

[11] Nelsen, above n 4, 23.

[12] Nelsen, above n 4, 8–9.

[13] Such as the requirement for objective criteria set out in ss 3–5 of the Petroleum Activities Act 1996 (Norway), and the criteria outlined in s 10 of the Petroleum Regulations 1997 (Norway).

[14] Norwegian Petroleum Directorate, 'Guidelines to Plan for Development and Operation of a Petroleum Deposit (PDO) and Plan for Installation and Operation of Facilities for Transport and Utilisation of Petroleum (PIO), 2000' (2000), accessed 14 July 2014 at www.npd.no/regelverk/r2002/frame_e.htm. See also Noreng, above n 8, 32.

The lower the points, the less likely a company will be allowed to participate in risky and complex areas.

Where companies are interested in applying for a license in Norway (either a standard licensing round, or a license in an 'award in predefined area' licensing round) they are required to be evaluated for prequalification. In order to successfully prequalify, applicants are required to prove they fulfil the technical, financial and conduct requirements for the grant of a license. However, it is important to realize that although a company may prequalify for a grant of license, there is no guarantee that a license will be awarded.

Similar to Norway, any company wanting to participate in petroleum operations in Nigeria is required to prequalify. Bidders are required to register and provide all required technical, financial and legal information. This information is evaluated, and if successful, the applicant qualifies for participation, and assigning with strategic partners. Such qualification is made public.

The use of WPB in Australia does not require prequalification. However, with the introduction of cash bidding in 2014, a system of prequalification for cash bidders has been introduced. Where an applicant wants to bid for acreage, they are first required to make an application, where their suitability to hold a license will be pre-assessed. If assessed as suitable the applicant will be invited to bid for the acreage.

## 3. ALLOCATION OF LICENSE – BID VERSUS DISCRETIONARY

### Use of Bid System

The allocation of licenses under the bid system is consistent with the North American system of petroleum licensing. Under this system, license allocation has traditionally been made utilizing the cash bid system. In the United States, under the Outer Continental Shelf Lands Act 1953 the Department of the Interior offers access to offshore petroleum resources utilizing the cash bid system. After the approval of the five year leasing plan, sealed cash bids are solicited from interested parties. The sale of leases is advertised in the Federal Register, which contains the terms and conditions of sale, bidding protocol, blocks available for lease and any other special information. The leases are awarded by competitive sealed bids for the acreage. The winner of each acreage block is announced live over the internet and radio.

This system of allocation by cash bidding is now being utilized by Nigeria. Previously, the allocation of petroleum licenses occurred through the arbitrary exercise of discretion before and after independence until the 2000s. Since this time, some level of open bidding has been introduced. Today most, but not all, grants and allocations of licenses are now based on a set of criteria made known to qualified and deserving companies at the time such blocks are open for bidding. Indeed, the Nigerian government has indicated its commitment to only allocate concessions or licenses to operators based on open competitive cash bidding rounds, where licenses can only be granted to a company incorporated in Nigeria and registered solely for petroleum activities.

Licenses in Russia are today also allocated through a cash auction. Applications for participation in an auction go through an examination process, assessing the applicant's financial and technical competence. Applications for participation in the competitive tender do not go through a preliminary examination. Rather the examination occurs throughout the tender process. The granted license, awarded to the winner of an auction or a tender, is then sent to the federal or the territorial geological funds for registration, which occurs within the period of one month from the time of its grant.

Prior to 1993 Australia utilized the cash bidding system for the allocation of exploration licenses, especially in highly prospective areas. However, in 1993 this method was abandoned in favour of the sole use of the WPB system for the allocation of petroleum licenses. WPB was implemented as a means of encouraging petroleum exploration, particularly given that the North West Shelf (NWS) area required detailed exploration. Under the Australian WPB system, the allocation of an exploration license is made to the highest bidder, where the applicant 'bids' how much money they will expend on exploration programs. The competitive WPB system is supplemented by *Guidance Notes for Applicants* that are released annually by the relevant government department. Under this competitive system, where there is more than one bidder under the WPB, the Joint Authority may grant the exploration license to the applicant which 'in the Authority's opinion, is most deserving of the grant having regard to criteria made publicly available by the Authority' under s 106(2) of the Offshore Petroleum and Greenhouse Gas Storage Act 2006 (Cth) (OPAGGSA). The assessment is made by considering which applicant is most likely to achieve the fullest assessment of the petroleum potential within the license area. Since 2014 Australia has embraced the use of cash bidding for mature areas, which is incorporated into the annual licensing rounds.

In most Canadian jurisdictions WPB is the sole criterion used for the allocation of licenses to access petroleum resources, in order to ensure that the process is fair and objective, and removes all form of discretion. Although, in Newfoundland and Labrador, bids for significant petroleum discoveries and production licenses can be made on the basis of cash bidding. However, the bid may not be accepted where it does not satisfy the terms and conditions of a license.

The Brazilian licensing system utilizes a complex bidding process for the award of petroleum licenses. The biddable factors that are assessed to define the winning proposal are threefold: the minimum exploration work program, the cash (signature) bonus and local content provisions.

## Discretionary Allocation

The two countries that are striking examples are the UK and Norway, which together developed the 'North Sea system' under the LCS.

The UK originally incorporated the use of discretionary criteria for the award of a petroleum license under the Petroleum Act 1934. The UK government initially debated the merits of using a discretionary versus an auction system in the 1960s, ultimately rejecting the auction system since it 'did not permit the same range of detailed control that discretionary licencing systems do'.[15] This choice of discretionary allocation enabled the UK Ministry of Power to allocate licenses to companies that were willing to carry out the wishes of the UK government for the rapid development of the resources on the UK Continental Shelf (UKCS).

In the early licensing rounds the UK government encouraged, through its discretionary award of licenses, the rapid and thorough exploitation of the UKCS, in order to exploit the petroleum resources for their economic worth.[16] The UK argued that the allocation of licenses using discretion with weighting on the work program was required because such emphasis promoted rapid and thorough exploration and exploitation of its North Sea oil resources.[17] Oil companies proposed a work program and entered into negotiations with the UK government regarding the required work program. When agreement was reached on the terms of the license, the companies committed to a minimum program of work on the UKCS in a

---

[15] Nelsen, above n 4, 19.

[16] Ibid.

[17] Select Committee in Energy, Third Report Session 1981–2, *North Sea Oil Depletion Policy* (House of Commons, London, 1982) [55], in Ursula M H Kretzer, 'Overcapitalisation in Licencing Systems Based on Size of Work Programme' (1993) 19(4) *Resources Policy* 299, 299.

specified period of time. Hence, there was not only an element of work program bidding, but also negotiation with the UK government to determine the most suitable applicant for that license area.

Under the Petroleum Act 1998 (the 1998 Act) the discretionary allocation of petroleum licenses continued. Section 3 of the 1998 Act empowers the Secretary of State to grant licenses to persons that they think fit, on such terms and conditions as they see fit, for such financial considerations as they, with the consent of the Treasurer, may determine. Such a system confers broad discretionary power upon the Secretary of State in the award of licenses.

The UK explored the use of the cash bid system for the allocation of some licenses in the fourth licensing round in 1971.[18] Fifteen blocks were auctioned, with a further 267 blocks allocated by the usual discretionary basis of license allocation.[19] Altogether, 31 bids were received for the 15 blocks, with winning bids providing £37 million in government revenue. In comparison, the other 267 blocks only realized revenue of £2.7 million.[20]

A number of academics have examined this licensing allocation experiment by the UK, with Kenneth Dam concluding that the 'discretionary system turned out to be the most expensive subsidy'.[21] Dam arrived at this conclusion based on an analysis of the income received.[22] However, this analysis compares only the initial payments made by the oil companies that bid for the licenses under the discretion and auction system. Whilst the auction system did provide an initial income of £37 million, compared to only £2.7 million for the discretionary licenses, the licenses awarded through discretion also contributed over £200 million in minimum work programs.[23] The income analysis by Dam fails to incorporate income realized from the calculation of resource rent tax and other special taxation on the income realized from the licenses awarded using discretion. As such, it is not possible to conclude that the income received was higher from the award of licenses by cash bid rather than

---

18 Walter Mead, 'Towards an Optimal Oil and Gas Leasing System' (1994) 15(4) *Energy Journal* 1, 5.

19 Kjell J Sunnevåg, 'Designing Auctions for Offshore Petroleum Lease Allocation' (2000) 26 *Resources Policy* 3, 5.

20 Walter Mead, in ibid, 5.

21 Dam, above n 10, 39.

22 See ibid, 38, for income received from the allocation of licenses using the bid and discretion methods in the licensing rounds.

23 House of Commons, First Report of the Committee of Public Accounts, North Sea Oil and Gas, Session 1972–3, at 45 (1973) in Dam, above n 10, 38.

through discretionary methods, since not all income from licenses awarded by discretion was included in the calculations, only initial income. Rather, it is only possible to conclude that bidding provided greater 'up front' revenue, an expected outcome from the auctioning of petroleum licenses.

The UK government reintroduced cash auctioning as one of the methods of allocating licenses in the seventh licensing round in 1980/81, whilst still retaining the right to allocate licenses using discretion, and to reject the highest cash bid.[24] The rationale for the reintroduction was that it would realize large sums of money without slowing offshore exploration, rather than an ideological argument relating to free markets.[25] During this period the UK economy was facing numerous structural problems, including high inflation, wage inflation, trade union strikes and high levels of government debt.[26] The UK government sought to reduce the government deficit by raising taxes, decreasing government spending and reducing interest rates. The effect of these actions was to cause economic recession in the UK.[27] Coupled with this were the effects of the appreciation of the pound sterling, particularly affecting the economy through income received from the export of oil. As part of the overall monetary strategy, the government sought a cash injection into the economy through the cash auctioning of these blocks. Forty-two blocks from the UKCS mature province were auctioned, realizing £210 million.[28] The eighth and ninth rounds each auctioned off 15 blocks (8 per cent of total number of blocks offered), with the return disappointing.[29] The eighth round raised only £33 million through auction, although the ninth round was more successful, raising £121 million.[30]

The tenth licensing round in 1987 saw a return to the discretionary allocation of both mature and frontier blocks. This coincided with a change in the fiscal policy of the Thatcher government, since the Treasury no longer required the injection of cash that would have been realized in the auctioning of the acreage.[31] In all, the auctioning of

---

[24] Nelsen, above n 4, 92.

[25] Ibid, 92.

[26] 'UK Economy under Mrs Thatcher 1979–1984', accessed 14 July 2014 at www.economicshelp.org/2007/03/uk-economy-under-mrs-thatcher-1979-1984.html.

[27] Ibid.

[28] Nelsen, above n 4, 92.

[29] Ibid, 92.

[30] Ibid, 92.

[31] Ibid, 92.

British acreage was not motivated by the economic desire to extract the greatest resource rent from the acreage, but to raise revenue. Using auction as a revenue raising exercise seriously undermines the value of the auction system, which is predicated on the notion of extracting the highest resource rent. The use of auction by the British as a revenue raising tool rather than as a tool to capture the greatest amount of economic rent undervalued the resources, and enabled bidders the opportunity to bid not for what the acreage was worth, but for what the British government was likely to accept for the acreage.

Norway has always granted petroleum production licenses on the basis of broad discretionary allocation, premised on the applicant's capacity. The capacity of the applicant is evaluated to determine which company is best able to combine effective exploration and production of petroleum with sound safety and environmental practices. The reasoning for the use of discretionary allocation is that the climatic challenges of the Nordic areas and the risk of damage in environmentally sensitive areas have made it necessary for licenses to be granted to companies with strong technical capability, large financial resources (given the strict liability regime) and a proven record of responsible conduct. In addition, Norway utilized the discretionary system to develop national oil companies and national industries for the fabrication of installations (including materials fabrication such as concrete and steel), and for services to the petroleum industry.

The discretionary scope of the allocation of petroleum licenses in the UK and Norway has been significantly curtailed with the introduction of the EU Hydrocarbon Directive,[32] which requires that states only utilize specific, non-discriminatory criteria when awarding a license. Such non-discriminatory criteria may include technical capacity, financial capability and the applicant's plan for exploration and production in the license area.[33]

However, both the UK and Norway still retain elements of discretion in awarding a license under the EU directive. In Norway, s 10 of the Petroleum Regulations 1997 (Norway) provides for state discretion in selecting a licensee where two or more applicants are regarded equal on the basis of the technical, financial and work criteria. In making the decision, the Norwegian state has regard to 'other relevant objective and

---

[32] Directive 94/22/EC on the conditions for granting and using authorizations for the prospection, exploration and production of hydrocarbons.

[33] For an example of such 'non-discriminatory criteria', see s 10 of the Petroleum Regulations 1997 (Norway).

non-discriminatory criteria' in order to make the final choice between the applications.

Whereas the allocation of licenses in Norway and the UK could be described as 'objective' discretionary, the grant of a license in Japan was 'discretionary' discretionary. Grants of a license under the Mining Act 1950 originally occurred on a first-come, first-served basis, where a license is granted to the applicant that makes the application for an area earlier than anyone else on the condition that the application satisfies certain required criteria. However, the 2011 amendments to the Mining Act excluded the first-come, first-served rule, instead opting for an 'objective' discretionary allocation, where the government invites competing applications through a licensing round and grants a license to the applicant with the best capability (assessed using defined criteria) to exploit petroleum in the area.

## 4. SPECIAL TREATMENT FOR MATURE AND FRONTIER AREAS

The allocation of licenses in mature areas has been of interest in the last ten years, as some petroleum provinces mature, whilst at the same time opening up new areas. In particular there has been special treatment in the UK and Norway, with Australia recently implementing special treatment for mature areas.

The UK has established two variations of the licensing system: the Promote and Frontier licenses. Promote licenses were introduced in 2003 in order to enable small companies with considerable geotechnical ability but limited financial resources to obtain licenses. Whilst these small companies may pose a risk (in that they may not be able to meet the obligations of the license), their knowledge of the geotechnical aspects of the license area and capability far outweighs this risk. It is recognized that such a license is crucial to assist the UK in meeting its objective of maximizing the economic recovery of petroleum from the UKCS. Frontier licenses, first introduced in 2004 and restructured in 2008, are designed to facilitate the exploration and development of the area to the West of Shetland. This is a frontier area with several technical difficulties including deep water, complex geology and minimal infrastructure. There are two variants of the infrastructure license:

- Atlantic Margin license: a license granted for a 6 year initial term, a 6 year second term and an 18 year production period; and

- Outer Hebrides license: for the Atlantic Margin west of the Outer Hebrides. Licenses granted for this area have a 9 year initial term, followed by a 6 year second term and an 18 year production period.

To date the Promote and Frontier licenses have been regarded as successful, measured by an increase in the number of new entrants to the North Sea and increased activity to the west of Shetland.

Similar to the UK, Norway has established a special licensing system for mature areas, particularly those licenses in the North Sea. In order to re-examine the geology in mature areas, an arrangement with award of licenses in predefined areas (APA) was introduced in 2003 parallel to the ordinary licensing rounds for new areas. The APA scheme gives companies the possibility to apply for licenses in mature areas of the continental shelf with a well-known geology.[34] In these areas the possibility of discoveries are high, but often the discoveries are small and dependent on the use of existing infrastructure. The APA scheme has proved to be important and has been the basis for new discoveries. In 2009, two license areas that were awarded APA licenses discovered significant oil reserves in the Avaldsnes and Aldous Major fields, which were later confirmed to be part of the same field. The fields were unitized and renamed the Johan Sverup field, and are estimated to contain 1.7–3.3 billion barrels of oil.

Unlike the UK, Norway does not provide for special conditions in frontier areas. Throughout Norwegian petroleum development, there has been a policy of releasing new areas for licensing on a sequential basis. The latest licensing round, the twenty-third, focused on new areas in the south-eastern Barents Sea, a new area that was previously the subject of a maritime border dispute between Norway and Russia. Since the delineation of the maritime border in 2010, the Norwegian government undertook a detailed seismic survey of the area prior to release of the area in the licensing round. Whilst not providing special conditions for acreage in frontier areas, this detailed seismic assessment of the geology of an area to be released provides an overview of the acreage for prospective applicants.

Australia recognizes that the allocation of licenses in mature areas based on exploration work program bidding is not practical, since the geology of the area is already well known and production has already

---

[34] See the announcement for TFO 2014, accessed 14 July 2014 at www.npd.no/en/Topics/Production-licences/Theme-articles/Licensing-rounds/APA-2014/APA-2014 – announcement-/. The predefined area was here extended with six blocks in the Norwegian Sea and three blocks in the Barents Sea.

occurred in the area. Rather, licenses in mature areas are allocated under the cash bid system. This is the complete antithesis of the UK and Norway, where licenses in mature areas are allocated by considering the technical capacity of the applicant to best explore and develop the license area. This approach has proved very successful, particularly in Norway, given that the Johan Sverup field was found under an APA license.

Frontier areas in Australia are not granted special treatment in the allocation licenses. Rather, this special treatment is provided for in the petroleum taxation system to encourage petroleum exploration in Australia's remote offshore areas in order to discover new petroleum provinces. Under the new tax concessions introduced in the mid 2000s,[35] expenditure associated with exploration in designated frontier areas could be claimed at a rate of 150 per cent, rather than the 100 per cent usually allowed for exploration expenditure in other areas.

The developing petroleum provinces that are considered in this book have not, at the time of publication, adopted any special treatment for the development of the petroleum areas. In addition, the countries considered have no mature areas to be given special consideration.

## 5. TRANSPARENCY AND STABILITY OF THE LICENSING SYSTEM

The allocation of licenses in the United States under the sealed cash bid system provide a confidential process of allocation of licenses. This process of allocation is transparent, as the allocation of licenses is televised live over the radio and the internet. This allocation occurs within a system that is extremely stable, having operated for over 60 years. However, the regulatory framework responsible for all petroleum activities has altered since the Deepwater Horizon accident in 2010. The existing regulatory body, the Minerals Management Service (MMS), has been disbanded and replaced with three equivalent bodies: the Bureau of Ocean Energy Management, the Bureau of Safety and Environmental Enforcement and the Office of Natural Resource Revenue.

The allocation of licenses in Brazil is made using set criteria that are transparent and apply equally to all participants. In addition, the contractual arrangements governing the license are outlined in the standardized Joint Operating Agreement (JOA). Although a young system, the

[35] Chapter 3, Schedule 5 of the Petroleum Resource Rent Tax Assessment Act 1987.

Brazilian licensing system appears to be stable. There is participation by many international companies, and the standardized JOA has been in use for over ten years. This stability, as well as vast offshore resources, continues to attract participation by international oil companies.

Even though Norway uses a discretionary system to award petroleum licenses, the system is transparent and stable. This transparency arises due to transparent criteria (set out in regulation 10 of the Petroleum Regulations) that are applied to all applicants. In addition, there is a standard, publically available JOA that all licensees are required to sign (similar to Brazil), which governs their conduct. The stability of the Norwegian licensing framework is imbued in the longevity of the system, which has been utilized since the early 1970s.

The allocation of licenses in the UK is similarly transparent, even though the UK also grants licenses under a discretionary framework. This transparency is in large part attributable to the EU non-discrimination requirements under the EU Hydrocarbon Directive, as well as the standard JOAs that are contained within the legislative provisions.

The licensing system in Nigeria has been plagued with issues relating to transparency. In particular, the system of allocation of licenses has been afflicted with concerns regarding the introduction of the confidential model PSC in 1991. As a consequence, since the early 2000s licenses have been allocated through the auction process, where the process can be seen to be more open and transparent.

The Japanese system of allocation of access to petroleum occurs under the Mining Act 1950, versions of which have been in existence for over 100 years. Under the Mining Act, prior to the 2011 amendments, the grant of a license could be perceived to be non-transparent, since licenses were granted on a first-come, first-served basis rather than through official licensing rounds. However, as Japan's interest in petroleum on the continental shelf increases, so has the transparency of the processes to award access to petroleum, with all licenses now awarded through a bid process.

## 6. HYBRIDIZATION – USE OF PSC AND LCS

Three of the developing petroleum regimes, Brazil, Nigeria and Russia, utilize a hybrid system of licensing, where both production sharing contracts and the licensing and concession system are used to award access to petroleum resources.

Production sharing contracts were used extensively in Nigeria since the 1970s. Although the contract used differed in some respects to the PSC pioneered in Indonesia. The experience of Nigeria in the use of the PSC has not been successful for two main reasons. Firstly, the PSC has not been successful in enabling the Nigerian state to achieve its objectives, particularly local content and Nigerian company participation. Secondly, PSCs have been marred by allegations of corruption, particularly with the introduction of the confidential model PSC in 1991. Ultimately, however, the PSC was seen as an unfavourable method of the allocation of access to petroleum and their use has been abandoned in favour of the LCS.

Unlike Nigeria, who have abandoned the use of PSCs, Brazil has introduced the use of PSCs alongside the LCS as a method to allocate access to petroleum resources. When petroleum activities were opened up to petroleum companies in 1995, the Brazilian government implemented a system of allocation using a licensing and concession system, with licenses granted through a bidding process and governed by a model JOA. However, the first model Brazilian concession agreement (BCA) was implemented in 1999, and has since been substantially altered. The BCA was introduced as an assurance for investors, as the agreement covers both exploration and production phases of petroleum activities. The PSC was introduced in 2012 to apply to the pre-salt areas where huge oil resources are located, in order to allow a greater government role and for greater government control. The PSC laws confer exclusive rights to Petrobras, and create a new petroleum company that will play a key role in the Brazilian PSC regime. The PSCs only apply in specific areas defined in the law, with the remaining parts of the country continuing to be governed by a concession agreement.

Since allowing non-Russian petroleum companies to participate in the extraction of petroleum, Russia has utilized a mixed system to enable access to petroleum resources. The Supreme Court Decree of 1992 outlines the requirements of the grant of a license through competitive tenders and auctions. However, high cost blocks not purchased at auction are made available under the production sharing agreement (PSA) regime. At any one time, blocks released through the PSA regime are not able to exceed 30 per cent of all allocated blocks.

## 7. CONCLUSION

Although all of the countries in this book utilize the LCS, this chapter has highlighted the differences in the application of an LCS. There are differences in many aspects. Some jurisdictions require applicants to be prequalified prior to applying for a license area. Other jurisdictions, particularly developed (or mature) jurisdictions, incorporate special treatment for mature areas. In addition, the UK has implemented a new license system for the areas west of the Shetlands.

Each of the countries in this book utilizes either a bid or discretionary method for the allocation of licenses. However, as seen in section 3 above, there are variations in these two main methods of allocation. It is the allocation of licenses that primarily accounts for the transparency and stability of the petroleum access regimes. In some jurisdictions, particularly Nigeria, the lack of transparency in PSCs has encouraged governments to utilize the LCS for the grant of licenses. However, in other jurisdictions, particularly Brazil, the PSC has been either retained or implemented as a method of the government retaining control in the petroleum area.

# Index